COLOR IN FOOD

Technological and Psychophysical Aspects

COLOR IN FOOD

Technological and Psychophysical Aspects

Edited by
José Luis Caivano
María del Pilar Buera

CRC Press
Taylor & Francis Group
Boca Raton London New York

CRC Press is an imprint of the
Taylor & Francis Group, an **informa** business

CRC Press
Taylor & Francis Group
6000 Broken Sound Parkway NW, Suite 300
Boca Raton, FL 33487-2742

First issued in paperback 2016

Version Date: 20120330

ISBN 13: 978-1-138-19964-4 (pbk)
ISBN 13: 978-1-4398-7693-0 (hbk)

Library of Congress Cataloging-in-Publication Data

Color in food : technological and psychophysical aspects / editors, José Luis Caivano, Maria del Pilar Buera.
p. cm.
Selections from a meeting of the International Color Association, held in Mar del Plata, Argentina, from October 12 to 15, 2010.
Includes bibliographical references and index.
ISBN 978-1-4398-7693-0 (hardback)
1. Color of food--Congresses. 2. Color--Psychological aspects--Congresses. I. Caivano, José Luis. II. Pilar Buera, María del. III. International Colour Association.

TP370.9.C64C64 2012
155.9'1145--dc23 2012005266

Visit the Taylor & Francis Web site at
http://www.taylorandfrancis.com

and the CRC Press Web site at
http://www.crcpress.com

Contents

Preface

Because most of our contact with the external world is channeled through vision, and color is the most salient visual aspect in this regard, the importance of controlling, measuring, and "designing" the color of food becomes evident.

This book includes a selection of keynote lectures and papers presented at AIC 2010 Color and Food: From the Farm to the Table, the interim meeting of the International Color Association, held in Mar del Plata, Argentina, from October 12 to 15, 2010.

The topic of color in food was not addressed in any previous AIC meeting and, being a wide interdisciplinary field, the readers of this book will find approaches from chemistry, physics, psychophysics, psychology, biology, engineering, and other disciplines.

The meeting featured 136 presentations divided into invited lectures, oral papers, and posters, which have been included in the proceedings, edited by José Luis Caivano and Mabel A. López, and published in Buenos Aires by the Grupo Argentino del Color in 2010. We made a selection from these papers, and their authors were invited to develop the articles further and submit them for this book. The members of the editorial committee reviewed the submissions, and this process resulted in the 44 chapters contained in the book, divided into 5 thematic parts: Part I (Food Color and Appearance), Part II (Food Colorimetry and Color Scales), Part III (Color Change as Quality Index of Food), Part IV (Color as an Index of Food Composition and Properties), and Part V (Food Environment: Color in Packaging, Sensory Evaluation, and Preferences).

The thematic scope comprises a variety of aspects focused on color research and application in various stages of food production, processing, marketing, purchasing, and consumption.

We hope to offer to the reader a rich panorama of the present state of color studies in relation to food, its industry, and the activities surrounding it.

For MATLAB® and Simulink® product information, please contact:

The MathWorks, Inc.
3 Apple Hill Drive
Natick, MA, 01760-2098 USA
Tel: 508-647-7000
Fax: 508-647-7001
E-mail: info@mathworks.com
Web: www.mathworks.com

Contributors

Nuria Acevedo
Department of Industries
Faculty of Exact and Natural Sciences
University of Buenos Aires
Buenos Aires, Argentina

Lina Marcela Agudelo Laverde
Department of Industries
Faculty of Exact and Natural Sciences
University of Buenos Aires
Buenos Aires, Argentina

María Victoria Agüero
Research Group on Food Engineering
Faculty of Engineering
National University of Mar del Plata, and Conicet
Mar del Plata, Argentina

Roberto Jorge Aguerre
Department of Technology
National University of Lujan
Lujan, Argentina

Mariano Aguilar
Faculty of Design Engineering
Polytechnic University of Valencia
Valencia, Spain

Terumi Aiba
Department of Health and Nutrition
Faculty of Health Science
Kyoto Koka Women's University
Kyoto, Japan

Guillermo Alcusón
Vegetable Research Laboratory
Department of Food Technology
Faculty of Veterinary
University of Zaragoza
Zaragoza, Spain

Johan Alferdinck
Perceptual and Cognitive Systems, TNO
Soesterberg, the Netherlands

Carlos Alberto Almada
Department of Technology
National University of Lujan
Lujan, Argentina

Ana Andrada
Laboratory of Apiculture
Department of Agronomy
National University of the South
Bahía Blanca, Argentina

Sebastià Balasch
Department of Statistics and Operations Research
School of Agricultural Engineering
Polytechnic University of Valencia
Valencia, Spain

Germán P. Balbarrey
Laboratory of Apiculture
Department of Agronomy
National University of the South
Bahía Blanca, Argentina

Dardo G. Bardier
Chajá and Guarapitá Street
Balneario Solís, Uruguay

Emilia Bejines Mejías
Food Color and Quality Laboratory
Department of Nutrition and Food Science
Faculty of Pharmacy
University of Seville
Seville, Spain

Marcelo Oscar Bello
Department of Industries
Faculty of Exact and Natural Sciences
University of Buenos Aires
Buenos Aires, Argentina

Cecilia Bernardi
Institute of Food Technology
Faculty of Chemical Engineering
National University of Litoral
Santa Fe, Argentina

Alicia Eva Bevilacqua
Center of Research and Development in Food Cryotechnology
National University of La Plata, and Conicet
La Plata, Argentina

Andrea Biolatto
Concepción del Uruguay Agricultural Experiment Station
National Institute of Agricultural Technology
Concepción del Uruguay, Entre Ríos, Argentina

Patrik Brandt
School of Computer Science, Physics and Mathematics
Linnaeus University
Kalmar, Sweden

María del Pilar Buera
Department of Industries
Faculty of Exact and Natural Sciences
University of Buenos Aires
Buenos Aires, Argentina

Elisa Colombo
Department of Lighting, Light and Vision
Faculty of Exact Sciences and Technology
Research Institute of Lighting, Environment and Vision
National University of Tucumán, and Conicet
Tucumán, Argentina

Eduardo Comerón
Rafaela Agricultural Experiment Station
National Institute of Agricultural Technology
Rafaela, Argentina

Paulo Felix Marcelino Conceição
Pró-Cor Association of Brazil
Sao Paulo, Brazil

Gabriela Cordon
Department of Inorganic, Analytical and Physical Chemistry
Faculty of Exact and Natural Sciences
Institute of Physical-Chemistry of Materials, Environment and Energy
University of Buenos Aires
Buenos Aires, Argentina

Osvaldo Da Pos
Department of Applied Psychology
Faculty of Psychology
University of Padua
Padua, Italy

Adriana Descalzo
Institute of Food Technology
National Institute of Agricultural Technology
Morón, Argentina

Juan Echazarreta
Laboratory of Apiculture
Department of Agronomy
National University of the South
Bahía Blanca, Argentina

Javier Enrione
Department of Science and Food Technology
Faculty of Technology
University of Santiago de Chile
Santiago, Chile

M. Luisa Escudero-Gilete
Food Color and Quality Laboratory
Department of Nutrition and Food Science
Faculty of Pharmacy
University of Seville
Seville, Spain

Rocío Fernández-Vázquez
Food Color and Quality Laboratory
Department of Nutrition and Food Science
Faculty of Pharmacy
University of Seville
Seville, Spain

Julia Inés Fossati
Gutenberg Foundation
Buenos Aires, Argentina

Lorena Franceschinis
Faculty of Engineering
Multidisciplinary Institute of Research and Development of Northern Patagonia
National University of Comahue, and Conicet
Neuquén, Argentina

Yoko Fukumoto
Faculty of Human Life and Science
Doshisha Women's College of Liberal Arts
Kyoto, Japan

Liliana M. Gallez
Laboratory of Apiculture
Department of Agronomy
National University of the South
Bahía Blanca, Argentina

Alicia del Valle Gallo
Department of Technology
National University of Lujan
Lujan, Province of Buenos Aires, Argentina

Juan M. Gisbert
Department of Crop Production
School of Agricultural Engineering
Polytechnic University of Valencia
Valencia, Spain

Claudia González
Institute of Food Technology
Agro-Industrial Research Center
National Institute of Agricultural Technology
Morón, Argentina

M. Lourdes González-Miret
Food Color and Quality Laboratory
Department of Nutrition and Food Science
Faculty of Pharmacy
University of Seville
Seville, Spain

Sergio R. Gor
Department of Lighting, Light and Vision
Faculty of Exact Sciences and Technology
Research Institute of Lighting, Environment and Vision
National University of Tucumán, and Conicet
Tucumán, Argentina

Belén Gordillo
Food Color and Quality Laboratory
Department of Nutrition and Food Science
Faculty of Pharmacy
University of Seville
Seville, Spain

Gabriela María Grigioni
Institute of Food Technology
Agro-Industrial Research Center
National Institute of Agricultural Technology
Morón, Argentina

Daniel Güemes
Institute of Food Technology
Faculty of Chemical Engineering
National University of Litoral
Santa Fe, Argentina

Leticia Guida
Department of Organic Chemistry
Faculty of Exact and Natural Sciences
University of Buenos Aires
Buenos Aires, Argentina

Silvina Guidi
Institute of Food Technology
Agro-Industrial Research Center
National Institute of Agricultural Technology
Morón, Argentina

Naoya Hara
Department of Architecture
Faculty of Environmental and Urban Engineering
Kansai University
Osaka, Japan

Francisco J. Heredia
Food Color and Quality Laboratory
Department of Nutrition and Food Science
Faculty of Pharmacy
University of Seville
Seville, Spain

Dolores Hernanz
Department of Analytical Chemistry
Faculty of Pharmacy
University of Seville
Seville, Spain

Robert Hirschler
Color Institute
National Service for Industrial Training
Technology Center for the Chemical and Textile Industry
Rio de Janeiro, Brazil

John B. Hutchings
Department of Colour Science
University of Leeds
Leeds, United Kingdom

Diego Iaconis
Laboratory of Apiculture
Department of Agronomy
National University of the South
Bahía Blanca, Argentina

Sara Ibáñez
Department of Crop Production
School of Agricultural Engineering
Polytechnic University of Valencia
Valencia, Spain

Martín Irurueta
Institute of Food Technology
National Institute of Agricultural Technology
Morón, Argentina

Hiroshi Iwade
Center for Civil Engineering and Architecture
The Kansai Electric Power Co., Inc.
Osaka, Japan

Wataru Iwai
Central Lighting Engineering Division
Panasonic Electric Works
Tokyo, Japan

M. José Jara-Palacios
Food Color and Quality Laboratory
Department of Nutrition and Food Science
Faculty of Pharmacy
University of Seville
Seville, Spain

Wei Ji
Department of Colour Science
University of Leeds
Leeds, United Kingdom

Päivi Jokela
School of Computer Science, Physics and Mathematics
Linnaeus University
Kalmar, Sweden

Ivar Jung
School of Design
Linnaeus University
Kalmar, Sweden

Jangmi Kang
Department of Nutrition
Hyogo NCC College
Nishinomiya, Japan

Saori Kitaguchi
Center for Fiber and Textile Science
Kyoto Institute of Technology
Kyoto, Japan

Yoji Kitani
Graduate School of Science and Technology
Kyoto Institute of Technology
Kyoto, Japan

María Gabriela Lagorio
Department of Inorganic, Analytical and Physical Chemistry
Faculty of Exact and Natural Sciences
Institute of Physical-Chemistry of Materials, Environment and Energy
University of Buenos Aires
Buenos Aires, Argentina

Leandro Langman
Institute of Food Technology
National Institute of Agricultural Technology
Morón, Argentina

Graciela Leiva
Department of Organic Chemistry
Faculty of Exact and Natural Sciences
University of Buenos Aires
Buenos Aires, Argentina

María Ana Loubes
Department of Industries
Faculty of Exact and Natural Sciences
University of Buenos Aires
Buenos Aires, Argentina

R. Daniel Lozano
Private Consultant
Florida, Argentina

Marcel Lucassen
Lucassen Colour Research
Landsmeer, the Netherlands

M. Ronnier Luo
Department of Colour Science
University of Leeds
Leeds, United Kingdom

Silvia B. Maidana
Faculty of Sciences and Food Technology
Multidisciplinary Institute of Research and Development of Northern Patagonia
National University of Comahue, and Conicet
Villa Regina, Argentina

Laura Malec
Department of Organic Chemistry
Faculty of Exact and Natural Sciences
University of Buenos Aires
Buenos Aires, Argentina

Alfredo Marconi
Laboratory of Apiculture
Department of Agronomy
National University of the South
Bahía Blanca, Argentina

Ángel Marqués-Mateu
Department of Cartographic Engineering, Geodesy and Photogrammetry
School of Civil Engineering
Polytechnic University of Valencia
Valencia, Spain

Lorena Martín
Institute of Food Technology
Faculty of Chemical Engineering
National University of Litoral
Santa Fe, Argentina

Silvia Matiacevich
Department of Science and Food Technology
Faculty of Technology
University of Santiago de Chile
Santiago, Chile

Motoko Matsui
Graduate School of Life and Environmental Science
Kyoto Prefectural University
Kyoto, Japan

Guillermo Meier
Concordia Agricultural Experiment Station
National Institute of Agricultural Technology
Concordia, Argentina

Antonio J. Meléndez-Martínez
Food Color and Quality Laboratory
Department of Nutrition and Food Science
Faculty of Pharmacy
University of Seville
Seville, Spain

Héctor Moreno
Department of Crop Production
School of Agricultural Engineering
Polytechnic University of Valencia
Valencia, Spain

María Luisa Musso
Faculty of Architecture, Design and Urbanism
University of Buenos Aires
Buenos Aires, Argentina

Gabriela Naranjo
Department of Organic Chemistry
Faculty of Exact and Natural Sciences
University of Buenos Aires
Buenos Aires, Argentina

Ángel I. Negueruela
Department of Applied Physics
Faculty of Veterinary and of Food Science and Technology
University of Zaragoza
Zaragoza, Spain

Kimiko Ohtani
Graduate School of Life and Environmental Science
Kyoto Prefectural University
Kyoto, Japan

Shino Okuda
Faculty of Human Life and Science
Doshisha Women's College of Liberal Arts
Kyoto, Japan

Rosa Oria
Vegetable Research Laboratory
Department of Food Technology
Faculty of Veterinary
University of Zaragoza
Zaragoza, Spain

Fernando Osorio
Department of Science and Food Technology
Faculty of Technology
University of Santiago de Chile
Santiago, Chile

Oranis Panyarjun
Faculty of Architecture
Chulalongkorn University
Bangkok, Thailand

Adriana Pazos
Institute of Food Technology
Agro-Industrial Research Center
National Institute of Agricultural Technology
Morón, Argentina

Claudio Petriella
Department of Technology
National University of Lujan
Lujan, Argentina

Andrea Piagentini
Institute of Food Technology
Faculty of Chemical Engineering
National University of Litoral
Santa Fe, Argentina

Leonor Pilatti
Catamarca Agricultural Experiment Station
National Institute of Agricultural Technology
Valle Viejo, Catamarca, Argentina

María Pirovani
Institute of Food Technology
Faculty of Chemical Engineering
National University of Litoral
Santa Fe, Argentina

Gustavo Polenta
Institute of Food Technology
Agro-Industrial Research Center
National Institute of Agricultural Technology
Morón, Argentina

Jaume Pujol
Centre for Sensors, Instruments and Systems Development
Polytechnic University of Catalonia
Barcelona, Spain

Jacqueline Ramallo
San Miguel Citrus Co.
Tucumán, Argentina

Cristina Rao
Department of Applied Psychology
Faculty of Psychology
University of Padua
Padua, Italy

Alessandro Rizzi
Department of Informatics and Communication
University of Milan
Milan, Italy

Francisco José Rodríguez-Pulido
Food Color and Quality Laboratory
Department of Nutrition and Food Science
Faculty of Pharmacy
University of Seville
Seville, Spain

Marcela María B. Rojas
Gutenberg Foundation
Buenos Aires, Argentina

Luciana Rossetti
Institute of Food Technology
National Institute of Agricultural Technology
Morón, Argentina

Maurizio Rossi
Department of Industrial Design, Arts, Communication and Fashion
Polytechnic of Milan
Milan, Italy

Sara Inés Roura
Research Group on Food Engineering
Faculty of Engineering
National University of Mar del Plata
Mar del Plata, Argentina

Ana María Ruiz de Castro
Vegetable Research Laboratory
Department of Food Technology
Faculty of Veterinary
University of Zaragoza
Zaragoza, Spain

Daniela M. Salvatori
Faculty of Engineering
Multidisciplinary Institute of Research and Development of Northern Patagonia
National University of Comahue, and Conicet
Neuquén, Argentina

Ana María Sancho
Institute of Food Technology
Agro-Industrial Research Center
National Institute of Agricultural Technology
Morón, Argentina

José D. Sandoval
Department of Lighting, Light and Vision
Faculty of Exact Sciences and Technology
Research Institute of Lighting, Environment and Vision
National University of Tucumán, and Conicet
Tucumán, Argentina

Tetsuya Sato
Graduate School of Science and Technology
Kyoto Institute of Technology
Kyoto, Japan

Carolina Schebor
Department of Industries
Faculty of Exact and Natural Sciences
University of Buenos Aires
Buenos Aires, Argentina

Ana Sfer
Department of Mathematics
Faculty of Exact Sciences and Technology
National University of Tucumán
Tucumán, Argentina

Patricia Silva
Department of Science and Food Technology
Faculty of Technology
University of Santiago de Chile
Santiago, Chile

Carla M. Stinco
Food Color and Quality Laboratory
Department of Nutrition and Food Science
Faculty of Pharmacy
University of Seville
Seville, Spain

Constantino Suárez
Faculty of Exact and Natural Sciences
University of Buenos Aires
Buenos Aires, Argentina

Suchitra Sueeprasan
Department of Imaging and Printing Technology
Faculty of Science
Chulalongkorn University
Bangkok, Thailand

Marcela Patricia Tolaba
Department of Industries
Faculty of Exact and Natural Sciences
University of Buenos Aires
Buenos Aires, Argentina

Alejandra Tomac
Research Group on Food Preservation and Quality
Department of Chemical Engineering
Faculty of Engineering
National University of Mar del Plata
Mar del Plata, Argentina

Keiko Tomita
Faculty of Agriculture
Department of Food Science and Nutrition
Kinki University
Nara, Japan

and

Graduate School of Life and Environmental Science
Kyoto Prefectural University
Kyoto, Japan

Elian Tourn
Scientific Research Commission of Buenos Aires Province
Department of Agronomy
National University of the South
Bahía Blanca, Argentina

Chawika Traisiwakul
Department of Imaging and Printing Technology
Faculty of Science
Chulalongkorn University
Bangkok, Thailand

María Concepción Urzola
Vegetable Research Laboratory
Department of Food Technology
Faculty of Veterinary
University of Zaragoza
Zaragoza, Spain

Franco Van de Velde
Institute of Food Technology
Faculty of Chemical Engineering
National University of Litoral
Santa Fe, Argentina

Ron van Megen
CoMore BV
Zeist, the Netherlands

Daniel Vázquez
Concordia Agricultural Experiment Station
National Institute of Agricultural Technology
Concordia, Argentina

Isabel M. Vicario
Food Color and Quality Laboratory
Department of Nutrition and Food Science
Faculty of Pharmacy
University of Seville
Seville, Spain

Ole Victor
School of Design
Linnaeus University
Kalmar, Sweden

Meritxell Vilaseca
Centre for Sensors, Instruments and Systems Development
Polytechnic University of Catalonia
Barcelona, Spain

Mabel B. Vullioud
Faculty of Sciences and Food Technology
Multidisciplinary Institute of Research and Development of Northern Patagonia
National University of Comahue, and Conicet
Villa Regina, Argentina

María Isabel Yeannes
Research Group on Food Preservation and Quality
Department of Chemical Engineering
Faculty of Engineering
National University of Mar del Plata, and Conicet
Mar del Plata, Argentina

Yuhi Yonemaru
Graduate School of Science and Technology
Kyoto Institute of Technology
Kyoto, Japan

Juan Manuel Zaldívar-Cruz
Program on Food Science and Engineering
Postgraduate College
Cárdenas, México

PART I

Food Color and Appearance

CHAPTER 1

Food Appearance and Expectations

JOHN B. HUTCHINGS, M. RONNIER LUO, and WEI JI

Contents

1.1 Introduction

The color of our natural foods has evolved first through optimization of plant life to prevailing geographical, geological, and climatological conditions and second, through coevolution with insect and animal visual characteristics. These early evolutionary links are so profound that we need to eat generous portions of pigment-containing fruit and vegetables to live a healthy life today. Most, if not all, plant pigments have a value to the plant, and very many pigments have a value to the herbivores and carnivores devouring them.

Color is important, but it is not the whole story of our visual link with food because we have evolved to respond to appearance as a whole. Appearance comprises visually perceived structure of the food, each element of which possesses color, translucency, gloss, and surface texture (roughness) properties. Each of these attributes behaves characteristically with time and processing, all contributing to our overall expectations of the food in front of us.

This chapter contains an outline of food appearance properties, the resulting expectations for food and the food environment, and gives examples of the halo effects that occur and the commercial exploitation arising. It will conclude with a brief consideration of specification and measurement of appearance attributes as well as of the expectations themselves.

1.2 Food Color and Appearance

Study of how foods look differs from that of all other mass-marketed materials. One of these differences arises from the fact that foods have a natural variation in appearance properties. Normally, solid foods are nonuniform in color; they are translucent and vary, often irregularly, in surface texture (i.e., roughness) and gloss. It is the complete package of these properties that leads to identification and preference for a particular food.

Many products, for example, meats, have a visual structure revealing different elements within the total product. We immediately receive information as to the number of muscles present, the degree of lean, fat, connective tissue, and gristle content.

Color is a characteristic of a particular food material and is vital to its identification and judgment of quality. Fish and red meats depend for their perceived quality on the balance between color and translucency. Cooking changes the absolute values and balance of visually perceived attributes. Ancient Egyptians clarified drinks by filtration, and today a reduction in clarity leads to rejection of many alcoholic drinks. Turbidity in fruit juices, such as apple, can be a positive or negative attribute depending on the expectations of the consumer.

Specific gloss characteristics are associated with different fruits and vegetables. High-quality chocolate normally has a high gloss, and light scattered from the surface is near mirrorlike specular reflection. When chocolate blooms, it loses gloss, the specular reflection changes to diffuse scatter, and the surface becomes dull. In the store, glossiness of moist surfaces such as fish reinforces perceptions of freshness, but only if there is sufficient directional light. Hence, for the store, apples may not only be coated with wax designed to reduce gas exchange, weight loss, and fungal growth but are also designed to be glossy. Gloss or glaze is achieved in the kitchen for vegetables and maintained for the dining room by coating with butter.

Surface texture is also a characteristic of foods. Meat cut along the grain reveals its fibrous structure, breakfast cereals have differing degrees and types of roughness, and some varieties of apple are rough skinned while some smooth.

These individual appearance attributes are not linked to quality in isolation. For example, the visually perceived quality of fish and meat depends on color and translucency; that of chocolate depends on color and gloss; breakfast cereals depend on color, color distribution, and surface texture; and drinks depend on color as well as translucency.

In summary, visual structure, color, translucency, gloss, and surface texture are attributes of appearance, and all affect the look of the product. Effects of processing, cooking, and consumer preference should be considered in terms of appearance as a whole because each element of appearance changes characteristically with time and processing and combines to affect our expectations.

1.3 Expectations, Halo Effects, and the Taste Panel

Appearance, as defined earlier, provides the visual stimulus to which we react. Total appearance is formed by interaction of these appearance properties with the human response. Resulting images and expectations control our response to the food along the whole supply chain from field to kitchen to the plate of food on the table. That is, the total appearance of a food comprises two parts—the scene and the viewer. The scene consists of the elements of the scene, the way in which the elements are presented (i.e., the design) and the illumination. The human response is governed by our individual visual characteristics, our upbringing, psychology, preferences, and our immediate environment, which includes our appetite, needs, and health. We respond to total appearance in terms of sensory, emotional, intellectual images, and expectations. The term total appearance, meaning a combination of assessable and measurable product attributes as well as our feelings about them, was first used with reference to foods (Hutchings 1999).

There are two main types of expectation that condition our subsequent responses and experiences. The first are those generated by what we believe, perhaps, from a religious knowledge. The second are the five general categories of expectation generated from our perceptions of a material or scene. This applies to everything we view, whether that is a landscape, store front, the waiter, a plate of food, or its packaging (Hutchings 2003).

Visually assessed safety involves safety of body and safety of mind, perhaps, this bread appears moldy.
Visual identification, for example, this food will taste sweet.
Visually assessed usefulness, for example, will this food contain the nutrients I need?
Visually assessed pleasantness, perhaps, this looks tasty.
Visually assessed satisfaction, for example, when I have finished eating this meal will it have been worth the money?

Expectations are fundamental to food marketing, and when a food pack, advertisement, dining room, or a particular food dish is being designed, it is helpful to consider the five types. An important consequence of expectations is that they generate halo effects. For example, it is relatively easy to confuse tasters by giving them inappropriately colored foods, for example, the yellow-colored, raspberry-flavored drink that may be identified as orange juice. Such effects have profound implications for the sensory testing of foods. Panel in-mouth scores can be influenced by sample appearance, the environment, panel organization, and panel organizer attitude. There are however two groups of subjects: the field independent, who attend to their taste and smell perceptions even in the presence of an inconsistent visual stimulus, and the field dependent, who make more mistakes when trying to identify flavors in the absence of visual cues as to their origin (Moskowitz 1983). Little account is often taken of the extreme influence of the brand. The investigative focus may concentrate too much on the product, while marketers may overfocus on the concept or on the potential customers.

Preconditioning influences taster beliefs. For example, knowledge of the fat content may affect consumer response to the product. Preferences of young children can be changed by identification of the product with hero figures. Product advertising affects panel members. For example, the British have been subjected to the persistent advertising claims that "smaller peas are sweeter" and that "larger peas are tougher peas." An unwary approach to the paneling of peas results in erroneous findings founded upon awareness of these claims.

Halo effects can be positive or negative. The color of an orange tells us that it contains a natural mix of healthy antioxidants that are good for us. But the color uniformity may tell us that it has been sprayed with fungicide and herbicide, and the gloss reveals that it has been waxed. We may therefore conclude that oranges are poisonous and therefore bad for us. Another consequence of expectations is that attitudes to specific foods can be regional. For example, in the United Kingdom, there are different preferences for tomato soup color. Dark, deep red soup is preferred by those used to tomato puree or tomato powder–based soup, but orange red is preferred by those raised on tins of Heinz cream of tomato soup.

Color is affected by lighting, and no single lighting regime is optimal for all foods. According to food folklore, diners eating in the dark can be made physiologically sick by switching on the lights, revealing that they are eating inappropriately colored food. Red-biased light is often used in the marketing of red meats. This conceals the brown specks in fresh beef that indicate pigment oxidation and the presence of metmyoglobin. Such meat may be perfectly edible, but the customer normally sees this as undesirable. Some regard such a use of store lighting as bordering on fraud, but is it unethical to display foods to their best possible advantage? Expectations can be optimized but can also be commercially exploited, sometimes to an unethical extent. Such examples include the orange-flavored, high-sugar, low-fruit-content drink marketed in a translucent container in the chill cabinet, the same used for high-fruit-content drinks. Such ethical examples are of concern to the food industry (Hutchings 2006).

During optimization of product flavor or texture, low illumination levels or colored lighting is often used in the tasting area. However, although the actual color of the product may be completely lost, conclusions may still be drawn about the sample from the light reflected. For example, the extent of baking of bread products can be detected even under low illumination levels. The halo effect is so powerful that variations in appearance should be entirely eliminated when judgments of flavor and texture are required.

1.4 Measurement

Measurement and specification of material properties (i.e., visual structure, color, translucency, gloss, surface texture, and changes with processing and time) are required for monitoring, specification, and communication, as well as a sensory panel aid, and as a tool for consumer understanding. Such measurements lead to an appreciation and knowledge of factors that underpin likeness, acceptability, and willingness to buy. Conventional color measurement methods and instruments were designed for use with paints, photography, textiles, and later, ceramics and plastics. They have severe limitations in that they are designed to measure samples that are flat and opaque and have uniformly colored surfaces. In contrast, most foods are translucent, have surface irregularity, and are nonuniformly colored. For example, meat is a translucent material, and a full understanding of its physiology, performance, and customer acceptability cannot be obtained until it is understood as a translucent material (MacDougall 1982). In many published papers, the lack of agreement between observers and measurements can be traced to this phenomenon. That is, in the use of conventional color specification instrumentation, there tend to be two major problems. These arise from the nature of the optics of the sample and its physical form where shape irregularity creates difficulties in the way the sample is presented to the instrument. Now, however, all attributes of appearance can be instrumentally specified with precision and accuracy using calibrated digital camera technology (Hutchings et al. 2002).

Even when a sound relationship is built between a measurement and the viewer, natural materials such as fruits and vegetables change in color from season to season. Climatological and geographical differences result in the development of different balances of pigments within the product. For example, for tomatoes given the same visual grade, the measured color changes between seasons (MacKinney and Little 1962). Rigorous experimental protocol must be used to obtain comparable results season to season. Hence, with all types of food materials, there is potential for significant error.

There are many applications for good color measurement practice within the food industry. These concern color specification, monitoring during storage and processing, in-field crop monitoring, color communication in what is a huge global industry, and understanding how the customer behaves when faced with food on the shelf and on the plate. Digital measurement techniques are rapidly overtaking conventional methods of color specification. These can be used for the rapid and nondestructive specification of all aspects of color and appearance. Successful color measurement investigations have been carried out, for example, on fruit and breakfast cereals (Hutchings et al. 2002), wines (González-Miret et al. 2007), fruit (Ji et al. 2012), vegetables (Hutchings et al. 2005), and fruit juices (Ji et al. 2005).

Color changes taking place during paneling can be specified using digital analysis. This enables panels to be set up, controlled, and monitored to take account of such changes. Digital imaging is also used to construct science-based food color order systems. Studies of the sensory aspects of food appearance are hampered by the three-dimensional nature of color itself. This has in the past prohibited the production of physical color scales that follow sequences of color changes occurring with natural pigments. However, using digital technology, a study of the color gamut of freshly cooked frozen peas led to the construction of a color science-based scale now used for screening and monitoring different varieties of the vegetable (Hutchings et al. 2005). First, the gamut of color space occupied by food products on the market is established. Second, the taste panel is used to indicate the area within this space that corresponds to the visual impact of the product. Third, using color-calibrated reproduction systems, a physical color fan can be created for routine panel use. The color science foundation of the scale enables panel scores to be interpolated accurately in color space.

Virtual products can be viewed on screen, thus avoiding the expense of sample manufacture (Wei et al. 2011). Effects on consumer expectations of changes in product as well as package color can be determined. For example, the volume of color space occupied by those drinks recognizable as orange juice can be determined. Within this volume, directions can then be plotted of increasing, for example, visually perceived sweetness, fruitiness, acidity, and liking. Such color-calibrated image analysis can be used in store, in the home, as well as in hospital.

Not only are color and appearance vital to judgments of food quality, but the eating environment also affects how we feel about a meal. This aspect of the eating experience is now also the subject of research. Studies involving quantification of impact and emotional response to scenes can reveal the extent to which the occupier of the space is affected by individual elements within the space, such as furniture, color, and lighting. This enables us to specify design elements necessary for maximum, for example, visually perceived comfort (Hutchings et al. 2011).

1.5 Summary

The story of food color and appearance involves the following:

- A story of vision, evolution, and of our continued survival as humans.
- A story of appearance, total appearance, expectations, halo effects, and population differences.
- A story that for completion and understanding requires disciplined assessment and measurement—we now have the means of accomplishing these.

References

González-Miret, M. L., W. Ji, M. R. Luo, J. B. Hutchings, and F. J. Heredia. 2007. Measuring colour appearance of red wines. *Food Quality and Preference* 18 (6): 862–871.

Hutchings, J. B. 1999. *Food Color and Appearance*, 2nd edn. Gaithersburg, MD: Aspen.

Hutchings, J. B. 2003. *Expectations and the Food Industry—The Impact of Color and Appearance*. New York: Kluwer/Plenum Publishers.

Hutchings, J. B. 2006. Talking about color?… and ethics. *Color Research and Application* 31 (2): 87–89.

Hutchings, J. B., R. Brown, B. Dias, K. Plater, and M. Singleton. 2005. Physical colour scales for sensory panels. *Food Science and Technology* 19 (2): 45–47.

Hutchings, J. B., M. R. Luo, and W. Ji. 2002. Calibrated colour imaging analysis of food. In *Colour in Food, Improving Quality*, ed. D. B. MacDougall. Cambridge, U.K.: Woodhead Publishing, pp. 352–366.

Hutchings, J. B., M. R. Luo, and L. C. Ou. 2011. Quantification of scene appearance—A valid design tool? *Color Research and Application*. Accepted, doi: 10.1002/col.20659.

Ji, W., M. R. Luo, and J. B. Hutchings. 2012. Measuring banana appearance aspects using spectrophotometer and digital camera. In *Color in Food: Technological and Psychophysical Aspects*, eds. J. Caivano and P. Buera. Boca Raton, FL: CRC Press, pp. 71–77. In this book.

Ji, W., M. R. Luo, J. B. Hutchings, and J. Dakin. 2005. Scaling translucency, opacity, apparent flavour strength and preference of orange juice. In *AIC Colour 2005, Proceedings of the 10th Congress of the International Colour Association*, eds. J. L. Nieves and J. Hernández-Andrés. Granada, Spain: Comité Español del Color, pp. 729–732.

MacDougall, D. B. 1982. Changes in colour and opacity of meat. *Food Chemistry* 9: 75–88.

MacKinney, G. and A. C. Little. 1962. *Color of Foods*. Westport, CT: Avi.

Moskowitz, H. R. 1983. *Product Testing and the Sensory Evaluation of Foods*. Westport, CT: Food and Nutrition Press.

Wei, S. T., L. C. Ou, M. R. Luo, and J. B. Hutchings. 2011. Optimisation of food flavour expectations using colour. *Food Quality and Preference*. Accepted.

CHAPTER 2

Color and Visual Appearance in Foods

R. DANIEL LOZANO

Contents

2.1 Introduction

Since almost a decade ago, I have been involved in the study and understanding of the phenomena called *visual appearance*, which was added to the experience, almost 40 years, in subjects related to color. Color of foods—salted anchovy, beans, corned beef, noodles, apple juice, orange juice, milk sweet (*dulce de leche*), corn, apples, margarine, honey, fish, sausages, tomatoes, wheat, wine, "yerba mate," etc.—has been one of the subjects I studied while working at INTI, the National Institute of Industrial Technology, in Argentina. The present work intends to describe something more than color. In fact, it tries to establish a reasonable ground to understand a very complex phenomenon, such as the whole visual appearance, which includes color but is not restricted to it. It is well known that only three primaries are necessary to see color. We shall forget here to define which primaries we are dealing with. Presently, we accept that there are three primaries: red, green, and blue. Also, we shall not specify which these colors were. Now, when we see a texture, such as the skin of an orange or a lemon and the shell of a walnut, a peach, or a strawberry, color indicates not only the product but also the morphological characteristics of its surface and its *visual appearance*. Simply, to imagine the complexity of the problem, it is sufficient to tell that up to six different orientation angles can be set up relative to the horizontal defined as 0° (90°, 60°, 30°, 0°, –30°, and −60°). To this, one must add what is called *spatial frequencies*, which are the lines of different widths that we recognize as bar code used to identify commercial products in the supermarket paying boxes. It is supposed that only eight of these spatial frequencies are needed to identify its appearance effect. Three colors, six orientation angles, and eight spatial frequencies make 144 variables, 144 possibilities. Then, the question produces gooseflesh. If we have three different detectors in the human retina, one for each of the three colors we see..., do we have 144 different detectors systems in the retina to see form, color, and texture? Which are they? How do they work?

The problem to evaluate color in food has been present since the human being appeared in the earth thousands of years ago. Since then, the political and economical changes in the society have introduced the mass production of foods. Primates choose the food to eat on the basis of visual experience; color and visual appearance in combination with other senses help them to choose their food. Even now, people choose the products in the food shops based on their experience.

The *total appearance*, as described by Hutchings (1999), involves all the senses in a whole perception mode including different aspects of the social life. He describes the total appearance as formed by three parts: *receptor mechanisms, inherited and learned responses to specific events,* and *immediate environment*. The *receptor mechanisms* are the inherited and acquired sensory characteristics such as color vision (adaptation, afterimages, constancy, discrimination, and metamerism), aging effects (cataract, glare, light intensity need, and yellowing), and other senses (hearing, smell, taste, and touch). The *inherited and learned responses* are the product of culture, memory, preference, fashion, physiology, and psychology. The *immediate environment* elements are geographical factors (climate, landscape, and seasonal changes), social factors (crowding, personal space, and degree of awareness), and medical factors (survival and need, state of well-being, and protection).

This chapter narrows its scope to the proposal of a new mode of classification of visual appearance limited only to what the people see, without any other consideration with respect to environmental, cultural, and historical background or surrounding.

2.2 New Classification of Visual Appearance

As the reader can see, this approach is much more than the "simple" perception of visual appearance (including color). For the "total appearance," the history and environment of the person who observes and evaluates the scene are pertinent. This chapter will only try to describe new information about a much more restricted view: the way people can see, recognize, and describe colors, forms, and objects, including surface finish and texture, without any relation to other external factors as in the case of Hutchings' total appearance.

In 1978, I published a book on color measurement (Lozano 1978). After 30 years, I tried to rewrite it, but when I went into appearance measurements, such as gloss, I found plenty of information about different aspects of visual appearance never mentioned before, particularly those related to the finishing of automotive paints. Simultaneously, the development of new characteristics of computers (such as memory and velocity) and the approach of new programming techniques such as the "graphic software" and the "inverse graphic rendering," the use of fractals and wavelets, the techniques used to create movie pictures—particularly animated films for children by the companies such as DreamWorks and Pixar—and the extensive use of Fourier maths, together with the advance and research on contrast sensitivity in human vision, have completely changed the approach to visual appearance.

In a meeting of the CIE Technical Committee 1–65 on visual appearance held in Paris (CIE 2006), I presented a work with the modified proposal of a previous work (Lozano 2006, 2007) (see Figure 2.1). It is important to stress that the circle is divided into three parts as is the whole visual appearance phenomena. They are color, cesia, and spatiality. Color is the best known aspect and is composed of luminosity and chromaticity, the latter being divided into two components: hue and saturation or chroma.

Luminosity or clarity allows going further to luminous reflectance and transmittance and, from there, to whiteness. This is a part of the circle shared with cesia, which is described as the perception of the spatial distribution of light (Caivano 1991, 1993, 1994, 1996, 1997, 1999, 2001; Caivano and Doria 1997). A new component is introduced, named spatiality, which is determined by the space appreciation or evaluation through the visual system.

There are visual appearance phenomena that connect spatiality with color, such as metallic, pearlescent, or iridescent appearances in which color changes as the angle of illumination or observation does. Therefore, they are dependent on the spatial distribution of the incident and observed light. We call this group spatial color. Cesia also has three components. One is shared with color: luminosity, luminous reflectance and transmittance, clarity, and whiteness. The connection of whiteness with color is the effect called yellowness, when whiteness is modified adding a color contamination, such as yellow, a product, in most cases, of aging or deterioration.

The other two components of cesia are permeability and diffusivity. The first is related to the capacity of the material or object to absorb light. Normally, the opposite of permeability is opacity. The second is related to diffusion of light by means of scattering. Gloss (or the contrary, matt or dullness), translucency, and transparency are visual appearances related to this property.

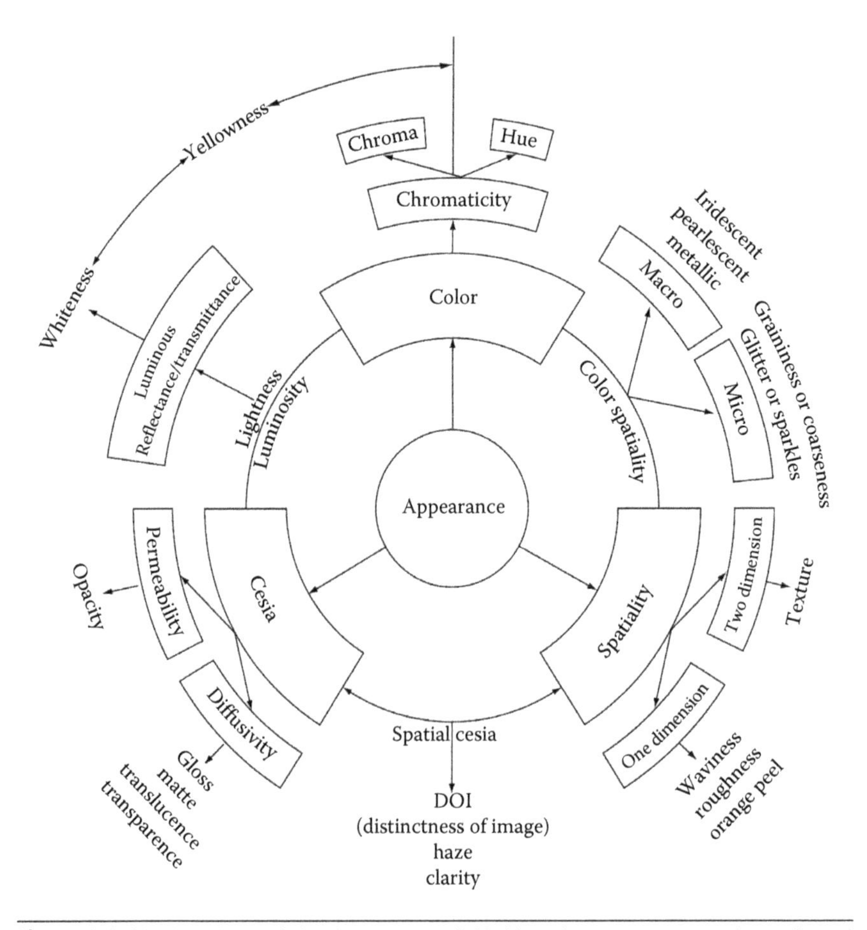

Figure 2.1 Circular scheme of visual appearance divided into three components: color, cesia, and spatiality. The intermediate components are spatial color, spatial cesia, and lightness or clarity between color and cesia.

Following down the circle, we find what is qualified as spatial cesia and which are properties of cesia based on the spatial perception of the appearance without color evaluation, such as definition of image (DOI), haze, and clarity (this is related to the perception of light diffusion in a transparent liquid). These appearances are related to cesia because they are independent of color perception and need a spatial evaluation of the visual effect. In the case of DOI, we need to observe the definition of images reflected on the surface, haze is perceived in the space around the light source reflection, and clarity is observed in the whole image of the liquid in the glass or bottle.

Then we can see the new proposal: spatiality of one and two dimensions. Why one and two dimensions? Why not three dimensions? Well, at first sight, it is difficult to explain it. When trying to catch an object, a human being uses both eyes to evaluate the distance. However, when looking at a scene, human beings are not able to calculate the distance directly; they depend on the elements that compose the scene to evaluate the size of the objects present in it, and on comparisons with previous experiences. We see a scene, as in a picture or a photograph, in two dimensions, and we guess the distance and the size of the objects with respect to the surroundings. We are then able to evaluate the visual appearance in two dimensions.

When an observer looks in a linear mode, for instance, at a footpath covered with paving tiles with lines transversal to his march, the lines are discriminated when they are near, but with distance, the transversal lines disappear, becoming a uniform surface. Instead, if the lines are parallel to the direction of walking, they converge in the horizon, normally into a vanishing point far away from the observer. In this case, the sight is evaluated in only one dimension. The same occurs when people judge the effect called orange peel.

The typical spatiality of two dimensions is texture. We need two visual dimensions to classify or evaluate texture. Within the classification of spatiality of one dimension are waviness, roughness, and orange peel.

The last item is what McCamy (1996, 1998) defined as micro-appearance. He also proposed to call macro-appearance what we defined as color or spatiality, composed by metallic appearance, pearlescence, and iridescence, already mentioned. Micro-appearance is something rather new. It appeared in the last years as a new form of finishing in automotive paints, with nothing similar in nature or in previous human manufacturing. The visual effects are glitter (or sparkling) and coarseness, which are produced by particles immersed in the paint and seen under specific modes of illumination. The first is seen under directional light sources reflected by the paint. The second needs an almost completely uniform diffuse illumination to evaluate its characteristics.

2.3 Food and Its Glossy Appearance

Normally, foods do not show too much gloss in their appearance, except some fruits that can be manually polished, as apples. But what is gloss? Are there several classes of gloss? Hunter and Harold (1987) have classified five types of gloss, but actually only three are accepted as different kinds of gloss, as one can see in Figure 2.2.

Normally, gloss is measured as specular reflectance, varying the incident angle and depending on the type of materials (usually 20°, 30°, 45°, 60°, 75°, and/or 85°), but as the reader will see later, the angularity does not take into consideration how the incident flux is spatially reflected but only how much is reflected with regard to an accepted standard. Sheen is classified as gloss when materials are observed at grazing angles, such as is done with matt papers. Surprisingly, the standards for measuring papers use 75° instead of 85°.

In the measurement of luster, the specular reflected flux measurement is compared with the diffuse reflected one. It is a way to evaluate this property,

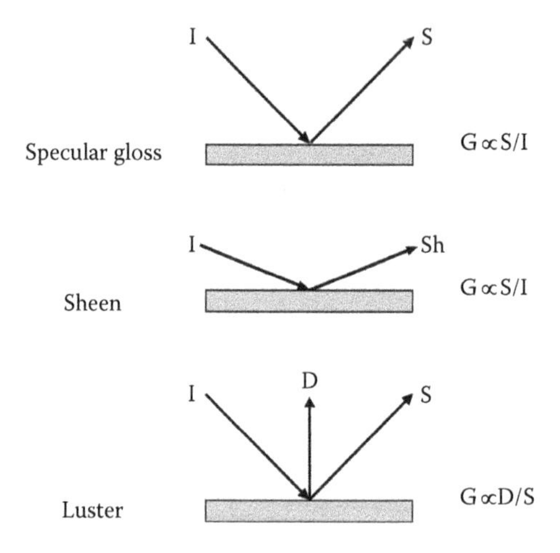

Figure 2.2 Three different types of gloss as classified by Hunter.

but on what grounds? There is little evidence of psychological evaluations of this type. If one revises different standards to evaluate other aspects of appearance referred to gloss and distribution of the light reflected near the specular angle, one can see Figure 2.3. The right scheme shows different angles near the specular direction to measure different types of appearance, at 18 min from the specular reflection angle is measured DOI. With respect to this definition, one must question the size of instrumental apertures.

The appearance called bloom is measured 2° apart from the specular direction. Finally, haze is measured at 5° apart. But … why these differences? Which psychological studies support these definitions? Figure 2.4 shows the shape of the specular reflectance curve for different glossy materials.

It is possible to see that curves of specular reflection can change the shape of the reflection peak. This represents surfaces more or less polished and how the reflected light is spatially distributed. A better understanding is possible by looking at the differences of goniophotometric reflectance curves in Figure 2.5. One can see the shape corresponding to a normal writing paper (above left) and to the most brilliant glossy coated paper (below right). The shapes shown change a lot, from an egg shape to a spine one, oriented to the specular reflectance direction. In conclusion, we can say that the spatial distribution of the reflected light must be known in order to really understand what the human eye sees and judges.

2.4 Texture, Spatial Frequencies, Fractals, Wavelets, Fourier Analysis, and Other Related Matters

Foods have texture. It is easy to define the difference between an orange, an apple, a plum, and a peach because their textures are quite different, but …

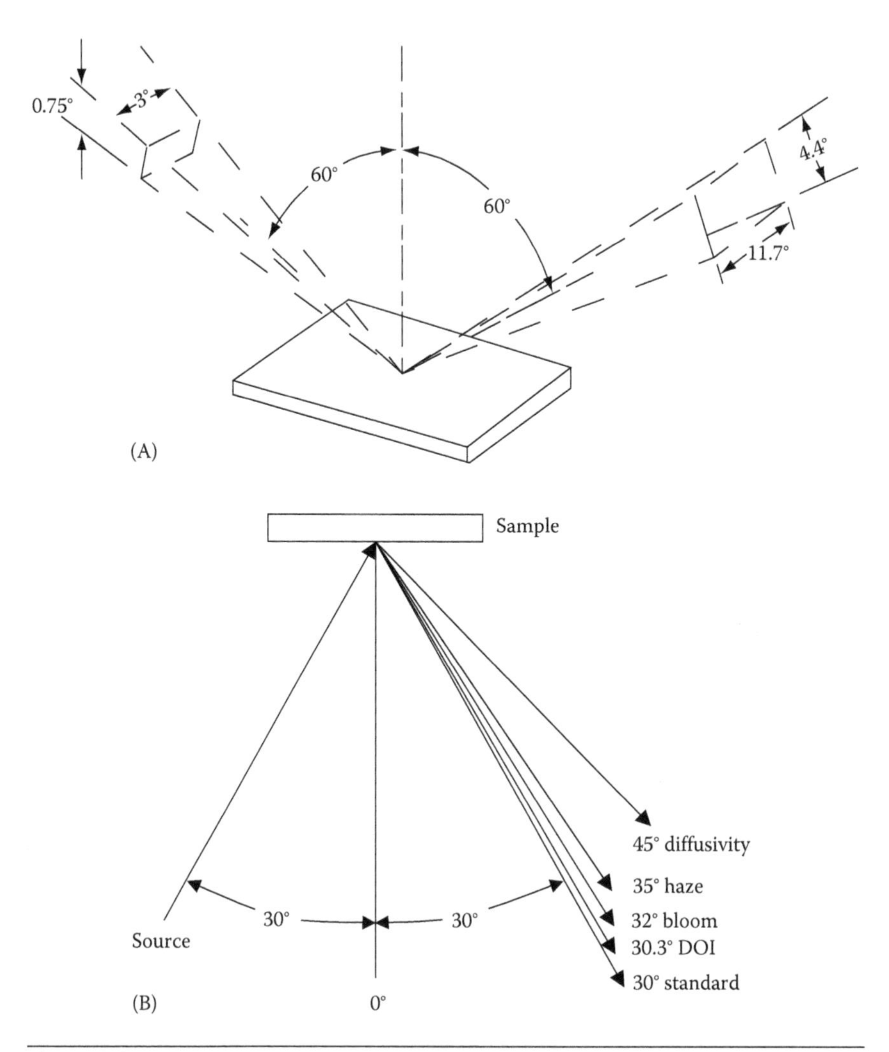

Figure 2.3 (A) Different illumination and measuring apertures. (B) With an incident angle of 30°, standard gloss is measured at 0°, DOI at 30.3°, bloom at 32°, and haze at 35°.

what about oranges, mandarins, and lemons? Their textures are similar; colors and shapes can help to differentiate them. What is really annoying is to give an answer when somebody asks us to describe a texture or to describe texture differences. To try to understand the problem we face, I introduce a new concept: the spatial frequencies (Figure 2.6). As one can see on the left drawing, what is defined as spatial frequencies are black and white stripes of the same width, and the number of lines in a unit of length is called spatial frequency. In the right part of the figure appears the drawing of an alphabet where a letter is displaced one position in each row, giving place to a formed order as one sees the figure at different distances.

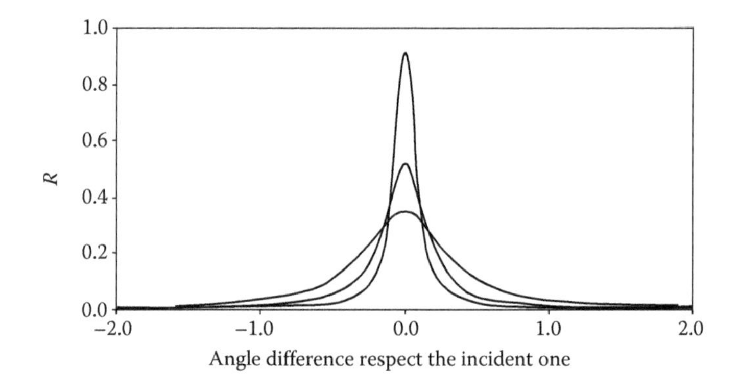

Figure 2.4 Different gonioreflectance curves of glossy materials.

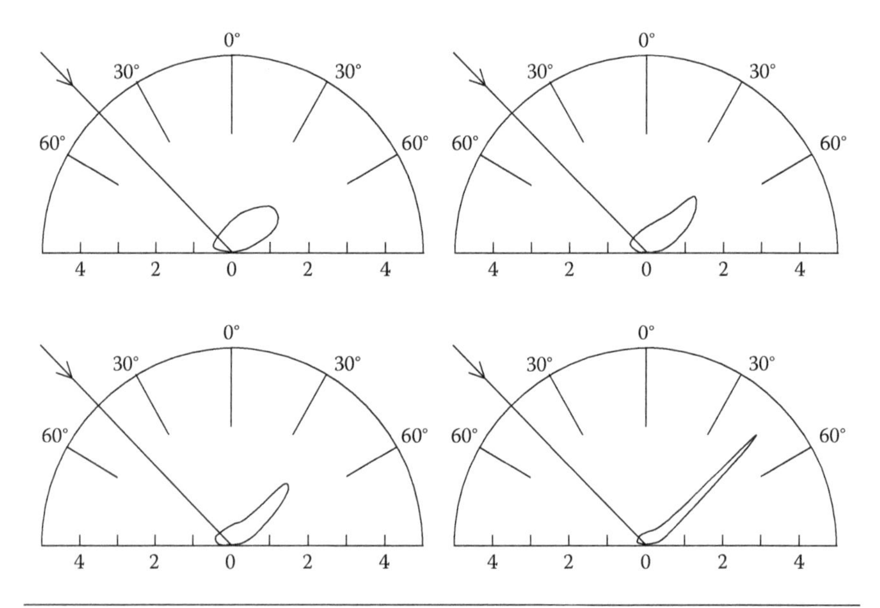

Figure 2.5 Goniophotometric curves of reflectance for different papers.

How do we deal with spatial frequencies? We deal with spatial frequencies through Fourier analysis, which are simple series that use sines or cosines to match periodical functions, as one can see in Figure 2.7 with the square function.

This effect is present in the observation of all textures and is evaluated through the spatial frequency analysis. But this is not the whole affair related to texture perception of the human being. Some other mathematical elements were introduced recently. One of them is fractal, created by Mandelbrot (1967) to explain why scientists cannot measure real images. We can take his own example: How to measure the length of Great Britain coast? Where do

Figure 2.6 (A) Different spatial frequencies. (B) The effect of distance of a geometric configuration.

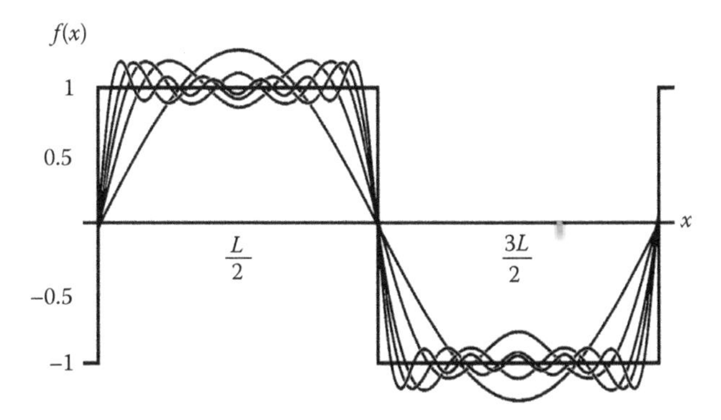

Figure 2.7 A square wave with five elements of the Fourier series approximating the square function.

we stop? In the green part? In the limit of the sand? Or the size of a grain of sand? Where?

Mathematical fractals cannot measure real images and surfaces; its existence must be defined in all scales. On the contrary, physical surfaces have a superior limit where the scale is applied. The inferior limit is defined by the size of the constituent particles. In his work on this subject, Pentland (1984) states that the surface is fractal if the fractal dimension is stable and can be approximated in just one scale of a wide range of them. Pentland suggested that fractals can be used as a model to describe textures.

In 1975, Julesz (1962, 1975, 1981a,b) presented the theory of textons as the basic mathematical compound of texture, element of which allows classifying texture structures mathematically. A hundred years ago, Haar (1910) created wavelets, which were rediscovered by Gabor (1946), but were really reused by Daubechies (1988), a little bit more than 20 years ago, which initiated the use of wavelet to compress information related to images needed to be transmitted.

To transmit images was, and is still, a problem. Twenty years ago, we had computers known as XT, succeeded by AT, with an operating memory of 640 kb. A diskette could have a memory from 720 kb to 1.2 Mb. Today, we talk about hundreds of gigabytes. The quality of a picture is consistent with a speed of a computer. Therefore, we have now enormous possibilities to process images that were impossible then. A consequence of this is the advancement of the techniques to develop what is called graphic software and inverse graphic rendering. Both techniques are able to show objects or drawings (such as comics or animated films) in different types of illumination while the animated personages of the film move in the scene. This can be seen in *Toy Story* and *Monsters Inc.*, which provide an incredible touch of reality to the scenes.

The software programmers tried hard and achieved a great goal, but let me now try to understand what happened. They have used new capacities of the computers available today, processed the images, and reproduced them in a reasonable way, delighting people with these films. Yet do they explain how the human mind processes that information and how a human being processes such complicated and enormous quantity of data that normally are processed by people who could have no education, especially primitive inhabitants of the earth?

2.5 Some Numbers to End

To measure the reflectance of a sample in all directions in the semi-sphere surrounding it, as one can see in Figure 2.8, it needs time and patience.

To measure such quantity of data, Takagi et al. (2005) needed 16 days during which they measured 48,139 points. This could be reduced to 1,485 points, which can be measured in around 4 h if a great simplification

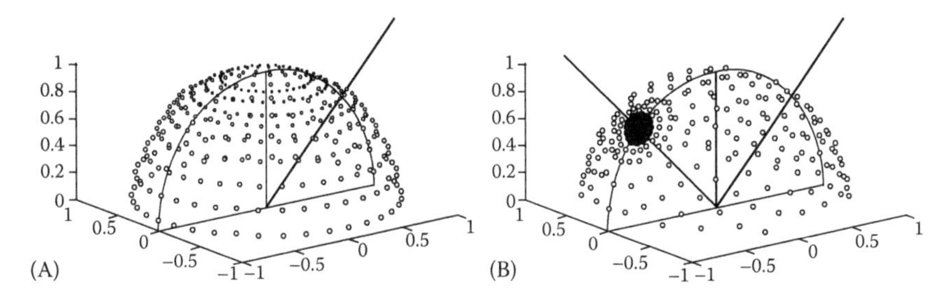

Figure 2.8 Measurement points with the requirements of the new techniques related to graphic software.

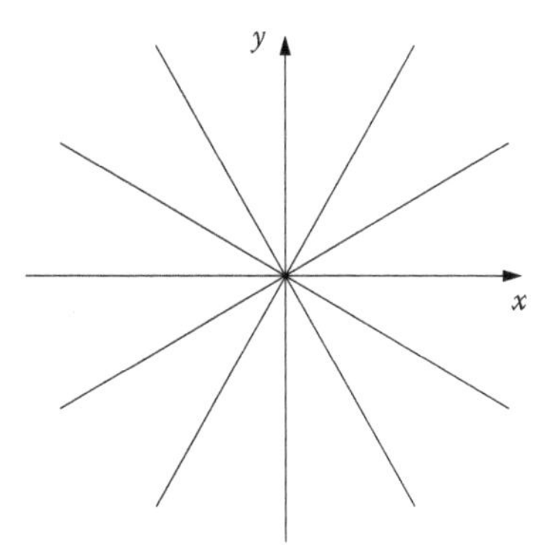

Figure 2.9 Detection angles for spatial frequencies in human vision.

is made. Nevertheless, each sample needs 4 h to characterize its spatial reflectance. These data are usually employed to evaluate automotive paint appearance. Once again, there is no answer to the following question: What is the correlation between the physical measurement and the psychophysical evaluation?

To finalize, I wish to mention that the study of textures allowed us to think that the human vision is capable to determine eight different spatial frequencies and six different angular positions with respect to the vertical or horizontal, as seen in Figure 2.9.

If it is accepted that to see colors the human being has three different cones in his retina, and to detect eight spatial frequencies and six angles of these frequencies with respect to the vertical or the horizontal, as shown in Figure 2.9, he would need additional detectors, then to evaluate a texture in color, he needs, at least, 144 simultaneous visual mechanisms to process the information. How is this achieved? As far as I know, there is no answer yet.

A final recommendation is as follows: Many manufacturers of color and appearance instruments claim about their marvelous properties. Be cautious; first we need to know how people make evaluation. An instrument can mean nothing if it is not really related to what people sec. If there is no psychological and psychophysical support of the scales used in the instrument, its validity is questionable.

References

Caivano, J. L. 1991. Cesia: A system of visual signs complementing color. *Color Research and Application* 16 (4): 258–267.

Caivano, J. L. 1993. Appearance (cesia): Variables, scales, solid. *Die Farbe* 39 (1–6): 115–125.

Caivano, J. L. 1994. Appearance (cesia): Construction of scales by mean of spinning disks. *Color Research and Application* 19 (5): 351–362.

Caivano, J. L. 1996. Cesia: Its relation to color in terms of the trichromatic theory. *Die Farbe* 42 (1–3): 51–63.

Caivano, J. L. 1997. Semiotics and cesia: Meanings of the spatial distribution of light. In *Colour and Psychology, Proceedings of the AIC Interim Meeting*, ed. L. Sivik. Stockholm, Sweden: Scandinavian Colour Institute, Colour Report F50, pp. 136–140.

Caivano, J. L. 1999. Evaluation of appearance by means of color and cesia: Visual estimation and comparison with atlas samples. In *Proceedings of the AIC Midterm Meeting*. Warsaw, Poland: Central Office of Measures, pp. 85–92.

Caivano, J. L. 2001. La investigación sobre los objetos visuales desde un punto de vista semiótico, con particular énfasis en los signos visuales producidos por la luz: color y cesía. *Cuadernos—FHYCS* (Univ. Nac. Jujuy, Argentina) 17: 85–99.

Caivano, J. L. and P. Doria. 1997. An atlas of cesia with physical samples. In *AIC 1997, Proceedings of the 7th Congress*, vol. 1. Kyoto, Japan: The Color Science Association of Japan, pp. 499–502.

CIE (Commission Internationale de l'Eclairage). 2006. Technical report: A framework for the measurement of visual appearance, CIE 175. Vienna, Austria: CIE Central Bureau.

Daubechies, I. 1988. Orthonormal basis of compactly supported wavelets. *Communications on Pure and Applied Mathematics* 41 (7): 909–996.

Gabor, D. 1946. Theory of communication. *Journal of the Institute for Electrical Engineering* 93: 429–457.

Haar, A. 1910. Zur Theorie der ortogonalen Funktionen-systeme. *Mathematische Annalen* 69: 331–371.

Hunter, R. S. and R. Harold. 1987. *The Measurement of Appearance*, 3rd edn. New York: Wiley.

Hutchings, J. B. 1999. *Food Color and Appearance*, 2nd edn. Gaithersburg, MD: Aspen.

Julesz, B. 1962. Visual pattern discrimination. *IRE Transactions on Information Theory*, IT-8: 84–92.

Julesz, B. 1975. Experiments in the visual perception of texture. *Scientific American* 232: 34–43.

Julesz, B. 1981a. Non-linear and cooperative processes in texture perception. In *Theoretical Approaches in Neurobiology*, eds. T. W. Werner and E. Reichardt. Cambridge, MA: MIT Press, pp. 93–108.

Julesz, B. 1981b. A theory of preattentive texture discrimination based on first-order statistics of textons. *Biological Cybernetics* 41: 131–138.

Lozano, R. D. 1978. *El color y su medición*. Buenos Aires, Argentina: Americalee.
Lozano, R. D. 2006. A new approach to appearance characterization. *Color Research and Application* 31 (3): 164–167.
Lozano, R. D. 2007. A new look into perception of appearance. In *Proceedings of the CIE Expert Symposium on Visual Appearance*. Paris, France: CIE, pp. 23–27.
Mandelbrot, B. 1967. How long is the coast of Britain? Statistical self-similarity and fractional dimension. *Science New Series* 156 (3775): 636–638.
McCamy, C. S. 1996. Observation and measurement of the appearance of metallic materials. Part I. Macro-appearance. *Color Research and Application* 21 (4): 292–303.
McCamy, C. S. 1998. Observation and measurement of the appearance of metallic materials. Part II. Micro-appearance. *Color Research and Application* 23 (6): 362–373.
Pentland, A. P. 1984. Fractal-based description of natural scenes. *IEEE Transactions on Pattern Analysis and Machine Intelligence*, PAMI-6: 661–674.
Takagi, A., A. Watanabe, and G. Baba. 2005. Prediction of spectral reflectance factor distribution of automotive paint finishes. *Color Research and Application* 30 (4): 275–282.

CHAPTER 3

Colors Seen through Transparent Objects

OSVALDO DA POS and CRISTINA RAO

Contents

3.1 Introduction

Both foods and drinks can be transparent or translucent (from now on, we do not differentiate the two terms), but not always they appear as such. The difference between transparency and translucency is rather controversial because it usually refers to physical properties of materials, like clear vs. diffuse light transmission. From a perceptual point of view, transparency is simply considered as the possibility of seeing something behind and through another object. The blurring of edges produced by translucent materials is a distinct effect from the contrast reduction that characterizes objects seen behind a transparent object. An impression of transparency can arise in simulations where contrast reduction occurs but margins are not blurred (Da Pos et al. 2007). The reason of the debate on the transparency/translucency terminology is that the definition of transparency is almost always framed in physical terms (transmittance or density). We need a psychological/perceptual definition which describes the visual appearance, for instance: "transparency is the property of an object through which we can see more or less distinctly what is behind." The degree of distinctness of the shape, color, texture, glossiness, and other features of the objects perceived in the background through a transparent object determines different kinds of transparency.

Research about transparency in the field of foods and drinks can be concentrated on two main characteristics, either on the transparent object or on the background seen behind and through it. We distinguish two main aspects, which can be studied in the field of perceptual transparency: first, the determination of the properties of the transparent object, like its chromatic characteristics (its color and its visual density), and secondly, the characteristics of the objects seen through the filter, like their colors (how they are modified by the transparent object) and other aspects of their surface (texture and glossiness). Metelli's model (1974) was essentially focused on the visual properties of the transparent object, while a number of other later studies (Masin 1998, D'Zmura et al. 2000, Ripamonti et al. 2004) were interested in explaining the color constancy, which, in stronger or weaker form, usually characterizes the objects seen behind and through the transparent object.

Most research on perceptual transparency shares the general concept proposed by Metelli (1974) that the color of the transparent object and the colors of the objects seen behind and through it arrive at the observer's eye in an additive mixture (Da Pos 1976), or in better formulation, in form of a partitive mixture (Brill 1976), while another view is maintained by Beck (1978) according to which the transparency phenomenon would be better described by a spectrally subtractive color mixture. Indications in favor of an additivity process derive also from the study by Gilchrist and Jacobsen (1983) who found substantial color constancy for colors seen through a veiling luminance. Moreover, Metelli's model is a convergent model according to the definition of D'Zmura et al. (1997), as in partitive mixtures, the uncovered backgrounds and the correspondent reduction colors converge in a unique color which characterizes the transparent filter (different models refer to different color space). It may happen that the convergence color is situated outside the color space, without greatly impairing the perception of transparency (Richards et al. 2009).

In normal vision, when perceptual transparency occurs, we see two colors, one in front and the other in the back in the area where the filter covers a specific region of the underlying Mondrian. But if we look at the same area through a hole made in a white opaque sheet, we see only one color, which is named reduction color because seen in a reduced context and can be described as a partitive mixture of radiations coming from the front filter and the back Mondrian. Some researchers (Metelli 1974, Anderson and Khang 2010) speak of a color scission when the reduction color, seen in the context, appears split in the two distinct colors, one belonging to the filter and the other to the background. The problem of the color adjustment procedure is that the observer can reproduce either the color of the Mondrian as she perceives it in the back or the reduction color, as a consequence of the focused attention in that area. There are different ways to increase the perceptive separation of the filter from the background, by time-to-time figural organization, depth, and movement. While D'Zmura et al. (2000) used to move backward and forward the transparent filter over a Mondrian, keeping the test area always completely covered, in this research, we used artificial stereovision to perceive the Mondrian and the filter in different depth planes. Both solutions have some limits. In the first case, the attention to the test area can easily induce a decontextualized perception and therefore facilitate the prominence of the reduction color. But, at the same time, it can ease the separation of the filter from the background as in many areas (around the test), one and the same color appears both covered and in plain view, indicating the effect of the filter. In the second case, the same reduction effect can be induced by individual difficulties in perceiving stereo depth, with the consequence that the two colors are not perceived, although all the test colors are perceived in a different plane from the filter and therefore well distinct.

As the present research deals with the perception of the colors seen through the transparent medium, our hypothesis can be formulated in terms of color constancy: "how much a surface color appears constant despite the modifications introduced in the proximal stimulus by a colored filter in front of it" (Foster and Nascimento 1994, Ripamonti et al. 2004). We expect that, contrary to what found by D'Zmura et al. (2000) and on the basis of previous works (Masin 1998), color constancy of surfaces completely covered by transparent objects is quite low.

3.2 Experiment

3.2.1 Material

To study this effect, we simulated in a calibrated 21″ Quato monitor a colored transparent veil completely covering a chromatic Mondrian (about 5 × 5 cm), about 10° at 30 cm of viewing distance, and protruding 0.5 cm from it over a white background (D65, 120 cd/m^2). The veil was perceived at a certain distance in front of the Mondrian (about 6 cm) by means of a simulated stereoscopic vision (two prismatic lenses were placed at 30 cm from the screen). The Mondrian consisted of nine nearly square regions of chromatic (B, G, Y, O, R, P) and achromatic colors (W, S, and A-gray) always presented in random order (Table 3.1, left). It appeared behind and covered by the filter as a result of the stratification in depth; the reduction colors (Katz 1935) were computed according to one of the most widely accepted physical models of phenomenal transparency (partitive mixtures: Metelli 1974, Da Pos 1989, D'Zmura et al. 2000), which can be considered a simple version of the Kubelka-Munk model

Table 3.1 CIELAB Specifications of the Colors of the Mondrian at Left and of the Filters at Right

Background	*L**	*a**	*b**	Filter	*L**	*a**	*b**
Red (R)	53.40	84.59	63.86	Y (60)	86.49	−1.00	66.86
Yellow (Y)	94.00	−10.27	90.98	R (60)	64.03	44.28	16.44
Blue (B)	46.85	−13.72	−37.46	B (60)	63.68	−17.38	-34.67
Purple (P)	37.32	41.83	−43.39	G (60)	68.85	−48.69	16.65
Green (G)	56.00	−51.69	25.64	Y (20)	92.29	−3.80	28.34
Orange (O)	63.06	61.40	69.62	R (20)	85.43	15.35	7.04
White (W)	100.00	0.00	0.00	B (20)	87.07	−8.29	-9.60
Gray (A)	53.98	0.00	0.00	G (20)	90.18	−16.31	7.63
Black (S)	0.00	0.00	0.00				

In brackets, color labels at left and the approximate NCS chromaticness at right.

(Brill 1976, Richards et al. 2009). The display was organized so that the adjustable Mondrian was perceived in front of a local background of the same color and size of the filter so as to keep the same contrast relationships with the background as the covered Mondrian, but seen in plain air. The filter could have one of the four unique hues (Y, R, B, G), at high or low saturation (20 and 60 NCS chromaticness) (Table 3.1, right) and at high or low density (0.50 and 0.1 achromatic transmittance).

3.2.2 *Method*

The task of four female observers was to reproduce in the adjustable Mondrian that was seen through the filter with its original colors (as if it were observed without the filter) the assumption being that they could perceive the original colors through the filter. A series of many small buttons displayed in the screen allowed easy changes in hue, lightness, chroma, and whiteness/blackness of each small squares forming the Mondrian. Observations were made in a dimly lit room (about 1 lx on the walls) after a suitable adaptation time. At the beginning, the observers had to get accustomed with the stereoscopic device and were asked where they perceived the veil to be sure that the depth perception was adequate to visually separate the veil from the Mondrian. Some training time was also assigned to the matching procedure. At the beginning of each trial, the modifiable Mondrian appeared with achromatic squares of different lightness, which had to be changed by the observers. Before closing each session, observers were invited to give an overall look both at the filtered and at the reproduced Mondrian to verify their global color correspondence and to adjust again the matches if necessary. There was no time limit to perform the task, and it could be interrupted as many times as the observers liked.

To check the ability of the observers in matching colors, a number of further control sessions were performed in which both Mondrians were uncovered while all other factors remained unchanged.

3.2.3 Results

All colors were transformed in the CAM02_UCS appearance color space to better deal with measures of color constancy (inconstancy index: Luo et al. 2006, Kim and Park 2010).

First of all, the performance of the four observers showed rather interesting differences. The observer 2 resulted significantly and also largely different from the other three (her median deviation $\Delta E' = 31.99$ compared with 16.68, 21.42, 13.14 of the other three observers, two tails Wilcoxon Z test $= \sim 8.2$ for the three comparisons, $p = 0.0000$) in that she reproduced with admirable care the "reduction" instead of the Mondrian colors (see an example in Figure 3.1).

This difference seems to depend either on the very poor stereoscopic vision of observer 2, not emerged in the initial training, or on her very strong selective attention which prevents spatial interactions between colors and consequently the appearance of complex phenomena like transparency, in which two colors are perceived at the same time in the same area, one in front and one on the back, along the same line of sight. All further statistical analyses have been therefore made on the data provided by three observers only.

On the basis of previous experiences, we expected that highly transparent filters would affect color constancy less than the more opaque ones. On the contrary, we found no significant difference between conditions in which transmittance was high (50%) and low (10%). Moreover, we also expected that more chromatic filters would distort more the back colors than little chromatic filters: results on the contrary indicate that slightly less deviations

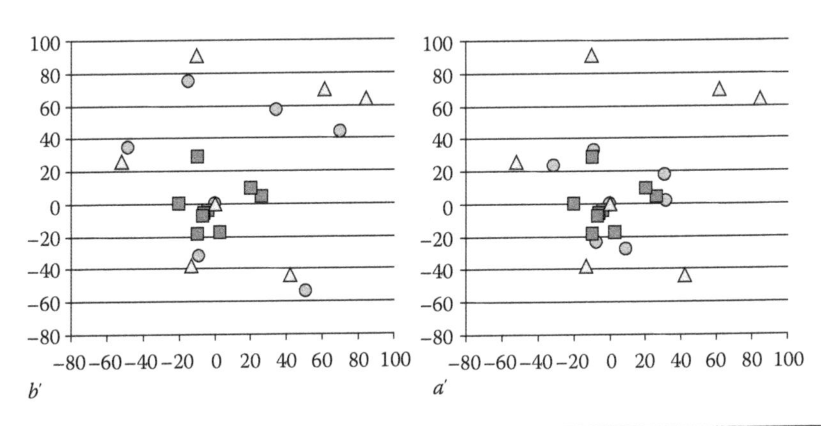

Figure 3.1 At left, the results plotted in a CAM02_UCS diagram of three observers when the filter was blue, little chromatic, and highly transparent. Squares: the chromatic colors, derived by calculations according to Metelli's model (1974) and presented in the screen; triangles: the colors of the Mondrian if seen in plain air; circles: the colors reproduced by the three observers (1, 3, 4). At right, the results of observer 2 in the same conditions. Perfect constancy is obtained if the reproduced colors are the same of the uncovered Mondrian. While rather good constancy is achieved by the three subjects at left, observer 2 carefully reproduces the reduction colors, failing to see the colors of the Mondrian.

(measured by $\Delta E'$) are derived when more chromatic filters are covering the Mondrians (median = 14.76 vs. 15.16). Although the difference is very small, it is significant (two tails Wilcoxon Z test = −8.07, p = 6.85E−16). Interesting results were obtained relative to the filter colors: increasing deviations from color constancy are produced by red, green, blue, and yellow filters in the order (median-R = 17.9, median-G = 18.7, median-B = 20, median-Y = 21.4; yellow is more deviating than the others, two-tail Wilcoxon Z test = −2.8, p = 0.005 in comparison with the blue filter; Z test = −2.9, p = 0.003 in comparison with the green filter; Z test = −2.6, p = 0.008 in comparison with the red filter). Also the colors of the background are more or less susceptible of being modified by the filter. Warm colors (Y, O, R) seem to be more constant than cold ones (B, G, P), irrespective of the kind of filters which cover them (median-Y = 21.9, median-O = 27.3, median-R = 32.3, median-B = 40.0, median-G = 40.1, median-P = 49.7). In the control trials, deviations were rather small: red, yellow, and blue colors are reproduced with small deviations (CAM02_UCS $\Delta E'$ = 3.1, $\Delta E'$ = 3.5, and $\Delta E'$ = 3.8, respectively) while slightly higher deviations are found in the case of orange, purple, and green (CAM02_UCS $\Delta E'$ = 7.1, $\Delta E'$ = 8.1, and $\Delta E'$ = 10.3, respectively). Lastly, lightness matching has been always very accurate.

3.3 Discussion and Conclusions

The first result concerns the rather remarkable individual differences found between our observers. As noted by Richards et al. (2009), observers may show significant individual differentiations in regard to their attitude, strategy, attention, and sensitivity, which can lead to unexpected experimental outcomes. Our observer number two was amazing in matching the reduction instead of the Mondrian colors, and therefore, it seems that she could not perceive any form of transparency during the adjustment procedure. We cannot decide whether her peculiar results derive from a weakness in perceiving stereo depth which would impair the separation of the Mondrian from the filter or from a particularly selective attention to the test area which could isolate it from the transparency context: in both cases, she could only match the reduction colors. This effect seems particularly linked to matching procedures which require very analytical attention to the area to be reproduced (Masin and Quarta 1984, Kingdom 2011). For this reason, color matching, although widely used in studies on transparency, seems not to be always suitable in this kind of research. We tried to prevent errors due to analytical observation by asking the observer to give an overall look both at the filtered and at the reproduced Mondrian to check their global color correspondence. This precaution, if actually taken by the observer, would justify the hypothesis that her stereoscopic vision, although sufficient to distinguish the two depth planes in the initial phase of the experiment, was actually weak during the process of color matching.

The second result regards the interaction between filter and Mondrian colors. On the one side, the lightness of colors was quite correctly preserved, and this result is in agreement with many previous researches (Da Pos 1976) which show that the achromatic variables are more relevant than the chromatic ones, at least for most observers (Richards et al. 2009). Nevertheless, the matched colors seem to be affected in various ways by the colors of both the Mondrian and the filter colors. An important interaction concerns the case where a specific Mondrian area and the filter have opponent or complementary colors: in these cases, the matched color is usually quite far from the actual color of the Mondrian. This result was already predicted by Hering (1920) who implicitly did not admit the possibility that a single color stimulation might originate the perception of two different and separate colors in the same area (the same direction of sight) unless they were two shades of a binary hue. For this reason, it should be impossible to see either opponent (that is incompatible) or complementary colors in the overlapping area. This point has relevant implications in a general color theory, as it seems that some cases of simultaneous presence of opponent colors in the same observed area actually occur, although very seldom, and probably at different perceptual depth (Mingolla, personal communication).

A third result relates to the unexpected negligibility of the density of the filter in distorting the colors beneath: lower transparency means that the background is much less visible in favor of the transparent medium, and therefore also its color would be easily distorted. But results showed that this was not the case: it seems that the color quality of the background seen through the transparent medium is preserved even though its presence in the overlapping area is very reduced. This result seems important in the light of a rather accepted theory that white cannot be transparent (Wittgenstein 1977). Note that for the same reason, also white backgrounds cannot be perceived through any translucent veil. The reason would be that an object reflecting light in the highest degree cannot, at the same time, transmit light from the background, as both colors of the object and of the background would mix in the overlapping area producing a color lighter than white. This theory found important criticisms (Westphal 1986), especially because it does not explain the common experience, but negates it. Both Metelli's model and common experience show that the perception of a color does not depend on how much stimulus is arriving at the eye, as a surface appears of the same color independently of the illumination level (at least inside a rather wide interval). The proportion of the two stimuli in the partitive mixture affects the variable "transmissibility" (Petrini and Logvinenko 2006) which correlates with the perceived extent of transparency, but does not affect the "reflectance" which correlates with perceived colors. Our results are therefore fully justified in the light of these reasons, although at first sight, they might appear contrary to commonsense.

One final comment relates to the type of filter we dealt with, that is, a simulated veil. The main characteristic of this filter is its nonselective transmission that means it can partially transmit radiation from the background by decreasing in intensity all wavelength by the same degree. Therefore, the spectral composition of the background radiation is not modified but in intensity, and this can be one reason for appearing the same color independently from the degree of transparency. Which objects behave like transparent veils? Those that are so thin that their material behaves like being holed (Da Pos 1989): for instance, slices of certain food are cut so thin to appear transparent. This definition of filter does not apply to many liquids, because most often, liquids behave like spectrally selective filters, and studying selective transparency appears particularly difficult as it involves subtractive mixtures, difficult to compute.

This work started from the assumption that perception of transparency can be studied as a case of color constancy: as resulted from the experiment, the variables which interfere with color constancy are not only relative to the structure of visual stimuli but also to the individual characteristics of the observers, as normally happens in all fields of human perception. This particular role of the observer has then to be further studied in future works, although its difficulty cannot be easily ignored (Richards et al. 2009).

Acknowledgments

We thank S. C. Masin for his precious advice. Research performed with the MURST grant PRIN 2007.

References

Anderson, B. L. and B. G. Khang. 2010. The role of scission in the perception of colour and opacity. *Journal of Vision* 10 (5): 1–16.

Beck, J. 1978. Additive and subtractive color mixture in color transparency. *Perception & Psychophysics* 23: 256–267.

Brill, M. H. 1976. Physical foundation of the perception of achromatic translucency. MIT Research Laboratory of Electronic Progress Reports No. 117, January, pp. 315–320.

Da Pos, O. 1976. Perceptual transparency: Additive or subtractive chromatic mixture? *Galilean Academy of Science, Letters and Art in Padua* (formerly *Accademia Patavina di Scienze Lettere ed Arti*) 88: 185–193.

Da Pos, O. 1989. *Trasparenze—Transparency*. Milan, Italy: Icone.

Da Pos, O., A. Devigili, F. Giaggio, and G. Trevisan. 2007. Color contrast and stratification of transparent figures. *Japanese Psychological Research* 49 (1): 68–78.

D'Zmura, M., P. Colantoni, K. Knoblauch, and B. Laget. 1997. Color transparency. *Perception* 26: 471–492.

D'Zmura, M., O. Rinner, and K. R. Gegenfurtner. 2000. The colors seen behind transparent filters. *Perception* 29 (8): 911–926.

Foster, D. H. and S. M. C. Nascimento. 1994. Relational colour constancy from invariant cone-excitation ratios. *Proceedings of the Royal Society, London, B*, 257: 115–121.

Gilchrist, A. and A. Jacobsen. 1983. Lightness constancy through a veiling luminance. *Journal of Experimental Psychology: Human Perception and Performance* 9 (6): 936–944.
Hering, E. 1920. *Grundzüge der Lehre vom Lichtsinn*. Berlin, Germany: Springer. English translation by L. M. Hurvich and D. Jameson, *Outlines of a Theory of the Light Sense*. Cambridge, MA: Harvard University Press, 1964.
Katz, D. 1935. *The World of Colour*. London, U.K.: Routledge.
Kim, Y. J. and S. Park. 2010. CIECAM02-UCS based evaluation of colorimetric characterization modeling for a liquid crystal display using a digital still camera. *Optical Review* 17 (3): 152–158.
Kingdom, F. A. A. 2011. Lightness, brightness and transparency: A quarter century of new ideas, captivating demonstrations and unrelenting controversy. *Vision Research* 51: 652–673.
Luo, M. R., G. Cui, and C. Li. 2006. Uniform colour spaces based on CIECAM02 colour appearance model. *Color Research and Application* 31 (4): 320–330.
Masin, S. C. 1998. The luminance conditions of Fuchs's transparency in two-dimensional patterns. *Perception* 27: 851–859.
Masin, S. C. and A. Quarta. 1984. Experimental demonstration that observers produce unbiased estimates of reduction lightness in transparent surfaces. *Bulletin of the Psychonomic Society* 22: 529–530.
Metelli, F. 1974. The perception of transparency. *Scientific American* 230 (4): 91–98.
Petrini, K. and A. D. Logvinenko. 2006. Multidimensional scaling (MDS) analysis of achromatic transparency. *Journal of Vision* 6 (6): article 394.
Richards, W., J. J. Koenderink, and A. van Doorn. 2009. Transparency and imaginary colors. *Journal of the Optical Society of America* 26 (5): 1119–1128.
Ripamonti, C., S. Westland, and O. Da Pos. 2004. Conditions for perceptual transparency. *Journal of Electronic Imaging* 13 (1): 29–35.
Westphal, J. 1986. White. *Mind*, New Series 95 (379): 311–328.
Wittgenstein, L. 1977. *Remarks on Colour*. Oxford, U.K.: Basil Blackwell.

CHAPTER 4

What Is the Color of a Glass of Wine?

IVAR JUNG, PÄIVI JOKELA, PATRIK BRANDT, and OLE VICTOR

Contents

4.1 Introduction

The area of color research is characterized by its multidisciplinary approach, including phenomenological, psychological, and physiological aspects of color perception (Billger 2000, Fridell Anter 2000, Hård and Sivik 2001); physical measurements of light spectra; and, most recently, also the color appearance in virtual reality (Stahre 2009). The main focus in perceptual color research has been on the surface colors, and the existing color systems, such as Natural Color System (NCS) and Munsell, are established using opaque colors (Hård et al. 1996a,b, Kuehni 2000, Nayatani 2004). When it comes to transparent materials, e.g., glass, plastics, and liquids, there is no available color atlas based on perception (Gladushko and Chesnokov 2007). Color sample collections Pantone and RAL (Pantone 2011, RAL Colours 2011) have a set of transparent plastic samples with specified color codes designated to these samples, but these samples are not organized systematically in a color space like NCS.

The lack of standards for measuring and representing the color perception in glass is a problem in the color generation process in glass industry. Consequently, it is often difficult for a glass designer to communicate the mental notion of the color so that the right chemical composition is used and the right conditions are maintained during the manufacturing process. In general, the industrial color generation process in glass artifacts is based on trial and error, which makes the procedure tedious and costly (Weyl 1990, Bamford 1997, Gladushko and Chesnokov 2007).

In this pilot study, the authors investigate the overall research question on how the color of transparent materials is perceived and matched with samples in visual color systems; the main focus is on the NCS and its potential to represent color in glass artifacts. The research group is multidisciplinary, and it comprises areas such as design, interaction design, and computer science as well as experts in the glass production processes. One of the partners in the project is Glass Research Institute, Glafo (Växjö, Sweden), and this partner has a unique collection of circa 3000 colored glass samples, which have been gathered during the last 60 years. The collection is frequently used by glass designers, but it does not cover the entire color range that can be produced in glass.

4.2 Research Questions Addressed

The overall research question is as follows: Is it possible to use the same perceptual representations (especially NCS) for transparent materials that are used for surface colors, and if this is possible, in which way should the existing color space be modified or extended to be able to incorporate transparent colors? Another research perspective is to study how the color appearance of transparent objects can be modeled in the virtual reality and how virtual models can be used to facilitate the communication in industrial color generation processes.

In the current study, the perceived color of glass objects was determined by trained observers in a standardized setting. The perceived color is expressed using NCS notation: hue (Φ) and nuance, i.e., the relationship between blackness (s), chromaticness (c), and whiteness (w). Similar method has been used by Fridell Anter (2000) in order to study the perceived color of painted house facades in different viewing situations. However, it is important to keep in mind that it is impossible to visually determine the "true" color of a transparent sample as the human color perception is always subjective. What is more, the perception also depends on external viewing conditions such as the type and position of the light source, viewing distance and angle, the thickness and shape of the sample, as well as surrounding colors.

4.3 Pilot Study

In the pilot study, a total of 420 color observations were conducted by 20 different observers. The observers were 18 students and lecturers from the School of Design (Linnaeus University, Sweden) and two glass experts from

Glafo. All the observers had prior experience of color matching using NCS atlas. The studied samples were seven glass sheets with the same dimensions ($5 \times 10 \times 4$ mm) and different colors; these samples will be referred to as samples A to G. The glass samples were matched with the NCS colors in three different ways, where the sample was placed:

1. On a metal frame 5 cm above the white paper, and the color was matched with the color samples in the NCS atlas
2. Directly on the computer screen, and the color was matched with screen images of the NCS colors
3. On a white paper, and the color was matched with the color samples in the NCS atlas

During all the observations, the light source, viewing distance, and angle as well as other significant conditions were according to Swedish Standard, SS 019104, color specifications with NCS. The experimental setting for matching techniques where the NCS atlas was used is shown in Figure 4.1; the color matching with screen images is shown in Figure 4.2.

The observers conducted the three different matching procedures for each glass sample and indicated in a questionnaire sheet which NCS color was the best match in each case. It was possible to choose two alternative colors if the observer did not find the perfect match in the NCS atlas or screen image. For each sample and for each observation technique, an arithmetic mean of hue (Φ) and nuance (s, c, and w) were calculated using the results from all 20 observers. If the observer indicated alternative colors, all the values were included in the mean value. The first matching technique is used as the reference value, and the results are reported as the differences ($\Delta\Phi$, Δs, Δc, and Δw) between the mean values of the reference color and the two other perception methods, respectively. As discussed earlier, there is no "true" perception of the color; therefore, the reference color is chosen in order to illustrate the

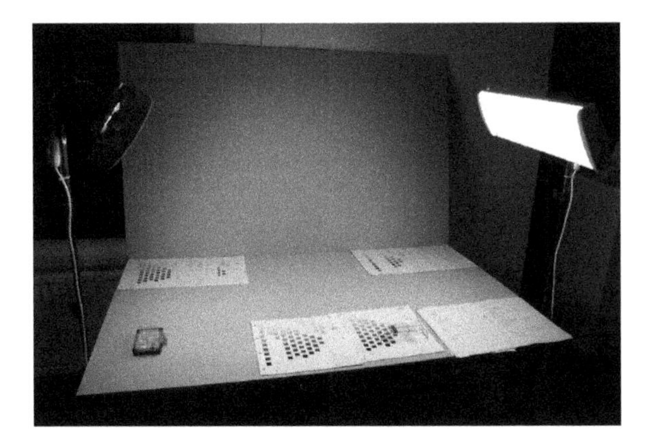

Figure 4.1 The experimental setting for color matching using NCS atlas.

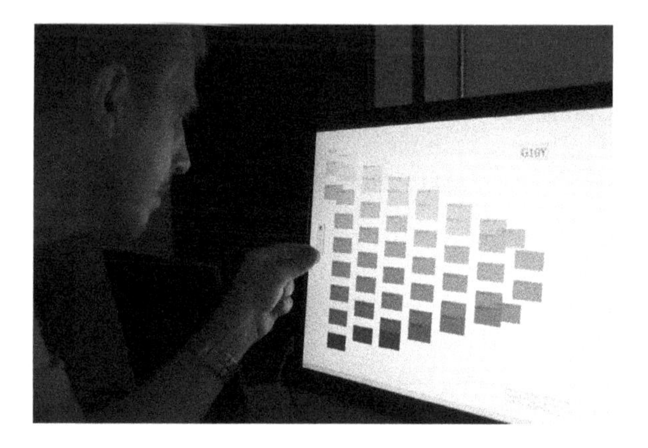

Figure 4.2 Color matching using screen images of NCS colors.

differences between the three observation techniques, but the choice per se is an arbitrary one.

4.4 Results and Discussion

The differences in nuance and hue between the mean values of the reference color and the two other perception methods are summarized in Tables 4.1 and 4.2. In these tables, 1 denotes the mean value when the sample is observed 5 cm above the white paper, 2 is the mean value when the sample is placed directly on the computer screen, and 3 is the mean value when the sample is placed directly on the white paper. The differences in nuance are calculated as follows:

$$\Delta s_2 = s_2 - s_1 \quad \text{and} \quad \Delta s_3 = s_3 - s_1 \tag{4.1}$$

$$\Delta c_2 = c_2 - c_1 \quad \text{and} \quad \Delta c_3 = c_3 - c_1 \tag{4.2}$$

$$\Delta w_2 = w_2 - w_1 \quad \text{and} \quad \Delta w_3 = w_3 - w_1 \tag{4.3}$$

Table 4.1 Summary of Differences in Nuance between the Three Observation Techniques

	s_1	Δs_2	Δs_3	c_1	Δc_2	Δc_3	w_1	Δw_2	Δw_3
A	34.00	−9.10	17.9	11.65	−3.50	4.55	54.40	12.6	−22.5
B	21.60	14.6	4.90	60.10	−14.0	3.80	18.55	−0.90	−8.75
C	5.50	2.00	0.40	27.25	−5.95	11.2	67.25	4.20	−11.5
D	19.05	−5.05	5.70	55.70	−0.95	4.50	25.70	5.65	−10.4
E	9.85	−2.95	−0.15	7.45	−1.15	4.30	83.00	3.90	−4.35
F	9.75	0.45	0.05	43.85	−2.20	26.6	46.55	1.75	−26.6
G	10.00	−0.70	0.70	17.10	−1.10	8.60	72.75	2.00	−9.15

Table 4.2 Summary of Differences in Hue between the Three Observation Techniques

	Quadrant	Φ_1	$\Delta\Phi_2$	$\Delta\Phi_3$	$\Delta\Phi_{c2}$	$\Delta\Phi_{c3}$
A	Yellow-red	Y(21.00)R	−7.58	2.67	−0.88	0.31
B	Red-blue	R(79.00)B	−1.25	0.08	−0.75	0.05
C	Blue-green	B(6.25)G	7.25	3.75	1.98	1.02
D	Blue-green	B(10.00)G	−0.58	−3.33	−0.32	−1.85
E	Blue-green	B(65.17)G	−20.9	14.1	−1.56	1.05
F	Green-yellow	G(29.50)Y	−7.5	−2.00	−3.29	−0.88
G	Green-yellow	G(62.00)Y	−16.5	1.75	−2.82	0.30

The differences are summarized in Table 4.1 and also illustrated in the NCS triangle as shown in Figure 4.3.

The differences in hue are calculated in the same manner:

$$\Delta\Phi_2 = \Phi_2 - \Phi_1 \quad \text{and} \quad \Delta\Phi_3 = \Phi_3 - \Phi_1 \tag{4.4}$$

Another way to estimate the difference in hue values also includes the chromaticness (the reference value) as follows:

$$\Delta\Phi_{c_2} = \frac{\Phi_2 - \Phi_1}{100} \times c_1 \quad \text{and} \quad \Delta\Phi_{c_3} = \frac{\Phi_3 - \Phi_1}{100} \times c_1 \tag{4.5}$$

The differences and the position of samples in the NCS circle are shown in Table 4.2.

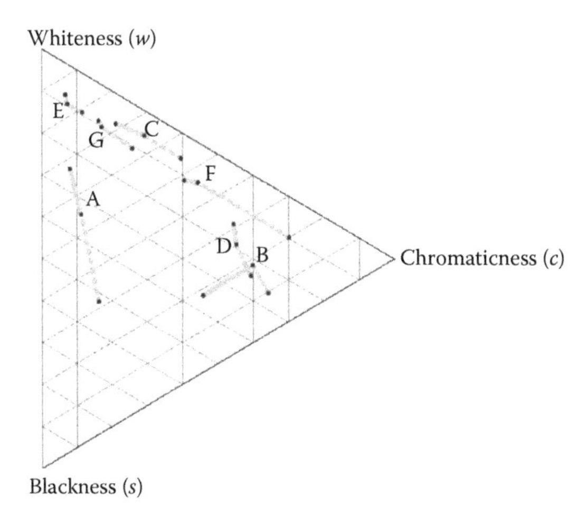

Figure 4.3 The sample coordinates are plotted in the NCS triangle. The midpoint represents matching technique 1 (5 cm above paper), the point at the end of the solid line is technique 2 (computer screen), and the point at the end of the dotted line is technique 3 (on the paper).

For all the samples, from A to G, the variation in nuance follows a similar pattern: samples observed 5 cm above the paper and on the computer screen are relatively similar, and they both show lower chromaticness as well as higher whiteness compared to the observation made directly on a white paper. Sample A and F have the largest overall variations, whereas E has the smallest variation. Comparing the observations made when sample was placed directly on a white paper with the reference technique, the following pattern can be used to describe variation in nuance:

- Samples C, F, and G, which all have low blackness values, show greater whiteness and lower chromaticness for the reference method. Sample E, which has the highest whiteness value, varies only slightly but could be classified in this pattern.
- Samples A and D, which have higher blackness values, show greater whiteness and lower blackness for the reference method.

Generally, the difference in hue values is greater when virtual images are matched with glass samples than when the NCS atlas is used. What is more, a general observation is that it seems to be more difficult to find the matching hue values when the sample has a low chromaticness value; e.g., the overall distribution of observed hue values is larger for the sample G than for sample B. Consequently, when $\Delta\Phi$ values are weighed against chromaticness, according to Equation 4.5, the obtained values show much less variation as the colors with the lower chromaticness are allowed to vary more in hue values.

4.5 Conclusion

The main conclusion from the pilot study is that it seems to be possible to use NCS representation also for transparent materials. However, the standard viewing conditions must be adapted to take account of the light that is transmitted through the object, which is illustrated by the difference in whiteness between the reference color and the color observed when the sample lies directly on a white paper. What is more, for transparent materials, the lightness value depends on the thickness of the sample. It is also indicated that the virtual images match the nuance values of reference method quite well, but further studies are needed to enhance the color appearance of virtual hue values.

Acknowledgments

This coproduction project is supported by an external grant (No. 2009/0217) from the Knowledge Foundation, Sweden. The project is conducted in cooperation with Glafo and Orrefors Kosta Boda AB.

References

Bamford, D. R. 1997. *Colour Generation and Control in Glass*. Amsterdam, the Netherlands: Elsevier.

Billger, M. 2000. Evaluation of a colour reference box as an aid for identification of colour appearance in rooms. *Color Research and Application* 25 (3): 214–225.

Fridell Anter, K. 2000. What colour is the red house? Perceived colour of painted facades. Stockholm, Sweden: Royal Institute of Technology, PhD dissertation.

Gladushko, O. A. and A. G. Chesnokov. 2007. Analysis of the color of glass products. *Glass and Ceramics* 64 (11/12): 24–27.

Hård, A. and L. Sivik. 2001. A theory of colors in combination—A descriptive model related to the NCS color-order system. *Color Research and Application* 26 (1): 5–28.

Hård, A., L. Sivik, and G. Tonnquist. 1996a. NCS, Natural Color System—From concept to research and applications, part I. *Color Research and Application* 21 (3): 180–205.

Hård, A., L. Sivik, and G. Tonnquist. 1996b. NCS, Natural Color System—From concept to research and applications, part II. *Color Research and Application* 21 (3): 206–220.

Kuehni, R. G. 2000. A comparison of five color order systems. *Color Research and Application* 25 (2): 123–131.

Nayatani, Y. 2004. Why two kinds of color order systems are necessary? *Color Research and Application* 30 (4): 295–303.

Pantone. 2011. *Pantone Plastics*. http://www.pantone.com. Accessed November 24, 2011.

RAL Colours. 2011. *RAL Plastics*. http://www.ral-farben.de. Accessed November 24, 2011.

Stahre, B. 2009. Defining reality in virtual reality. Exploring visual appearance and spatial experience focusing on colour. Gothenburg, Sweden: Chalmers University of Technology, PhD dissertation.

Weyl, W. A. 1990. *Coloured Glasses*. Sheffield, U.K.: Society of Glass Technology.

CHAPTER 5

Color as a Code in Food Packaging

An Argentine Case

MARÍA LUISA MUSSO

Contents

5.1 Introduction

Intellectual and emotional aspects of the product's image raise the concerns about its qualities, concerns about nutrient content, the ingredients, the amount of sugar, salt, and fat. The product appearance induces expectations. "Expectations govern our attitude to food and the food scene. We deduce from the appearance of the food in front of us whether it will harm us or be good for us," says Hutchings (2003). The consumer searches for the attributes he or she considers most suitable according to his or her internal needs, in the products he or she wants to buy. It is necessary to define a target segment to propose the product that is effectively closer to the ideal of the buyer and to communicate its benefits.

This is why the packaging has a big responsibility. Emotions, memory, and social patterns are behavioral areas in which color plays an important part. There is a message to remember, and the packaging is a message in itself;

it helps to guide, motivate, and encourage consumers in their purchase decision. It is imperative to choose the potential customer for the message; each target has its language, its expectations.

5.2 Segmentation

It is significant to consider two types of segmentation: functional segmentation and psychological one. In the functional segmentation, the idea is to group consumers according to the functional advantages the consumer is looking for. When products are addressed to health conscious consumers, the packaging, associated to the product concept, can produce a functional segmentation. Packaging has a decisive influence on the consumers' perception and, therefore, in the purchasing decision.

Psychological segmentation is based on the characteristics of the social class of the consumers, their lifestyle, reference models, and their personality, involved in the emotional satisfaction they obtain in the purchase. The psychological differential advantages are often more sustainable than the functional. Packaging must show specific signs and build confidence as much as unambiguous product identity.

5.3 Changes in Consumer

Consumers, increasingly informed and demanding, have taken the lead. They show new desires, seeking harmony between quality and wellness. In the twenty-first century, the will to live in a more human and less frivolous world emerged. Banality is being outdated. The importance of healthy living, honesty, and revaluation of emotions are values that flourished and increasingly permeate many aspects of life. Packaging design is one of them. With information and consciousness accompanying the suitable healthier urban living, many products are incorporating functional values as a vital part of their communication systems. Nutritional and functional information is relevant when choosing a brand or a particular variety.

5.4 Visual Identification

Colors are created in our brain as a perceptual tool to facilitate our visual–cognitive and visual–emotional functions. Colors are more than a physical process: they work as a sign system, a source of information decoding the world around us (Caivano 1998). In this world, the products we buy every day are present. The consumers develop their opinion about the products they see in less than 90 s from their first interaction with them. Between 62% and 90% of that assessment is based on the color of the product (Institute for Color Research, Color Communications, Inc., Chicago, USA).

The communicative properties of a color can be defined by two categories: natural associations and psychological or cultural associations.

Research conducted by the secretariat of the Seoul International Color Expo 2004 documented the following relationships between color and marketing: 92.6% said that they put most importance on visual factors when purchasing products. Only 5.6% said that the physical feel via the sense of touch was most important. Hearing and smell each drew 0.9%.

Vision is the primary source for all our experiences. Current marketing research has reported that approximately 80% of what we assimilate through the senses is visual. Color addresses one of our basic neurological needs for stimulation.

A complex semiotic process enables the understanding of product differentiation on the market. The impact of color on the decisions about what product to buy is due to the fact that it is a symbol that reflects the image we have of ourselves, our personality.

5.5 Color of the Product

Color is an essential element used as a sign to represent desirable product attributes. Consumers respond to the "total product" that also includes their image. Successful design requires an awareness of how colors communicate meaning. Color can provide information about the quality of a product and can also show a strong association with certain product categories. Green, for instance, is associated with natural products (Figure 5.1).

Figure 5.1 Argentina: green = natural.

The marketing function of the package includes the location, because the consumer must be able to identify the type of product (dairy desserts, breakfast cereals, detergents, etc.) from a distance, and also the identification, because once the product has been located, the consumer must clearly identify the products they really want to buy within a family or brand. The packaging must show specific signs, the information that helps build confidence in the product and strengthens consumer's purchasing decisions.

The graphics and colors used in the package must be consistent with the status, image, or expectations it wants to satisfy and must serve to identify and locate the product. Color improves readership. Color can be used as a referent code system for the product. Color coding helps to clearly identify the desired product.

In package design, some actions apply to the expected typology change in order to produce a strong identification with the brand. We also find the opposite strategy, which is to favor the association with the category identity. Breaking the category code can be a key to differentiate a new product. In spite of the fact that some brands traditionally use green as a strong identifier (Figure 5.2), green color is used in association with the green countryside and healthy products in most countries. Green is related to nature, freshness,

Figure 5.2 Green as brand identifier.

Figure 5.3 Germany: green = bio.

fertility, peace, hope, humidity, regeneration, growth, and relaxation and is calming, curative, and balsamic, in its positive meanings (Tornquist 1999). In many countries, green is used to identify bio products (Figure 5.3). In some others, green is also a visual attribute related to low fat, so are pink and light blue (Figure 5.4).

5.6 Diet or Light Products

Important changes in consumption values caused a typological substitution in the colors expected for certain products. The irruption of diet and light products and the explosive growth in value reached by the concept "low calorie" produced an unexpected change in the color paradigm.

The first experiences in color for diet products focused on white, silver, and pastel colors as pink. Finally, green, associated with nature, became a strong identifier for this type of products. Then, color is not talking about product attributes but on their feature of being "light" instead.

Color meanings can also have a regional value, given by a mixture of cultural interpretations associated with some colors and their historical use (Gage 1993).

(a)

(b)

Figure 5.4 (a) Chile: light blue = low fat and green = no fat; (b) Italy: pink = low fat and light blue = low fat.

In Spain, for instance, light products began to appear in the 1980s proposed as healthy products. Begoña Hernández Salueña, from the Department of Physics of the Public University of Navarra, after consulting several dairy companies, says that there is no official code for the colors of the milk pack. The use of color by type of milk (blue for the whole, green for semi-skimmed and skimmed) has to do with the organization in the supermarket shelves and with the communication for consumers. These colors appear having been selected by the first brand that sold these products, followed by the other brands. But several brands have recently decided to break that tradition and begun using colors more for identifying the brand than the category. Central Dairy Asturiana, for example, has decided to use red for whole milk (instead of blue), blue for the semi-skimmed, and green for the skimmed. Pascual uses dark blue for whole milk and light blue and pink for semi-skimmed and skimmed.

5.7 Light Products in Argentina

It is interesting to see how the green color has definitely been adopted as a category code in Argentina, especially in dairy products. The low-fat dairy-product area at the supermarket is easily recognizable from far away in a green spot.

Gonzalo Petracchi, packaging designer for Sancor, says that the green code emerged in the Argentine market around the 1990s, when the changes from diet products to light ones came out in order to clarify what was being offered to consumers. Diet products sought a cleaner and pure image, in association with reduced-calorie diets choosing a color as blue/cyan (Figure 5.5). At the

Figure 5.5 Argentina: blue = diet products.

beginning of the change, diet and light were virtually synonymous, but light category products wanted to find their own individuality in a codified meaning, an identity charged with emotional values as care and health, without giving up flavor (Figure 5.6).

The official Argentine food code and the one of Mercosur (2003) include the requirements for food labeling, in order to give the information that builds confidence, but no rule appears mentioning the use of color as an identifier (Código Alimentario Argentino 2007).

In the Argentine food code, for instance, food labeling included in chapter XVII—food or dietary regime—specifies the words to use, but not the color: food with low lipid content will be labeled with the name of the product and the indication "diet, reduced lipid value" or "diet of low fat" and also may bear the legends "reduced calorie or low calorie."

In spite that in most cases the color code for whole-fat products is blue, red is also used in Argentina and in other countries (Figure 5.7).

Green is definitely the color code for low fat and fat free in Argentina. Some brands are recently incorporated two greens, a light green and a darker one, to show a different identification for "no fat" and "low fat."

(a)

(b)

Figure 5.6 (a) Argentina: green = low fat zone in supermarkets; (b) Argentina: green = low or zero fat: Ser, Las Tres Niñas and other brands.

Figure 5.7 Argentina: green = no fat, red = full fat, red and blue = full fat.

There are a few exceptions. Nestlé uses light blue for 0% fat powder milk, and Sancor have introduced recently pink to identify 0% fat milk.

Green is the color of security. It is also the color of permission. Green packaging assures us to eat healthy, preventing us from getting fat, with safety, confidence, and certainty. Big companies of massive consumer products invested heavily in communication to encourage the establishing of an expressive symbolic code. In the last 2 years, this code has widely spread to other categories in Argentina (Figure 5.8).

Celebrating the power of color as a code in food packaging, I would like to remember what Charles Riley wrote in his book *Color Codes*: "completely mastering color is impossible, but the power it imparts to those who dare to handle it is as profound as that of light itself." He says also, "color is a third Promethean gift, like language and fire" (Riley 1995).

P.S. Though samples from a few countries are shown, research has also been conducted in Brazil, Paraguay, Peru, Mexico, United States, England, Spain, Sweden, France, Australia, Tahiti, and New Zealand, among others.

Figure 5.8 (a) Argentina: cereal bar and crackers, low fat = green; (b) Argentina: special bread and diet bread, low fat = green; (c) Argentina: sweet jelly, no fat = green; and (d) low fat = green, pizza and pie dough.

References

Caivano, J. L. 1998. Color and semiotics: A two-way street. *Color Research and Application* 23 (6): 390–401.
Código Alimentario Argentino. 2007. *Ley 18.284 18/07/1969. Decreto 2126/1971. Alimentos de régimen o dietéticos*, updated April 2007.
Gage, J. 1993. *Colour and Culture*. London, U.K.: Thames and Hudson.
Hutchings, J. B. 2003. *Expectations and the Food Industry*. New York: Kluwer Academic, Plenum Publishers.
Mercosur. 2003. *Reglamento técnico Mercosur para rotulación de alimentos envasados GMC, Resolución 26/03*.
Riley II, Ch. A. 1995. *Color Codes*. Hanover, NH: University Press of New England.
Tornquist, J. 1999. *Colore e luce*. Milan, Italy: Istituto del Colore.

CHAPTER 6

Effect of Illuminance and Correlated Color Temperature on Visibility of Food Color in Making Meals

SHINO OKUDA, YOKO FUKUMOTO, NAOYA HARA, HIROSHI IWADE, and WATARU IWAI

Contents

6.1 Introduction

The color appearance along with visual texture and flavor is one of the major factors affecting food eating quality. The degree of pigmentation and the extent of light scatter are very important in distinguishing the degree of freshness (MacDougall and Sansom 2007). In the case of food preparation, color appearance, glaze, and asperity are important in checking the freshness of the food and in detecting perished, spoiled, or damaged portions.

In cooking, meat color keeps changing with heat, so the color appearance of meat is significant in understanding the degree to which it has been cooked.

In previous studies (McGuiness et al. 1983), subjective evaluation under the condition of an incandescent lamp was carried out for four tasks involved in food preparation: finding details of a recipe, weighing packages, slicing cucumbers, and inspecting utensils. In another study on the lighting conditions of a kitchen (Fukumoto et al. 2009a,b), the task performance in cutting, peeling, and cooking was studied under different light sources, illuminances, light colors, and light directions. In the general rules of recommended lighting levels issued within the JIS Z-9110 (JIS 2010), the value of the horizontal illuminance for task lighting on the worktop in the kitchen is 300 lx, Ra = 80.

This chapter aims to determine the lighting conditions in food preparation and cooking. In this chapter, we conducted subjective evaluation on the visibility of two color charts for the food ingredients and cooked meat under different conditions of light source, illuminance, and correlated color temperature.

6.2 Methods

We carried out a subjective experiment of the visibility of food ingredients and cooked meat using light of different sources, illuminances, and colors. Figure 6.1 illustrates the plan and the section of the experimental space. We created 36 different lighting conditions with a combination of illuminance and light color using 10 fluorescent lamps (D65 lamps [Toshiba FLR40S D-EDL-D65/M] and 2700 K lamps [Panasonic FLR40S L-EDL-M]) on a

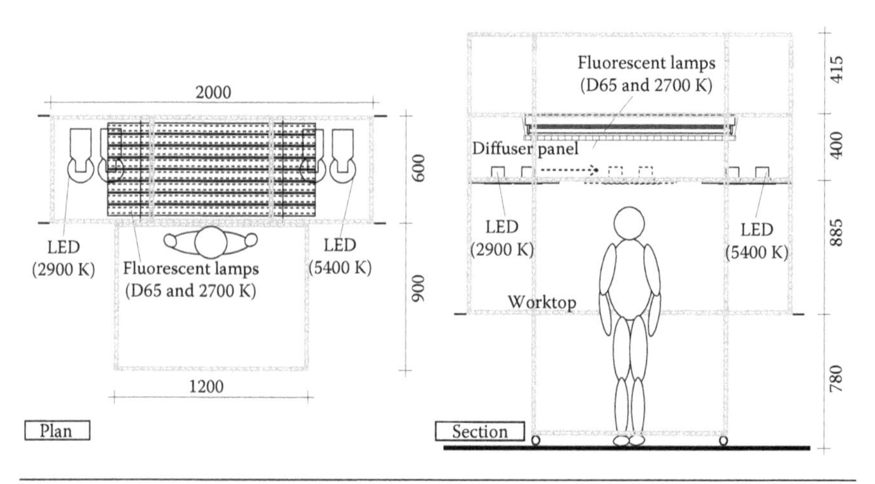

Figure 6.1 Experimental space.

Table 6.1 Chromaticity under Each Lighting Condition

Light Source	CCT (K)	x	y	Light Source	CCT (K)	x	y
Fluorescent	3000	0.43	0.40	Fluorescent	4500	0.36	0.36
Fluorescent	3500	0.40	0.38	Fluorescent	5000	0.35	0.35
Fluorescent	4000	0.38	0.37	Fluorescent	5500	0.33	0.35
LED	2900	0.45	0.42	LED	5400	0.34	0.36

ceiling in the experimental space, and 12 lighting conditions with six steps of the illuminance, using two types of 2900 K LED [Panasonic NDNN 21938] and 5400 K LED [Panasonic NDNN 21937]. And we could manage the range of illuminance from 50 to 1200 lx and that of correlated color temperature from 3000 to 5500 K. Table 6.1 shows the chromaticity under each lighting condition.

Subjects observed two color charts for the food ingredients and cooked meat according to the measuring results of the luminance and chromaticity (Fukumoto et al. 2010).

The color chart for the food ingredients consisted of nine color chips: 7.5GY 4/4, 5GY 5/6, 5GY 7/4, 7.5R 4/9, 2.5R 4/4, 10Y 4/3, 7.5YR 6/8, 2.5Y 8/12, and 5Y 6/4, and the color chart for the cooked meat consisted of nine color chips: 7.5YR 5/6, 7.5YR 5/4, 10YR 6/4, 5YR 4/4, 7.5YR 4/5, 10YR 5/4, 5YR 3/4, 7.5YR 4/3, and 7.5YR 4/4.

And they evaluated the visibility for the food ingredients and the visibility for cooked meat according to the four steps of categorical scales: "clearly visible," "visible," "barely visible," and "invisible." They also evaluated according to a numerical score with the method of magnitude estimation as the basis for the lighting condition scored 100, 1200 lx in illuminance with D65 lamps. The subjects were 22 female university students in their twenties. And they were accustomed to cooking because they belonged to the department of food science and nutrition.

6.3 Results and Discussion

6.3.1 Evaluation Results for the Visibility of Color Charts for Food Ingredients and Cooked Meat

Figure 6.2 shows the evaluation results for the visibility of two kinds of color charts for food ingredients and cooked meat. Twenty percentile values mean the evaluation values of 80% of the subjects, and 50 percentile values mean the evaluation values of half of the subjects. It was found that the visibility was higher as the level of illuminance on the worktop increased. It was also found that the visibility slightly increased as the correlated color temperature was higher.

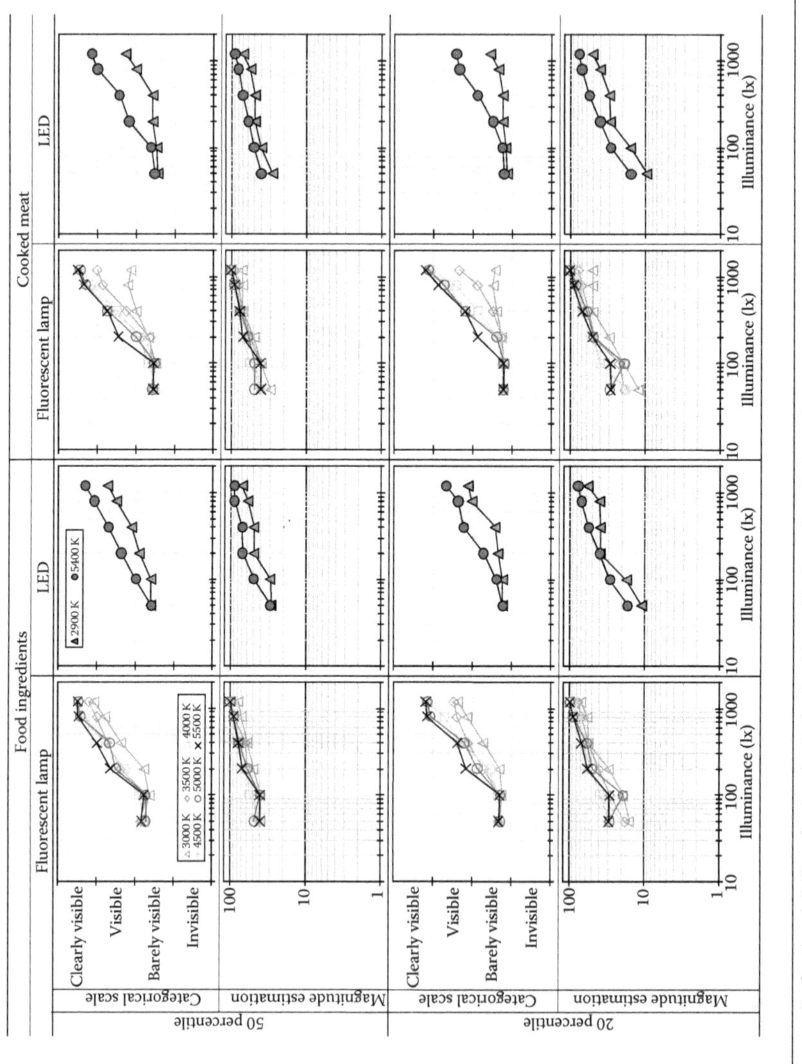

Figure 6.2 Evaluation results for the visibility of color charts for food ingredients and cooked meat.

Eighty percent of the subjects evaluated "visible" or "clearly visible" for the food ingredients under the lighting conditions of fluorescent lamps in 800 lx/3000 K and in 400 lx/5000 K. But they did not mark "clearly visible" under every lighting condition of LED lamps.

Eighty percent of the subjects evaluated "visible" or "clearly visible" for cooked meat under the lighting condition of fluorescent lamps in 400 lx/5000 K, but in the case of 3000 K, they answered "barely visible" under all illuminance conditions. Additionally, 50% of the subjects marked "barely visible" or "invisible" under LED lamps in 800 lx/2900 K.

6.3.2 Charts Indicating the Visibility Level under the Lighting Condition with a Combination of the Illuminance and Correlated Color Temperature

Figure 6.3 shows the charts that could indicate the visibility level under the lighting conditions with a combination of the illuminance and correlated color temperature under lighting conditions of fluorescent lamps. According to these charts, 80% of the subjects would evaluate "clearly visible" for food ingredients in 1200 lx/4000–5500 K and in 800–1200 lx/4500–5500 K. And they would evaluate "clearly visible" for cooked meat in 1200 lx/4000–5500 K.

These charts show the range of the illuminance and the correlated color temperature for the required visibility level. We can use these charts to evaluate the visibility level under some lighting conditions, illuminance, and correlated color temperature.

6.4 Conclusion

In this chapter, a subjective experiment was carried out to study the visibility of food ingredients and cooked meat under different conditions of illuminance and correlated color temperature. It was shown that a higher illuminance was required for the visibility of the food ingredients and cooked meat when the correlated color temperature was higher. And we made the charts that could indicate the visibility level under the lighting condition with a combination of the illuminance and correlated color temperature under lighting conditions of fluorescent lamps.

Acknowledgments

We thank all the people who helped us in carrying out the experiment. This study was supported by a grant from Kansai Electric Power Company to promote the sponsored research, 2009–2010, and the Ministry of Education, Culture, Sports, Science, and Technology to promote multidisciplinary research projects, 2006–2011.

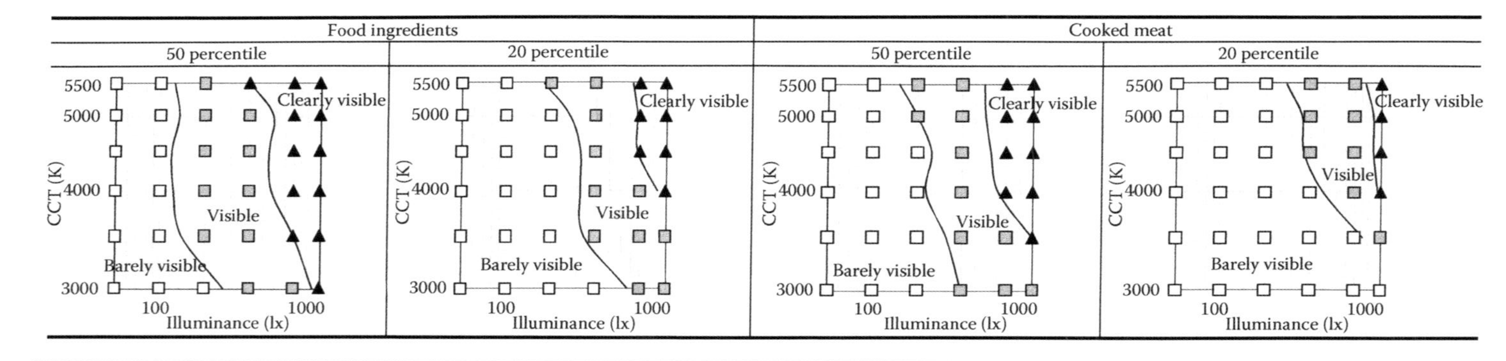

Figure 6.3 Charts indicating the visibility level under the lighting condition with a combination of the illuminance and correlated color temperature.

References

Fukumoto, Y., S. Okuda, N. Hara, A. Ishiguro, and W. Iwai. 2009a. Study on lighting environment for cooking in the kitchen—Effect of lighting direction on estimation. In *Proceedings of 2009 Annual Conference of the Illuminating Engineering Institute of Japan*, Tokyo: IEIJ, pp. 91–92.

Fukumoto, Y., S. Okuda, N. Hara, and W. Iwai. 2009b. Study on appropriate lighting environment for the kitchen. In *Proceedings of the 6th Lux Pacifica*, ed. W. Julian. Bangkok, Thailand, pp. 145–148.

Fukumoto, Y., S. Okuda, and K. Okajima. 2010. Colour references for estimating actual conditions of food material and cooked meat. In *AIC 2010 Color and Food, Proceedings*, eds. J. Caivano and M. López. Buenos Aires, Argentina: Grupo Argentino del Color, pp. 556–559.

JIS (Japanese Industrial Standard). January 20, 2010. General rules of recommended lighting levels. In Japanese Industrial Standard, Z-9110. Revised May 9, 2011.

MacDougall, D. B. and H. Sansom. 2007. Translucency measurement in foods using the Kubelka-Munk analysis with special reference to cured meat. In *Proceedings of the CIE Expert Symposium on Visual Appearance*. Paris, France: CIE, pp. 45–52.

McGuiness, P. J., P. R. Boyce, and S. D. P. Harker. 1983. The effects of illuminance on tasks performed in domestic kitchens. *Lighting Research & Technology* 15 (1): 9–24.

CHAPTER 7

Effect of Hydrothermal Conditions on Translucence of Milled Rice

MARCELO OSCAR BELLO, ROBERTO JORGE AGUERRE, MARCELA PATRICIA TOLABA, and CONSTANTINO SUÁREZ

Contents

7.1 Introduction

Rice parboiling involves hydration, steaming, drying, and milling in order to obtain milled parboiled rice. Traditional method using pressure steam has been widely used mainly due to the economic benefits from improved milling yields and less loss of solids during cooking (Bhattacharya 2004). Parboiling has a marked impact on the organoleptic properties of cooked rice. Starch gelatinization occurring during the steaming step and the complexation of rice lipids with the amylose fraction affect the rice constituents and give their own characteristics to the product (Deryckea et al. 2005). Gelatinization produces translucent grain with greater plasticity and resistance to the breakage. However, the application of high temperatures above gelatinization temperature increases the cost of parboiling equipment and affects the quality of parboiled rice in terms of hydration and cooking characteristics (Pillaiyar and Mohandoss 1981), tenderness and color (Pillaiyar 1984), and yellowness of the grain (Fellers and Deissinger 1983).

In alternative hydrothermal methods, the pressurized steam is replaced by water vapor at 15°C–20°C above the gelatinization temperature, and tempering stage is set among hydration and steaming steps. They lead to complete or partial gelatinization providing, as well as the traditional method, a better head rice yield (HRY) than nonparboiled rice. Characteristics of parboiled rice are more firmness, lowest stickiness of cooked rice, higher thiamine content, longer cooking time, and its own organoleptic properties (Tuley 1992, Champagne et al. 2004).

The influence of vapor pressure parboiling on milled rice properties has been widely studied (Champagne et al. 1998, Mohapatra and Bal 2006). However, the information related to the effect of hydrothermal conditions on cooked rice properties is very poor (Miah et al. 2002).

The objective was to study the hydrothermal treatment of rough rice including four steps: soaking, tempering of soaked grain, grain heating using hot water, and drying of parboiled rice. The purpose was to evaluate the effect of process conditions on the translucence of milled rice among other rice properties such as HRY and gelatinization degree (GD). Hydrothermal conditions were set based on process temperature, steaming time, and tempering time. Correlations between translucence index (TI%) and other quality indexes were also evaluated in this work.

7.2 Materials and Methods

7.2.1 Samples

A local variety of long rice paddy (Don Juan) provided by the National Institute of Agricultural Technology (INTA, Concepción del Uruguay, Entre Ríos, Argentina) was used.

The samples with initial moisture content of about 13% dry basis were stored at 4°C before use. The amylose content, determined by ISO 6647/87 iodine colorimetric method, was 26 g amylose/100 g dry matter. A gelatinization temperature of 70°C was determined for the milled rice grain by differential scanning calorimetry (DSC).

7.2.2 Parboiling Method

An experimental cooking chamber with water circulation was used. Rough rice samples of 100 g were placed in aluminum container with screen bottom and lid. The samples were soaked in a well-stirred and sealed water tank with controlled temperature (±0.5°C) during hydration step. After that, the rice container was elevated above the water level, and the rice was heated in the vapor space of the sealed tank (steaming step). Rice was finally oven-dried at 25°C until 12% dry basis and was stored at ambient temperature.

A multilevel factorial design consisting of 16 runs was performed (Table 7.1). Variables studied in this process were temperature, steaming time, and tempering time (x_1, x_2, and x_3, respectively). Each one of 16 runs of experimental design was duplicated.

In this procedure, the soaking and heating steps were performed at the same temperature. In order to have at least 15°C above gelatinization temperature and considering that 100°C is a limit for liquid water at atmospheric pressure, two temperatures were studied, 85°C and 95°C.

Soaking times were 10 and 5 min, respectively, to reach a saturation moisture content of 43% dry basis (Elbert et al. 2001).

The criteria to set the steaming time range were to have significant gelatinization avoiding grain splitting. The heating ranges were 90–180 min for 85°C and 45–75 min for 95°C.

Tempering before steaming step was performed in half of the experimental design points. The soaked grains were quickly withdrawn and transferred to an adiabatic container which was sealed and maintained at 25°C during 24 h. In these conditions, the mean moisture content of the grain remained constant. After that, the rice was returned to the heating chamber to steaming.

Table 7.1 Average Values of TI%, HRY, and GD Obtained for Different Hydrothermal Process Conditions: Temperature (x_1), Steaming Time (x_2), and Tempering Time (x_3)

Test No.	x_1 (°C)	x_2 (min)	x_3 (h)	TI (%)	HRY (%)	GD (%)
1	85	90	0	42.6	77.9	23.4
2	85	120	0	53.0	81.6	26.3
3	85	150	0	49.0	80.8	25.2
4	85	180	0	72.8	83.6	32.9
5	95	45	0	5.0	81.5	14.4
6	95	55	0	56.0	79.7	27.2
7	95	65	0	47.9	78.1	24.9
8	95	75	0	54.0	77.5	26.6
9	85	90	24	71.5	84.2	32.4
10	85	120	24	80.6	84.8	36.1
11	85	150	24	89.6	85.4	40.6
12	85	180	24	98.3	86.0	46.7
13	95	45	24	86.9	86.1	39.1
14	95	55	24	87.0	85.9	39.2
15	95	65	24	87.2	85.7	39.3
16	95	75	24	87.6	85.60	39.5

7.2.3 Head Rice Yield

Approximately 1 week after drying, rice was milled in a Suzuki MT-95 laboratory mill, and HRY was calculated as percentage of whole milled grains with respect to brown rice. The average value of duplicated was calculated.

The degree of milling calculated as percentage of rice bran with respect to the brown rice was approximately 13% dry basis in all experiments.

7.2.4 Gelatinization Degree

GD was calculated by relating the gelatinization enthalpy of the sample, ΔH, with $\Delta H_{control}$ corresponding to nonprocessed rice. Previously, samples were milled and rice flour was obtained. In sample pans, approximately 4 mg of rice flour was used and water was added to reach a 3:1 water/flour ratio.

Enthalpies were measured in PL-DSC calorimeter (Polymer Laboratories Ltd., United Kingdom), calibrated using pure indium standard ($\Delta H_{melting}$ = 28.41 J/g, 156.66°C). Tests were performed in triplicate in pans hermetically sealed heated from 30°C to 120°C (10°C/min).

7.2.5 Translucence Index

A Hunterlab Labscan Spectrocolorimeter (LS-5000, Hunter Associates Laboratory Inc., Reston, Virginia) was used to measure the TI% of milled rice. The instrument was standardized with the illuminant D65 (CIE [Commission Internationale de l'Eclairage] 1964, 10° Standard Observer) and a white standard plate LS-12084 (X = 79.01, Y = 83.96, Z = 86.76). A layer of black or white modeling paste (area, 28 cm^2) was used to support a monolayer of kernels. The rice grains covered the entire surface of mastic layer. Based on significant change of CIE chromaticity of translucent sample if white or black contrast was used (Hutchings and Gordon 1981), CIE chromaticity coordinate (x) was determined from eight different areas (diameter of illuminated area, 44 mm), and mean values were calculated for each background color. Following, the difference $\Delta x_S = x_S(\text{white}) - x_S(\text{black})$ was obtained. Taking as references the Δx_S values of unprocessed and traditional parboiled rice samples, the TI% for milled rice was calculated as

$$\text{TI\%} = \left(\frac{\Delta x_S - 2.72 \times 10^{-2}}{5.65 \times 10^{-2} - 2.72 \times 10^{-2}} \right) \times 100 \tag{7.1}$$

where

Δx_S is the sample change

2.72×10^{-2} and 5.65×10^{-2} are the Δx_S values of unprocessed rice (TI% = 0) and traditional parboiled rice (TI% = 100), respectively

7.2.6 Response Surface Method

A response surface method (RSM) was applied to analyze the effect of process conditions on quality variables. In such method, the responses studied

(Y_k, $k = 1, \ldots, p$) are matched to the factors (x_i, $i = 1, \ldots, n$) by the polynomial model associated with the experimental design (Khuri and Cornell 1987). The Statgraphics software package (Statistical Graphics Corporation, Warrenton, Virginia) was used for statistical analysis of the experimental results.

7.3 Results and Discussion

Processing conditions and experimental results of translucence are shown in Table 7.1. The comparison with the control sample (TI% = 0) evidences the convenient effect of hydrothermal treatment, particularly for tempered samples, resulting in increments of 98% for TI%. From a visual observation, it was found that all samples with TI% < 70% presented an undesirable "white belly." This defect has neither been observed in tempered samples which aspect was translucent. TI% was correlated with HRY ($r^2 = 0.76$) and also with GD ($r^2 = 0.98$).

The nonlinear relationship between HRY and TI% ($r^2 = 0.89$) or TI% and GD ($r^2 = 0.93$) are shown as follows:

$$\text{HRY} = 92.9 - 31.1\exp\left[-0.018\,\text{TI\%}\right] \tag{7.2}$$

and

$$\text{TI\%} = -0.0562\,\text{GD}^2 + 6.33\,\text{GD} - 74.7 \tag{7.3}$$

A similar relationship has been reported between HRY and GD determined by DSC (Marshall et al. 1993).

The effect of hydrothermal conditions on TI%, performed by RSM method, was satisfactorily modeled ($r^2 = 0.97$) by means of the following expression in terms of process variables:

$$\text{TI\%} = -58.87 + 1.27x_1 - 0.27x_2 + 1.35x_3 + 0.002x_2^2 \tag{7.4}$$

Process variables were both significant (p value < 0.05), and quadratic effect of steaming time was evidenced. Figures 7.1 and 7.2 show the estimated response surfaces of TI% as function of steaming time (x_2) and tempering time (x_3) to be 85°C and 95°C, respectively. It can be appreciated that TI% values increased with time and tempering. From optimization analysis, it was found that the conditions to optimize the process include 24 h of tempering at 25°C after hydration step. The steaming can be performed using 85°C (174.4 min) or 93.7°C (45 min). Both conditions provide a more firm and translucent grain with satisfactory values of TI%.

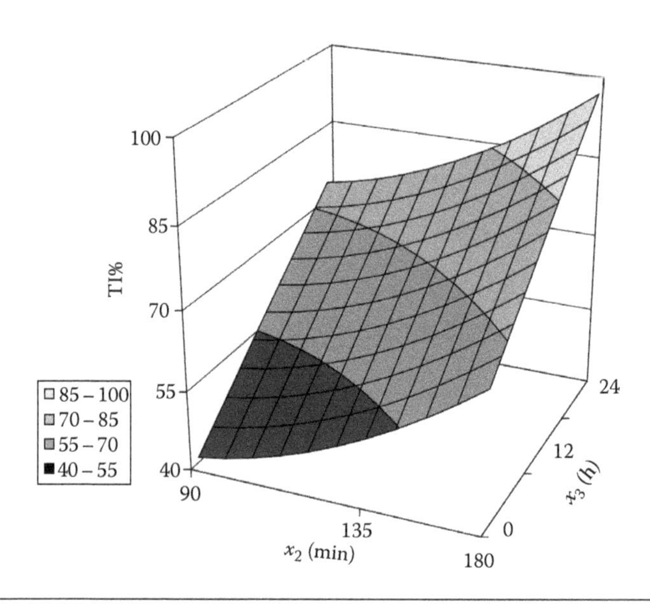

Figure 7.1 Effect of steaming time (x_2) and tempering time (x_3) on TI% to 85°C.

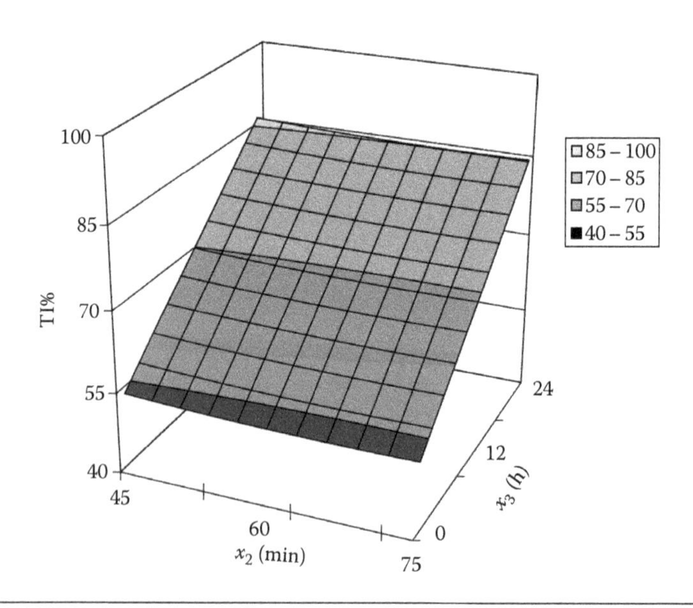

Figure 7.2 Effect of steaming time (x_2) and tempering time (x_3) on TI% to 95°C.

7.4 Conclusion

Hydrothermal conditions significantly affected TI%, which values resulted between 5% and 98%. TI% was correlated with GD ($r^2 = 0.98$). A nonlinear relationship between TI% and HRY was found with $r^2 = 0.89$. The significances of these results lie in the possibility to quantitatively assess the translucence of rice and also the fact of having mathematical expressions to connect TI% with other quality parameter characteristics of parboiled rice.

Acknowledgment

The authors acknowledge the financial assistance of Secretaría de Ciencia y Técnica, Universidad de Buenos Aires (FCEN-UBA), Argentina.

References

Bhattacharya, K. R. 2004. Parboiling of rice. In *Rice: Chemistry and technology*, ed. E. T. Champagne, 3rd edn. St. Paul, MN: American Association of Cereal Chemists, pp. 329–404.

Champagne, E., B. Lyon, B. Min, B. Vinyard, and K. Bett. 1998. Effects of postharvest processing on texture profile analysis of cooked rice. *Cereal Chemistry* 75 (2): 181–186.

Champagne, E. T., D. F. Wood, B. O. Juliano, and D. B. Bechtel. 2004. The rice grain and its gross composition. In *Rice: Chemistry and technology*, ed. E. T. Champagne, 3rd edn. St. Paul, MN: American Association of Cereal Chemists, pp. 77–107.

Deryckea, V., G. Vandeputtea, R. Vermeylena, W. De Manb, B. Goderisc, M. Kochd, and J. Delcoura. 2005. Starch gelatinization and amylose-lipid interactions during rice parboiling investigated by temperature resolved wide angle X-ray scattering and differential scanning calorimetry. *Journal of Cereal Science* 42 (3): 334–343.

Elbert, G., M. Tolaba, and C. Suárez. 2001. Model application: Hydration and gelatinization during rice parboiling. *Drying Technology* 19: 571–581.

Fellers, D. and A. Deissinger. 1983. Preliminary study on the effect of steam treatment of paddy on milling properties and rice stickiness. *Journal of Cereal Science* 1: 147–157.

Hutchings, J. B. and C. J. Gordon. 1981. Translucency specification and its application to a model food system. In *AIC Color 81, Proceedings of the 4th Congress*, ed. M. Ritcher, vol. 1. Berlin, Germany: Deutsche Farbwissenschaftliche Gesellschaft.

Khuri, A. and J. Cornell. 1987. *Response Surfaces: Designs and Analyses*. New York: Marcel Dekker.

Marshall, W., J. Wadsworth, L. Verma, and L. Velupillai. 1993. Determining the degree of gelatinization in parboiled rice. *Cereal Chemistry* 70 (2): 226–230.

Miah, M., A. Haque, M. Douglass, and B. Clarke. 2002. Parboiling of rice. Part II: Effect of hot soaking time on the degree of starch gelatinization. *International Journal of Food Science and Technology* 37: 539–545.

Mohapatra, D. and S. Bal. 2006. Cooking quality and instrumental textural attributes of cooked rice for different milling fractions. *Journal of Food Engineering* 73 (3): 253–259.

Pillaiyar, P. 1984. Applicability of the rapid gel test for indicating the texture of commercial parboiled rices. *Cereal Chemistry* 61: 255–256.

Pillaiyar, P. and R. Mohandoss. 1981. Cooking qualities of parboiled rices produces at low and high temperatures. *Journal of Science Food Agriculture* 32: 475–480.

Tuley, L. 1992. The rice revolution. *Food Review* 19 (5): 13–14.

CHAPTER 8

Measuring Banana Appearance Aspects Using Spectrophotometer and Digital Camera

WEI JI, M. RONNIER LUO, and JOHN B. HUTCHINGS

Contents

8.1 Introduction

An unripe banana fruit is always perceived as hard, green, sour, mealy (starchy), and odorless. This status ends when ethylene is produced. Ethylene is a simple hydrocarbon gas ($H_2C{=}CH_2$) that causes developmental changes resulting in fruit ripening. Ethylene apparently "turns on" the genes that are then transcribed and translated to make new enzymes. The enzymes then catalyze reactions to alter the characteristics of the fruit. These include hydrolases to help break down chemicals inside the fruits, amylases to accelerate hydrolysis of starch into sugar, and pectinases to catalyze degradation of pectin. During this process, chlorophyll is broken down, and sometimes new pigments are synthesized so that the fruit skin changes color from green to red, yellow, or blue. Acids are broken down so that the fruit changes from sour to neutral. The degradation of starch by amylase produces sugar which reduces the mealy (floury) quality and increases juiciness by osmosis. The breakdown of pectin between the fruit cells unglues them, so they can slip past each other, resulting in a softer fruit. Also, enzymes break down large organic molecules into smaller ones that can be volatile, and this gives an aroma sensation.

Visual color appearance of banana fruit dominates the consumer's decision during purchase (Hutchings 1999). Consumers use color as an index for quality when purchasing foods, and the color has to be "right" (Wrolstan and Culver 2008). Leading industries are using banana color charts to manage the fruit quality through the whole supply chain globally. The method is subjective, and the scale is coarse. Others apply spectrophotometers for the color measurement on a small area of the peeled skin (Ward and Nussinovitch 1996). Previous studies show that digital imaging methods can be applied for fruit color measurement (Park et al. 2005, Balaban 2008). A more reliable instrumental method is needed to describe ripeness of banana using color appearance terms and freckle percentage. This would benefit decision making in the supply processes (such as harvest, transportation, pack house, and on the market shelf), speed up the communication between supply stages, and reduce waste.

8.2 Experimental Conditions

In two boxes, one of green and one of ripe, Fyffe® bananas were purchased, each containing more than nine banana hands. Each hand included five or six fingers which were individually marked. Each hand (H1–H9) was investigated every day for a period of 9 days (D1–D9), by which time, a large proportion of skins had turned brown/black. Samples were kept in a laboratory at a temperature of 14 ± 1°C (see Figure 8.1). The daily procedure for the experiment was as follows: (1) image capturing of hands, (2) detaching and numbering fingers, (3) image capturing of fingers, (4) peeling the skin, and (5) spectrophotometer measurement of the peeled skin.

A conventional bench-top sphere-based spectrophotometer (Macbeth® CE7000A) was used for measuring the skin color as indicated in Figure 8.2. Three points were selected, each 25.4 mm circular shape. The results were averaged for each banana.

Banana images were captured in a neutral gray interior standard viewing cabinet (VeriVide®, Enderby, United Kingdom). The light source was a diffused D65 simulator. The luminance level of a barium sulfate white tile placed in the middle of the floor was 441.36 cd/m^2 with chromaticity coordinates $x = 0.3196$ and $y = 0.3348$.

Figure 8.1 Banana samples are grouped and stored in a temperature-controlled laboratory (dark bananas are green, and lighter bananas are yellow).

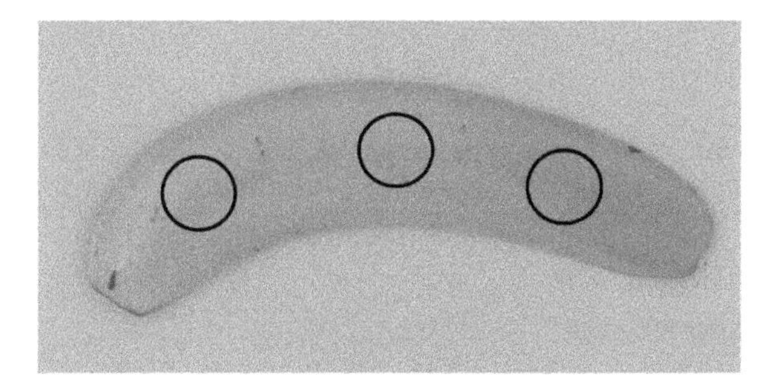

Figure 8.2 Spectrophotometer measurements on peeled banana skin at three positions.

A Nikon D80 digital camera was used for image capturing. The reproduced color of the camera was calibrated using an X-rite® reference color target of 240 color patches. The CIE (Commission Internationale de l'Eclairage) tristimulus values of those color patches were measured on a GretagMacbeth® CE7000A spectrophotometer. Before conducting each experiment, all devices were given 30 min warm-up time.

8.3 Results

When comparing the results, the technique of MCDM (mean color difference calculated from individual color values and the mean values of the group) was applied. This is used to calculate the mean color difference from a set of color differences between pairs of samples. The smaller the color difference number, the more repeatable or more reproducible the results are. Typically, about $3\Delta E^{*}_{ab}$ units can be visually perceived for banana samples.

8.3.1 Uncertainty of the Spectral Measurement

The following analysis was applied to evaluate the uncertainty of the spectrophotometer measurement. For each hand on each day, each banana skin was measured at the three positions as shown in Figure 8.2. The color difference was calculated between each point and the mean of three points. The average color differences for all fingers for the particular day were used to indicate variation for that banana hand as listed in Table 8.1. It can be seen that the green banana had a color variation ranging from 1 to 1.9□E^*_{ab} units. This is smaller than for the yellow bananas, which ranged from 1.7 to 3.0□E^*_{ab} units. The average color difference MCDM was 1.4□E^*_{ab} units for green and 2.1□E^*_{ab} units for yellow fruit.

Figure 8.3 is a plot of banana colors for the 9 days on CIE a^*b^* and L^*C_{ab} color charts. These two charts clearly show that the green bananas had much smaller color variation remaining green throughout the experiment. The yellow bananas, however, increased in redness, while lightness and chroma decreased. That is, color changed from bright colorful yellow to dark pale brown.

Table 8.1 The Uncertainty of the Spectrophotometer Measurement on Each Day for Each Banana Type in Terms of Color Differences

	No. of Fingers	L^*	a^*	b^*	C^*	h	DE^*_{ab}
Hand/day for green							
H1	5	53.80	−12.27	34.57	36.69	109.56	1.0
H2	6	59.07	−10.93	35.90	37.54	106.95	1.1
H3	5	58.47	−11.34	35.36	37.14	107.80	1.3
H4	5	56.95	−10.92	35.20	36.86	107.23	1.5
H5	6	60.19	−10.14	34.99	36.44	106.15	1.3
H6	5	57.46	−11.05	34.76	36.48	107.67	1.3
H7	6	56.24	−11.53	34.66	36.53	108.41	1.7
H8	6	58.42	−10.86	36.23	37.83	106.69	1.9
H9	5	55.25	−11.65	35.68	37.53	108.10	1.4
Hand/day for yellow							
H1	6	70.94	7.63	48.60	49.22	80.99	2.0
H2	5	69.39	9.43	46.66	47.61	78.56	1.7
H3	6	70.49	9.49	45.61	46.60	78.24	2.0
H4	6	58.19	13.32	39.91	42.18	71.08	2.9
H5	5	62.91	12.41	43.27	45.04	73.91	1.8
H6	6	63.46	12.18	43.15	44.94	74.00	1.7
H7	6	56.67	14.50	37.67	40.41	68.86	3.0
H8	6	53.39	16.29	37.66	41.15	66.14	2.1
H9	6	44.79	14.52	26.83	30.52	61.38	1.7

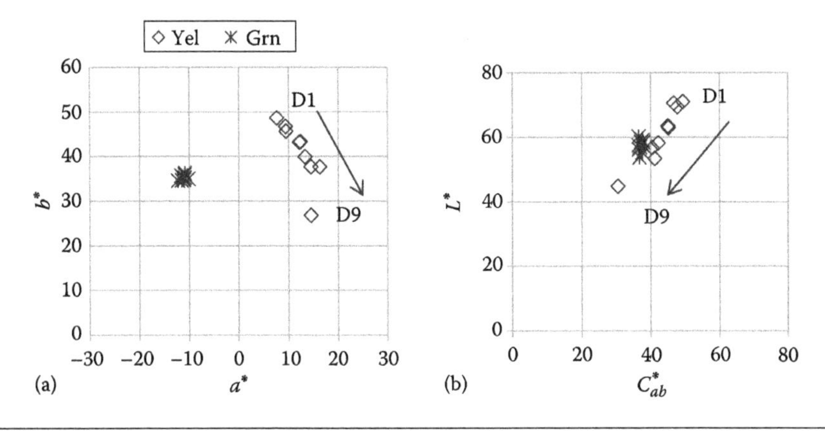

Figure 8.3 Green and yellow banana colors for 9 days plotted on a CIE a^*b^* (a) and L^*C_{ab} (b) diagrams.

8.3.2 Uncertainty of the Digital Image Method

A similar analysis process was used for the digital imaging methods. Uncertainty of the digital image system was quantified on whole hand images and each individual finger.

For whole hand images, color differences at three areas were calculated for all fingers in each hand against the mean values of the hand, and between the face and back positions (see Table 8.2). From this table, it can be seen that the variation for green bananas is smaller than that for yellow fruit. Variations between face and back are 1.9□E^*_{ab} for green and 2.7□E^*_{ab} for yellow.

In summary, the color results using the digital image method had □E^*_{ab} values ranging between 0.2 and 4.1 for green bananas and between 0.4 and 5.9 for yellow, with mean values less than 1.6 for green and less than 2.8 for yellow fruit. These results indicate that (1) images between hands or between fingers are repeatable, (2) images from hands and figures give repeatable results,

Table 8.2 Banana Hand Variations (□E^*_{ab})

DE^*_{ab}		H1	H2	H3	H4	H5	H6	H7	H8	H9	Mean
Green	Back	1.2	0.5	2.0	0.5	4.1	0.8	0.2	0.9	3.0	1.5
	Face	1.8	0.7	0.3	3.4	1.8	1.1	1.3	1.9	1.7	1.6
Yellow	Back	4.1	0.5	0.7	0.4	4.9	0.6	3.8	5.9	4.0	2.8
	Face	2.6	1.7	1.2	1.2	3.1	0.7	0.4	5.7	3.8	2.3
Green	Face vs. back	2.2	0.8	2.0	3.4	2.1	1.2	1.1	2.0	2.7	1.9
Yellow	Face vs. back	3.8	2.1	1.3	0.9	1.7	1.7	4.2	1.6	7.5	2.7

and (3) images from face and back are not so repeatable as some □E^*_{ab} values can be very large.

8.3.3 Comparing the Digital Imaging and Spectrophotometer Methods

The spectrophotometer measurements on the peeled skins were compared with color results from the digital image method (see Figure 8.4). It can be seen that hue values gave good agreement, but some discrepancies were found for the lightness and chroma values. This was caused by differences in the sampling positions when using different methods, especially as bananas ripened.

8.3.4 Banana Color Changes with Time

The CIECAM02 hue composition values were plotted with days to reveal the ripening process (see Figure 8.5). In Figure 8.5a diagram, green bananas changed little in hue as they remained consistently unripe. Figure 8.5b shows clear systematic pattern for yellow samples, i.e., the riper the banana, the redder the hue. Hue composition values were utilized here as they are closely linked with human color perception (CIE 2004). Figure 8.5c indicates similar trends: the freckle percentage for green banana not changing, while as the yellow fruit ripened, the percentages of freckling increased.

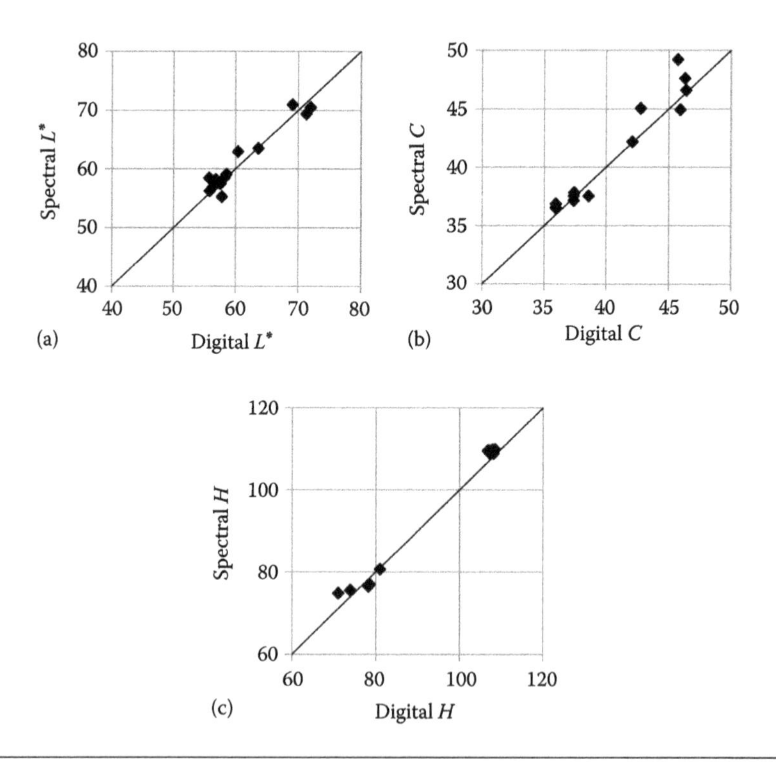

Figure 8.4 Comparing digital image color and spectrophotometer results for lightness (a), chroma (b), and hue (c).

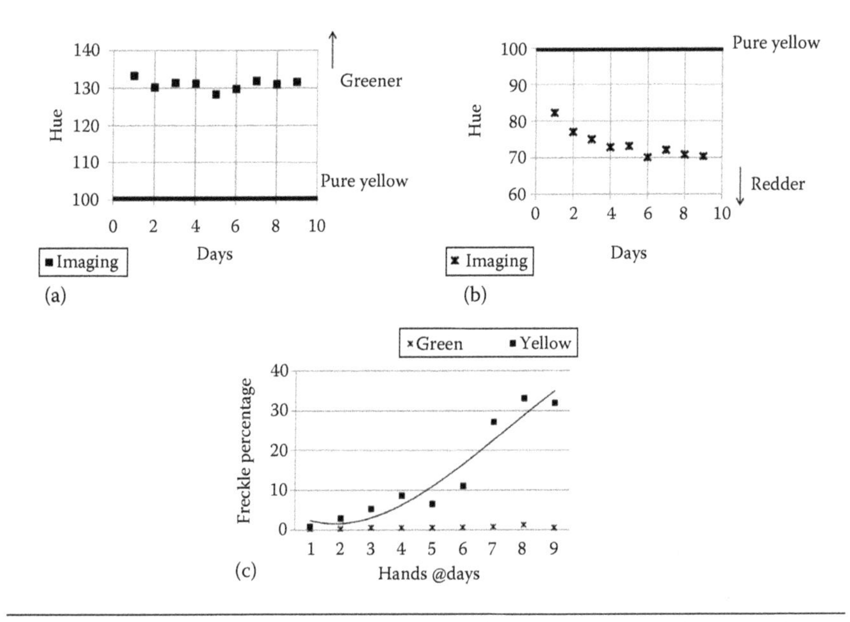

Figure 8.5 CIECAM02 hue composition changes with time for green bananas (a), yellow (b), and freckle percentage changes for green and yellow bananas (c).

8.4 Conclusions and Discussions

A digital image method has been developed for automatic scaling of banana images of color and freckle percentage to define the ripeness of the fruit. This method is repeatable, reproducible, and nondestructive. The technique can be further developed for quality control purposes along the whole fruit supply chain. The industry will benefit from this objective technique by reducing decision time and waste.

References

Balaban, M. O. 2008. Quantifying nonhomogeneous colors in agricultural materials. Part I: Method development. Part II: Comparison of machine vision and sensory panel evaluations. *Journal of Food Science* 73 (9): S431–S442.

CIE (Commission Internationale de l'Eclairage). 2004. *A Color Appearance Model for Color Management Systems: CIECAM02.* Vienna, Austria: CIE Central Bureau, publication 159.

Hutchings, J. B. 1999. *Food Color and Appearance*, 2nd edn. Gaithersburg, MD: Aspen.

Park, Y. K., J. B. Hutchings, W. Ji, J. W. Wu, and M. R. Luo. 2005. Specification of food colour and appearance using digital cameras. In *AIC Colour 2005, Proceedings of the 10th Congress of the International Colour Association,* eds. J. L. Nieves and J. Hernández-Andrés. Granada, Spain: Comité Español del Color, pp. 847–850.

Ward, G. and A. Nussinovitch, 1996. Peel gloss as a potential indicator of banana ripeness. *Lebensmittel Wissenschaft und Technologie* 29: 289–294.

Wrolstan, R. E. and C. A. Culver. 2008. *Color Quality of Fresh and Processed Foods.* Washington, DC: American Chemical Society.

PART II
Food Colorimetry and Color Scales

CHAPTER 9

Is the Color Measured in Food the Color That We See?

ÁNGEL I. NEGUERUELA

Contents

9.1 Introduction

In the 30 years that I have been measuring food color, I have had a doubt: Are food colors well measured? Or rather, do the color coordinates that have been found in the laboratory correspond to the food colors seen by customers or experts?

On numerous occasions, I have found that the answer is negative and I have found four groups of problems responsible for this: First, official methods to measure the color of certain foods are not related to the actual observation methods of expert color assessment. Second, some authors (and, consequently, referees) do not know the significance of color coordinates. Third, some authors use measuring instruments that are not suitable for the optical properties of the food to be measured, producing unacceptable results. Finally, foods whose colors are to be determined are not homogeneous.

This chapter will discuss these problems, based on, first, the cases I have found in the different kinds of foods whose color I have measured and, second, on those in the literature I have had to study and discuss with experts in other areas of food technology. I shall give priority to wine and fruit as I have worked most and have found the most important problems with them.

9.2 First Group: Official Methods

Official methods to measure the color of certain foods are not related to the actual observation methods of expert color assessment, for example, the color of red wines.

The sensory evaluation of wine is performed in standard glasses (OIV 2009). If we observe the color of a red wine in this glass, it is black with dark reddish tints at the edges (Figure 9.1). This is why the expert tilts the glass to observe the color of different thicknesses of wine, where the blue tones of young wines can be appreciated. If we perform spectrophotometric measurements of the light transmitted, we will see that the blue zone of the spectrum disappears as the path length of the cell increases (Figure 9.2). This disappearance of light transmitted in the blue zone indicates changes in the color of the wine (Table 9.1). If the wine is dark, the blue tones disappear at 10 mm thickness and the measurement does not correspond to the color that is observed by the wine taster.

Figure 9.1 Young red wine in wine-tasting glass (ISO 3591:1977).

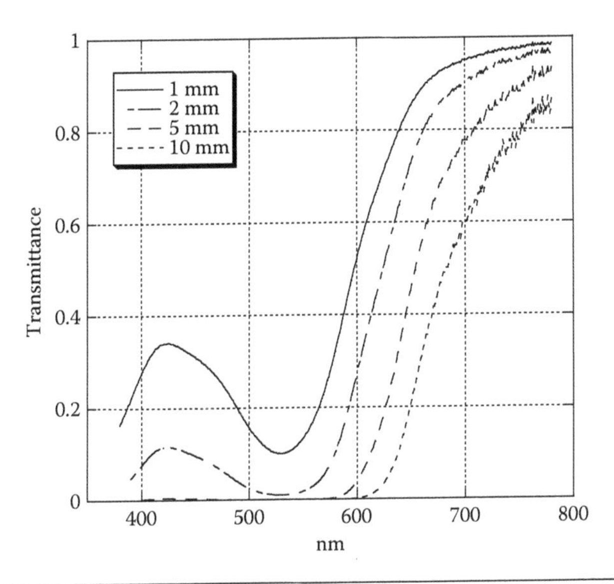

Figure 9.2 Young red wine spectra measured at different path length cells.

Table 9.1 Color Changes with Path Length of Cell in a Rioja Young Red Wine

Path Length of Cell (mm)	L^*	a^*	b^*	C^*_{ab}	h_{ab}
1	59.0	49.13	−3.55	49.25	336.90°
2	40.1	60.09	7.63	60.57	7.24°
5	19.5	50.88	29.31	58.82	30.11°
10	7.5	37.13	12.97	39.33	19.21°

The new official method for determining the wine color (OIV 2011) establishes that the red wine color must refer to a 10 mm path length cell. My work group suggested that the color should refer to a 2 mm path length cell, due to the blue tones of dark young red wines (Pérez-Caballero et al. 2003). But the proposal was not accepted by some wine experts as they preferred to maintain the same norm as before because of "tradition."

However, we decided to keep the 2 mm path length cell for determining the color of red and rose wines in our MSCV® program (Simplified Method for Wine Color), which is available as freeware on the Web (Ayala et al. 2001). For white wines and brandies, we have kept the measurements in 10 mm path length cell since at 2 mm they would be practically colorless.

This program allows us to calculate the CIELAB color coordinates of a wine from four absorbance measurements (which is the spectrophotometric unit most used by oenologists). In this version, measurements are accepted and results are displayed for any path length cell. Furthermore, we have provided for an approximated representation of the color calculated which allows us to see how the color of the wine changes in relation to its thickness.

The mean error calculated from more than one thousand three hundreds wines and brandies is lower than 1 CIELAB unit of color difference.

9.3 Second Group: Misinterpretation of Color Coordinates

Some authors (and, consequently, referees) do not know the significance of CIELAB color coordinates. These coordinates correspond to related colors, and they are always expressed in function of the reference white.

For example, the L^* coordinate, which some authors call "luminosity" instead of its correct name "lightness," is defined as "the brightness of an area judged relative to the brightness of a similar illuminated area that appears to be white or highly transmitting" (CEI and CIE 1987a).

Some authors misinterpret the a^* and b^* coordinates as the magnitude of color red or green and yellow or blue, respectively, of any color. This implies that the color is a phenomenon of four stimuli (or five, if we included the brightness) rather than three stimuli as in reality. They do not bear in mind that the h_{ab} coordinate is a function of the quotient of both, b^*/a^* and is related to the color hue. Furthermore, they do not take into consideration that the positive axis of the a^* coordinate ($h_{ab} = 0°$) corresponds to a red–purple color, and not to a spectral red color.

Finally, many authors do not use the C^*_{ab} and h_{ab} coordinates, not bearing in mind that these two coordinates, together with L^*, are those that correspond to the visual perception of related colors.

The C^*_{ab} coordinate is related to the perceived chroma, which is "the colorfulness of an area judged relative to the brightness of a similar illuminated area that appears to be white or highly transmitting" (CEI and CIE 1987b).

The h_{ab} coordinate is related to the perceived hue: red, yellow, green, blue, or combinations of two of these consecutive colors (CEI and CIE 1987c). The spectrum locus commences with values of h_{ab} around 30° for the red hues and ends at h_{ab} around 270° for the spectral blues, the rest of the circle being taken up by the purple hues.

If we want to compare colorimetric results, we must use the color differences between the colors measured, such as the old CIELAB color difference: ΔE^*_{ab}, or, even better, one of the later color differences derived from this color space, such as the recent CIEDE2000 difference, recommended by the CIE (2004).

This lack of knowledge concerning the meaning of the CIELAB color coordinates and color differences led to a problem when performing a comparative study of spectrophotometric measurements of the color of wine by the International Organisation of Vine and Wine (OIV), in which several laboratories took part (2002).

The color of a series of red, rose, and white wines was measured, and the results were sent to the coordinating laboratory of the study where the statistical analysis of each coordinate was performed separately. This analysis showed that all the measurements were within acceptable deviation if the measurements were made in 10 mm path length cells. However, when I carried out the study of color differences between the coordinates of samples and their statistical mean values, there were some color differences ΔE^*_{ab} higher than 3 CIELAB units, which indicated that they would be distinguished by an observer (Martínez et al. 2001).

I warned the group in charge of the comparative study about this, but they just withdrew the samples from the study and work continued as before.

I should mention that the "bad" samples were of dark red wines.

After thinking about this for a long time, I have reached the conclusion (which I cannot prove) that the problem was that the spectrophotometric measurements referred to 10 mm path length cells are not good in the blue zones, where the absorbances of the samples are higher (Figure 9.3).

These measurement errors, together with the different calibration of the instruments, cause the small deviation in each coordinate that, added together when calculating the color differences, lead to their being distinguishable to the naked eye. This did not occur with wines that were not dark (or if the color of the wine was referred to 2 mm path length cell).

Finally, I should like to recommend that when data are statistically analyzed, no conclusions should be made about each coordinate separately when comparing results. What is important are the color differences, especially if they are visible to the naked eye.

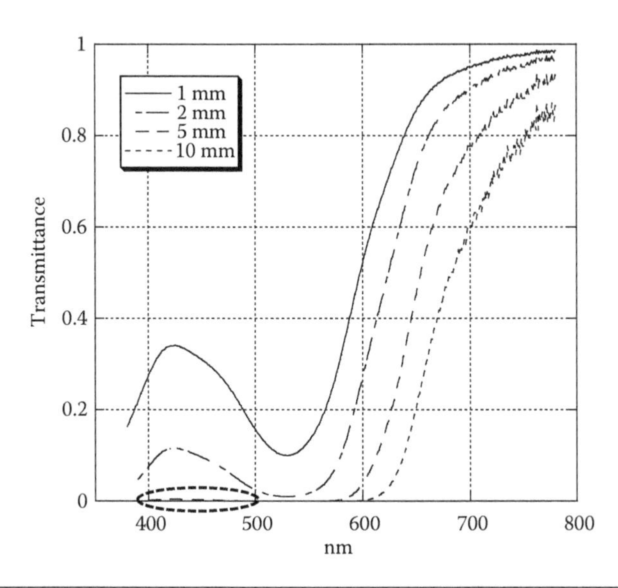

Figure 9.3 Spectrophotometric measurements in the blue zones are not good for young red wine spectra measured at 5 and 10 mm path length cells.

9.4 Third Group: Measuring Instruments

Some authors use measuring instruments that are not suitable for the optical properties of the food to be measured, producing unacceptable results.

According to the classical and the Kubelka–Munk theories (Judd and Wyszecki 1975), when white light from medium 1 hits a body (medium 2), various processes may occur (Figure 9.4).

a. Partial reflection of the light in all wavelengths of the visible spectrum on the first surface, owing to the change of refraction index between the two media.
b. Transmission of the remaining light to medium 2.
c. Partial absorption of the light in those wavelengths characteristic of the material of medium 2.
d. If the body is not transparent, the nonabsorbed light undergoes scattering inside the body until it reaches the second surface of the body.
e. Partial reflection on the second surface, analogous to that produced on the first surface due to refraction index change.
f. Transmission of the remaining light outside the body (medium 3).
g. The light reflected on the second surface returns inside the body, undergoing the same processes of absorption and scattering as in (c) and (d).
h. Partial reflection on the first surface, analogous to (a) and (e), and repetitions of processes from (a) to (h).
i. Transmission to medium 1 of the remaining light in the wavelengths that produce the characteristic color of the body. This color is similar to that corresponding to the transmitted light in (f). For example, a leaf is green because of transmission and reflection, even though these may not be the same green.

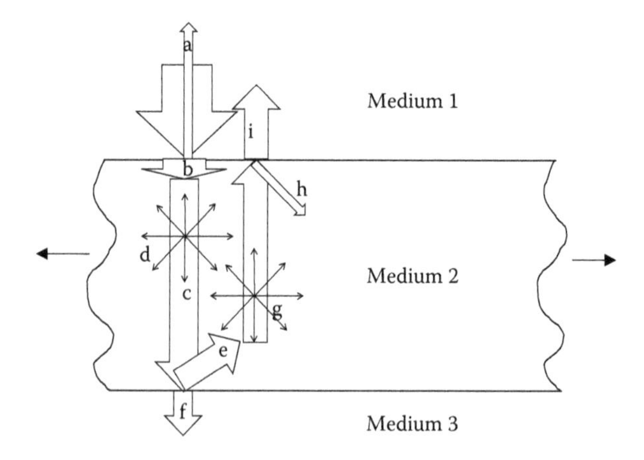

Figure 9.4 Processes in light hitting a turbid medium 2.

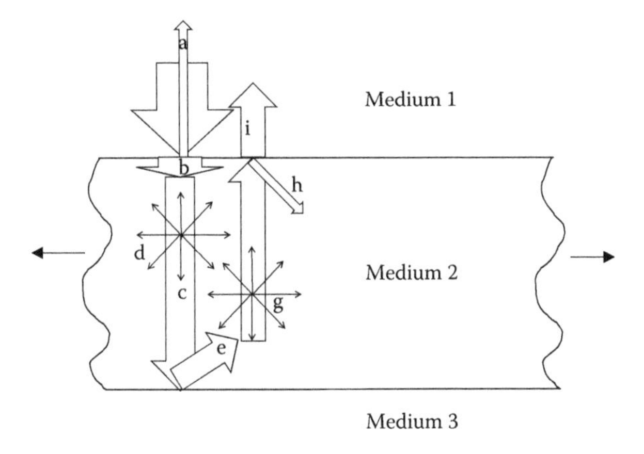

Figure 9.5 Processes in light hitting an opaque medium 2.

If the illuminated body is perfectly opaque, there is no light transmitted (lost) on the second surface and, therefore, the quantity of light that returns to medium 1 is greater than in the previous case (Figure 9.5).

If the body is not perfectly opaque but is sufficiently thick, light is not transmitted to medium 3, either. The reflectance of this body is called "reflectivity."

In normal conditions of natural or artificial illumination, many foods are opaque, and the light that indicates their color has no losses on the second surface.

This process is similar to the vision of the color of paintings or other opaque objects and most color-measuring instruments are designed to measure opaque objects.

Many color-measuring instruments, such as portable colorimeters and spectrocolorimeters, use a flash as a system of illumination, which compensates the short period of illumination with the high power of light emitted. Those of us who have worked with these devices know that, each time we take measurements, it is necessary to calibrate them with a perfectly opaque white reference surface. This indicates to the device the maximum quantity of light it is going to receive and must interpret as white color.

The problem of this illumination is that it has sufficient power to go through many food samples and loses part of its light through the second surface. The first time I observed this phenomenon was when I was measuring the color of veal with a thickness of 2.5 cm. I was on the computer and my colleague was managing the portable spectrocolorimeter with one hand, while holding the recipient containing the sample with the other (Figure 9.6). I saw the light of the flash after it had gone through the sample and the white recipient.

Obviously, this affects color measurements since the instrument receives less light than it should if the food to be measured were opaque as the

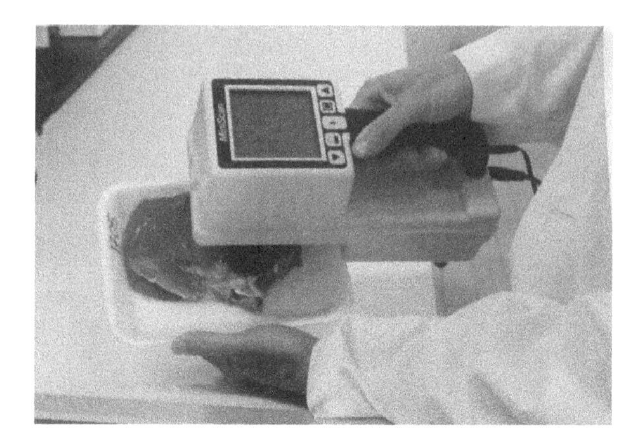

Figure 9.6 Measuring the color of veal with a portable spectrocolorimeter.

instrument "supposes" in its calibration. Consequently, the measurement corresponds to a darker object than what the tested food really is.

The same thing has happened with other, vegetable, foods.

On other occasions, I find it very difficult to understand the results presented as color measurements. The most extraordinary case I have found so far is the color measurement of different varieties of peach. One member of my team, a food technologist, showed me a paper by Drogoudi and Tsipouridis (2007) when we were measuring the color of another variety of peach. He told me that the measurements did not agree with ours. I looked at the paper and the measurements and I told him to find me those peaches of blue–white color, some of which have L^* values higher than 100 (Table 9.2).

Table 9.2 Color Coordinates of Several Clistone Peaches

L^*									
GF677	100.1	101.6	97.6	99.7	96.5	97.1	96.4	95.1	93.2
KID1	96.4	96.4	92.0	102.0	99.5	97.0	95.6	97.4	95.1
PR204	99.4	97.2	93.9	101.0	101.0	98.5	97.8	96.2	98.1
a^*									
GF677	4.2	3.4	10.9	3.7	8.1	7.0	7.9	8.7	11.9
KID1	5.4	4.0	9.6	5.3	9.1	6.6	6.8	8.1	11.9
PR204	5.6	4.4	9.9	6.7	8.0	6.3	6.3	9.1	12.0
b^*									
GF677	−18.3	−19.1	−26.8	−17.1	−18.6	−30.1	−42.0	−49.2	−19.6
KID1	−22.2	−10.4	−18.9	−26.6	−19.6	−23.6	−26.6	−51.0	−18.8
PR204	−18.7	−20.1	−16.3	−21.9	−24.0	−17.0	−21.9	−46.0	−21.2

Source: Data from Drogoudi, P.D. and Tsipouridis, C.G. *Sci. Hort.*, 115, 34, 2007.

I believe that, behind these results, there are two problems:

1. A poorly calibrated instrument
2. Corroborating the previous section, the most absolute ignorance of the authors and referees of the paper regarding the meaning of the CIELAB color coordinates

9.5 Fourth Group: Fruits without a Homogeneous Color

Finally, I should like to talk about the color measurement of fruits without a homogeneous color, mainly because of the ripening process. As is well-known, a fruit changes color while ripening and this change is usually from green to yellow and then to a lighter or darker red. The area of the fruit which first turns red is that which is directly facing the sun. When they are picked, these fruits usually have green and red areas, such as some apples, or yellow and red, such as some peaches.

Some authors (Drake and Elfving 1999, Benavides et al. 2002, Echevarría et al. 2002) studied the postharvest color of these fruits by measuring the color of both areas separately, but it is clear that the panel of experts who evaluated these fruits could not appreciate them separately since they had to give their color evaluation as a single parameter.

We faced this problem when we had to measure the color of Fuji apples that have alternating stripes of red and green (Figure 9.7). Following an old colorimetric method, we decided to set up the equipment in such a way that the apple was rotating (Francis and Clydesdale 1975: Chapter 38, p. 424) while a spectroradiometer measured the reflectance spectrum of one band all the way round the apple for one complete turn, integrating the result. The spectra obtained should differ according to the red and green areas of the fruits.

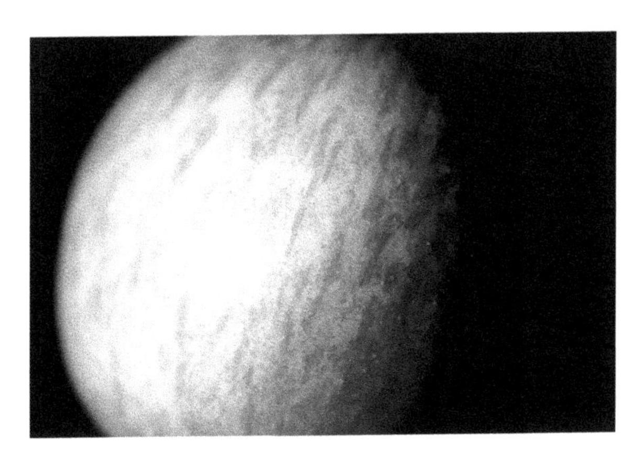

Figure 9.7 Fuji apple with alternating stripes of red and green colors.

To verify these results, we measured the reflectance spectrum of a completely green area and that of a completely red area and we calculated the spectra that would be obtained by adding both spectra in different proportions: point nine times red plus point one time green; point eight times red plus point two times green, and so on. The results are similar to the measured spectra of some apples (Marquina et al. 2004). This confirmed that the spectra measured on the fruits were a sum of the spectra of the red and green areas weighted with the areas occupied by each color.

When we compared our results with the experts' response, we saw that there was a good correlation between the a^* and h_{ab} coordinates obtained from the reflectance spectra measured and the panel's evaluation of the level of ripeness of the apples.

It seems that the panel of experts evaluated the quantity of the red area of the fruit to establish its level of ripeness and the method of spinning the fruit in front of the spectroradiometer did something similar.

In the case of measuring fruits of different colors, I would recommend this method, but I realize that it is impossible with a portable colorimeter or spectrocolorimeter. The use of digital photography to establish its color might be able to use this idea, although it should be taken into account that the image of the fruit will be of just one part and not of the whole fruit.

To end this point, I would like to suggest that when they are measuring the CIELAB coordinates of food, researchers should use some computer program that is able to reconstruct the color on the screen. They should have a clear idea that the color they are going to see will only be similar to that indicated by their color coordinates, unless they have correctly calibrated the screen response, which is not normal. However, if the color they see on the screen is not similar to the one they see on the fruit, something is wrong with the measurement method and should be corrected: the color measured by the device should be approximately the one we observe in the food.

9.6 Conclusion

The foundations of colorimetry establish that the measured colors must correspond to the color seen by the observer. If the correspondence is not good, the response of observers is the correct one, and the experimental conditions of color measurement must be revised.

References

Ayala, F., J. F. Echávarri, and A. I. Negueruela. 2001. MSCV©, Método simplificado para el color del vino/Simplified method for wine colour. Available as freeware on the Web sites: http://www.unizar.es/negueruela and http://www.unirioja.es/dptos/dq/fa/color/color.html. Accessed on April 25, 2011.

Benavides, A., I. Recasens, T. Casero, Y. Soria, and J. Puy. 2002. Multivariate analysis of quality and mineral parameters on Golden Smoothee apples treated before harvest with calcium and stored in controlled atmosphere. *Food Science and Technology International* 8(3): 139–146.

CEI and CIE (Commission Electrotechnique Internationale and Commission Internationale de l'Eclairage). 1987. *International Lighting Vocabulary*. Geneva, Switzerland: CEI Publ. 50 (845), CIE Publ. 17.4. (a) 845-02-31.1; lightness; (b) 845-02-35; hue; (c) 845-02-42 chroma.

CIE (Commission Internationale de l'Eclairage). 2004. *Colorimetry*, 3rd edn. Vienna, Austria: CIE Central Bureau, Publ. 15:2004.

Drake, S. R. and D. C. Elfving. 1999. Quality of 'Fuji' apples after regular and controlled atmosphere storage. *Fruit Varieties Journal* 53(3): 193–198.

Drogoudi, P. D. and C. G. Tsipouridis. 2007. Effects of cultivar and rootstock on the antioxidant content and physical characters of clingstone peaches. *Scientia Horticulturae* 115: 34–39.

Echevarría, G., J. Graell, and M. L. López. 2002. Effect of harvest date and storage conditions on quality and aroma production of 'Fuji' apples. *Food Science and Technology International* 8(6): 351–360.

Francis, F. J. and F. M. Clydesdale. 1975. *Food Colorimetry: Theory and Applications*. Westport, CN: Avi Publishing.

Judd, D. B. and G. Wyszecki. 1975. Chapter 3: Physics and psychophysics of colorant layers. *Color in Business, Science and Industry*. New York: John Wiley & Sons.

Marquina, P., M. E. Venturini, R. Oria, and A. I. Negueruela. 2004. Monitoring colour evolution during maturity in Fuji apples. *Food Science and Technology International* 10(5): 315–321.

Martínez, J. A., M. Melgosa, M. M. Pérez, E. Hita, and A. I. Negueruela. 2001. Visual and instrumental color evaluation in red wines. *Food Science and Technology International* 7(5): 439–444.

OIV (International Organisation of Vine and Wine). 2011. Compendium of international methods of wine and must analysis, vol. 1. Method OIV-MA-AS2-11. http://news.reseau-concept.net/pls/news/pk_recherche3.article?i_sid=&i_article_id=20490&i_app_id=19433&i_modele=http://www.oiv.int/uk/recherche/modele.php. Accessed April 25, 2011.

OIV (International Organisation of Vine and Wine). 2009. Resolutions of the 7th General Assembly. Resolution OIV/Concours 332B/2009. Guidelines for granting OIV patronage of international wine and spirituous beverages of vitivinicultural origin competitions. http://news.reseau-concept.net/pls/news/pk_recherche3.article?i_sid=&i_article_id=20490&i_app_id=19433&i_modele=http://www.oiv.int/uk/recherche/modele.php. Accessed April 25, 2011.

Pérez-Caballero, V., F. Ayala, J. F. Echávarri, and A. I. Negueruela. 2003. Proposal for a new Standard OIV method for determination of chromatic characteristics of wine. *American Journal of Enology and Viticulture* 54(1): 59–62.

CHAPTER **10**

Whiteness, Yellowness, and Browning in Food Colorimetry

A Critical Review

ROBERT HIRSCHLER

Contents

10.1 Introduction

Whiteness is an important characteristic of many food products from milk and rice to surimi and pasta. In many cases whiteness is desirable, in others it is not. Deviation from whiteness may be perceived as yellowness or browning, and there are hundreds of articles (and some pages in a few textbooks) describing this phenomenon using whiteness, browning, and—less frequently—yellowness indices. In a review of over 200 articles published in more than 30 journals dedicated to food science and technology, we have found ample references to the application of these formulae, most of them related to the description of the change in some kind of technological or process parameter, rather than the perceptual change of the white, yellowish, or brownish color of the product.

10.2 Whiteness, Yellowness, and Browning Indices

10.2.1 Whiteness Indices

In the food industry, the most frequently used whiteness indices are L^* (often erroneously called "whiteness" instead of "lightness") and

$$\mathrm{WI}_{\mathrm{JUDD}} = 100 - [(100 - L^*)^2 + (a^*)^2 + (b^*)^2]^{1/2} \tag{10.1}$$

(or the equivalents with Hunter coordinates), first suggested by Judd and Wyszecki (1963). In many publications, the

$$\mathrm{WI}_{\mathrm{HUNTER}} = L^* - 3b^* \tag{10.2}$$

formula with CIELAB coordinates (originally proposed by Hunter 1960, for Hunter L, a, b coordinates) is used, but to the current CIE whiteness formula (CIE 2004):

$$\mathrm{WI}_{\mathrm{CIE}} = Y + 800\,(x_n - x) + 1700\,(y_n - y) \tag{10.3}$$

we found reference only in two cases (Kotwaliwale et al. 2007, Zapotoczny et al. 2006). The reason for neglecting the only internationally recognized formula lies in the fact that natural or processed food products are very rarely white enough to fall within the limits of its validity. In the two cases cited, the $\mathrm{WI}_{\mathrm{CIE}}$ values are negative, which is, of course, nonsense.

10.2.2 Yellowness Indices

The Hunter b or CIELAB b^* coordinate is often used for the characterization of yellowness. Yellowness indices are unduly neglected in the publications reviewed; they report only in a few cases the application of the

$$\mathrm{YI}_{\mathrm{E313}} = \frac{100(C_X X - C_Z Z)}{Y} \tag{10.4}$$

according to ASTM (2005), where C_X and C_Z are illuminant- and observer-specific constants, or the

$$YI_{FC} = 142.86b^*/L^* \quad (10.5)$$

formula often referenced to Francis and Clydesdale (1975).

10.2.3 Browning Indices

Browning index in the literature may mean one of two things: a simple indicator of a chemical change (often characterized by the optical density at a given wavelength or the ratio of the reflectance at 570 and 650 nm) or the color change due to oxidation of a freshly cut fruit or vegetable surface, during storage or drying, or the baking of bread. The simplest (and probably least adequate) indicator of the color change is the L^* coordinate (or $100 - L^*$ or $100/L^*$). The best known and most often quoted browning index is a form of excitation purity that following the suggestion of Buera et al. (1985) is expressed as

$$BR_{BUERA} = 100(x_c - 0.31)/0.172 \quad (\text{for } C/2°) \quad (10.6)$$

where x_c is the CIE chromaticity coordinate and the constants were determined for given limiting conditions.

Feillet et al. (2000) calculated the yellowness and the brownness of pasta discs from the equations: $BI_{FEILLET} = 100 - R_{550}$ and $YI_{FEILLET} = 100\ (R_{480} - R_{550})$, where the R values are reflectance factors for selected wavelengths.

10.3 Case Studies

Illustrating some of the concepts of using whiteness, yellowness, and browning indices, we have taken colorimetric data from the literature. Where not enough details were given in the published work, the authors have kindly provided them in private communications.

10.3.1 Sensory Characteristics of Yoghurt

Vargas et al. (2008) used the WI_{JUDD} formula (D65/10°) to characterize the whiteness of yogurts prepared from mixtures of fresh raw caprine and fresh raw bovine milk.

As illustrated in Figure 10.1, L^* and WI_{JUDD} (very similar in behavior) are very little sensitive to the change in milk composition. In this case, WI_{HUNTER} or WI_{CIE} shows higher sensitivity, but the color of the milk mixtures (except for pure GM) is too yellowish; thus, WI_{CIE} is not supposed to be applied. The most adequate would be either YI_{E313} or YI_{FC} (which are in this case strongly correlated, $R^2 = 0.9983$) or even b^*, which in this case behaves very similarly to the two yellowness indices.

10.3.2 Whiteness of Surimi Gels

Xiong et al. (2009) investigated the effect of different levels of konjac glucomannan (KGM) on the whiteness ($WI_{JUDD} - C/2°$) of grass carp surimi gels (Figure 10.2).

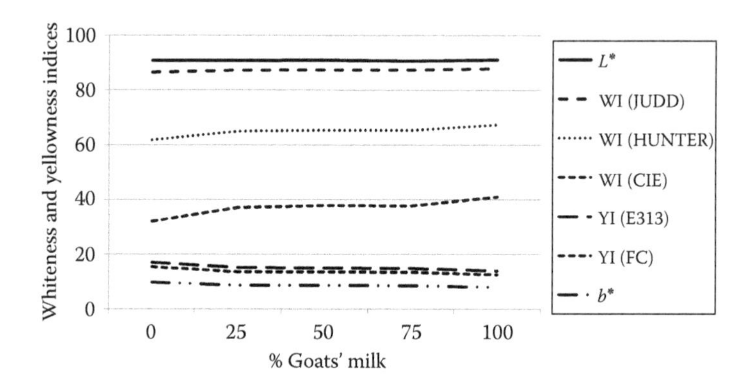

Figure 10.1 Whiteness and yellowness indices for yogurts made of different milk compositions. (Data kindly provided by Vargas, M. et al., *Int. Dairy J.*, 18, 1146, 2008.)

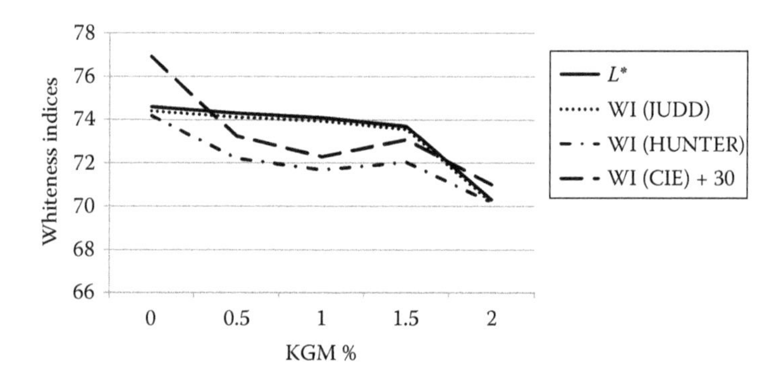

Figure 10.2 Effect of KGM concentration on the whiteness of surimi gels. (Based on data by Xiong, G. et al., *Food Chem.*, 116, 413, 2009.)

As there is very little change in a^* and b^* (and very low chroma for all samples), the L^* and WI_{JUDD} values are very close. Other whiteness indices, however, show a somewhat different tendency. It must be emphasized that the differences are very small, on the border of perceptible differences.

10.3.3 Yellowness of Mashed Potatoes

Fernández et al. (2008) measured the effect of biopolymer concentration and freezing and thawing processes on the color parameters of fresh and frozen/thawed mashed potatoes and calculated the YI_{FC} (D65/10°).

For their sample sets, the correlation between YI_{FC} and the CIELAB coordinate b^* is $R^2 = 0.83$, while that between YI_{E313} and b^* is as high as $R^2 = 0.95$ as illustrated in Figure 10.3. We shall see later that these correlations are very highly dependent on the sample set in question.

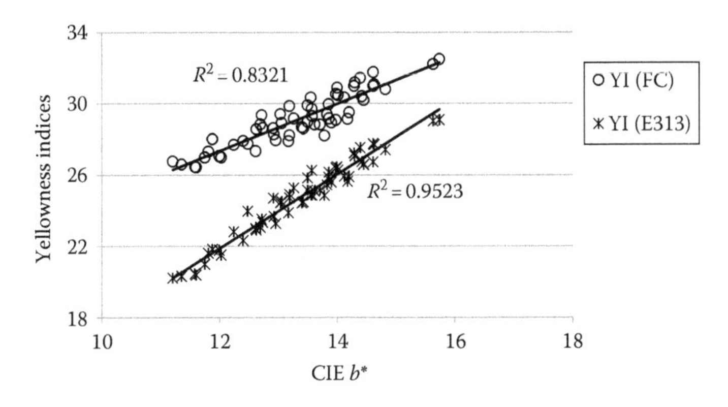

Figure 10.3 Correlation between b^* and yellowness indices for differently treated frozen/thawed mashed potatoes. (Based on data by Fernández, C. et al., *Food Hydrocolloids*, 22, 1381, 2008.)

10.3.4 Drying of Onion Slices

Arslan and Özcan (2010) studied the effect of sun, oven, and microwave drying on quality of onion slices. They measured CIELAB (*C*/2°) coordinates and discussed the color change in terms of the individual coordinates. If we use one of the well-known yellowness or browning indices, interesting further conclusions may be drawn from their data: they rank the color effect of different drying methods differently than the L^*, a^*, or b^* coordinates individually.

Figure 10.4 shows that there is very high ($R^2 > 0.99$) correlation between the yellowness and browning indices, but very low between them and b^*. This goes to show that b^* cannot always be used as a yellowness index, depending on the sample set it may or may not correlate well with yellowness/brownness.

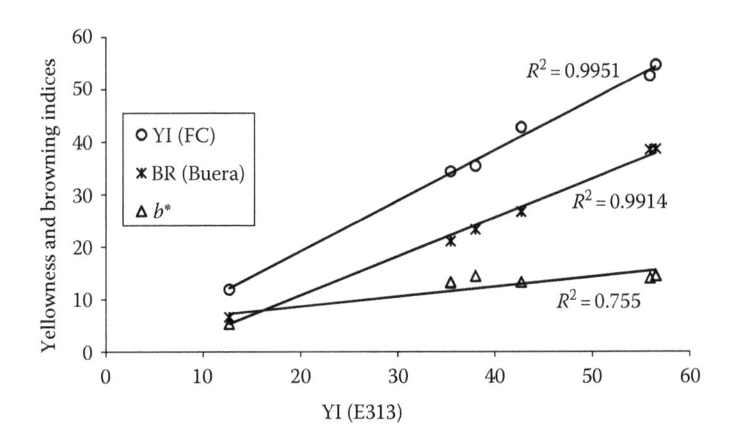

Figure 10.4 Correlation of different yellowness and browning indices and CIELAB b^*. (Based on data kindly provided by Arslan, D. and Özcan, M.M., *LWT Food Sci. Technol.*, 43, 1121, 2010.)

10.3.5 Color Improvement in Surimi

Taskaya et al. (2010) investigated the color improvement by titanium dioxide of proteins recovered from whole fish and arrived at the conclusion (based on WI_{JUDD} values) that the whiteness of restructured fish products based on proteins recovered from whole fish via isoelectric solubilization/precipitation can be similar to the whiteness of surimi seafood. The same conclusion could be drawn based on L^* values. However, if we compare the WI_{HUNTER} (Figure 10.5a) or the YI_{FC} or YI_{E313} (Figure 10.5b) values, we can see that surimi is in fact much whiter (less yellow) than recovered proteins.

This is an interesting example of how selecting the wrong index (in this case WI_{JUDD} or L^*) may lead to completely erroneous conclusions.

10.3.6 Cheese Color

Sheehan et al. (2009) studied the effect of partial or total substitution of bovine for caprine milk on the color of semi-hard cheeses. Their data (*C*/2°)

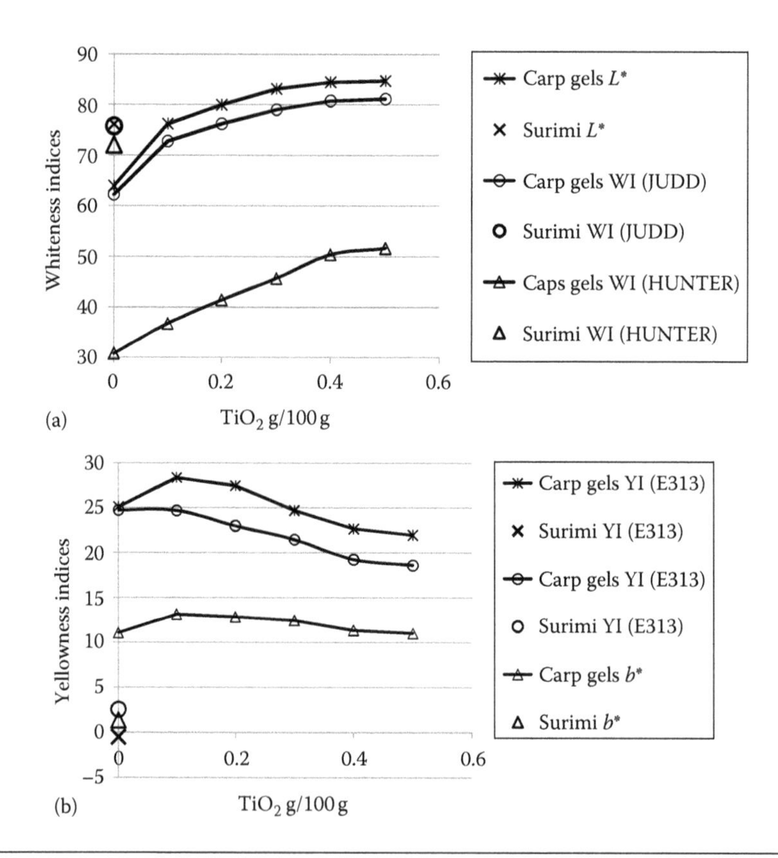

Figure 10.5 (a) Whiteness indices in function of titanium dioxide concentration in proteins recovered from whole fish. (b) Yellowness indices in function of titanium dioxide concentration in proteins recovered from whole fish. (Based on data by Taskaya, L., et al., *LWT Food Sci. Technol.*, 43, 401, 2010.)

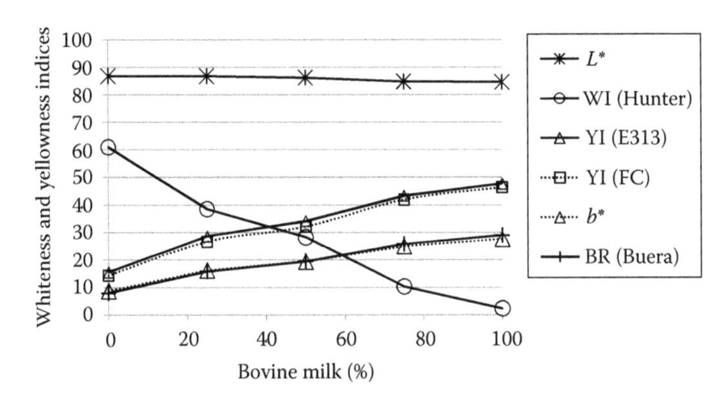

Figure 10.6 Whiteness and yellowness indices of semi-hard cheeses made of different compositions of bovine and caprine milk. (Based on data by Sheehan, J.J. et al., *Int. Dairy J.*, 19, 498, 2009.)

show that only the b^* coordinate changes significantly; there are only very small changes in L^* and a^*. For such a sample set, the YI_{FC} and YI_{E313} are very strongly correlated as are the BR_{BUERA} and b^* values (Figure 10.6).

Based on Figure 10.6, we may conclude that for this type of change the WI_{HUNTER} is the most sensitive and describes the same tendency as any of the yellowness or browning indices (i.e., yellowness increases and whiteness decreases with increasing bovine milk concentration). L^* is not an adequate descriptor of the changes in cheese color due to changes in milk composition.

10.3.7 Color Evaluation of Pasta Samples

Švec et al. (2008) evaluated the color of different pasta samples made of three types of flour: bright M1 as well as semi-bright M2 were milled from common wheat, while M3 from durum wheat. As shown in Figure 10.7, the number of

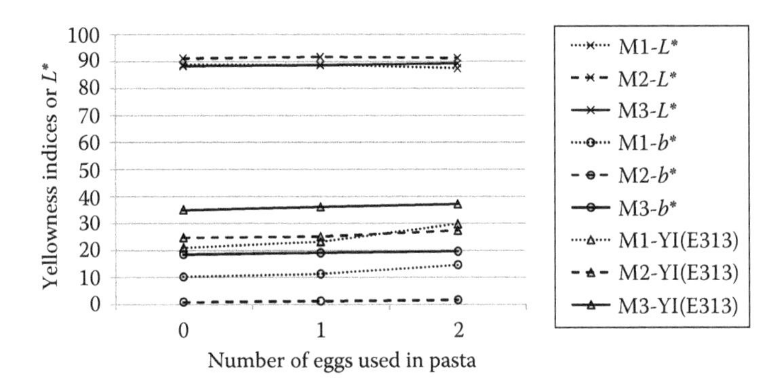

Figure 10.7 Yellowness indices and L^* values of different pasta compositions as a function of the number of eggs used in the pasta. (Based on data by Švec, I. et al., *Czech J Food Sci.*, 26, 421, 2008.)

eggs used in the pasta had the most significant influence on the yellowness for M1 (bright), less for M2 and M3. Adding two eggs reverses the yellowness of M1 and M2 pastas as measured by YI_{E313} (and the strongly correlated YI_{FC}, not shown in the figure) but not by b^*. The L^* (lightness) values show little differences between the different pasta compositions, and practically no influence of the number of eggs, so they are not really useful in measuring this effect.

10.3.8 Color of Fried, Battered Squid Rings

Baixauli et al. (2002) investigated the effect of the addition of corn flour and colorants on the color of fried, battered squid rings. Increasing the amount of corn flour seems to increase the yellowness indices (YI_{FC} and YI_{E313}), primarily due to the increase in the a^* coordinate, while the results for the b^* coordinate are rather erratic (Figure 10.8a).

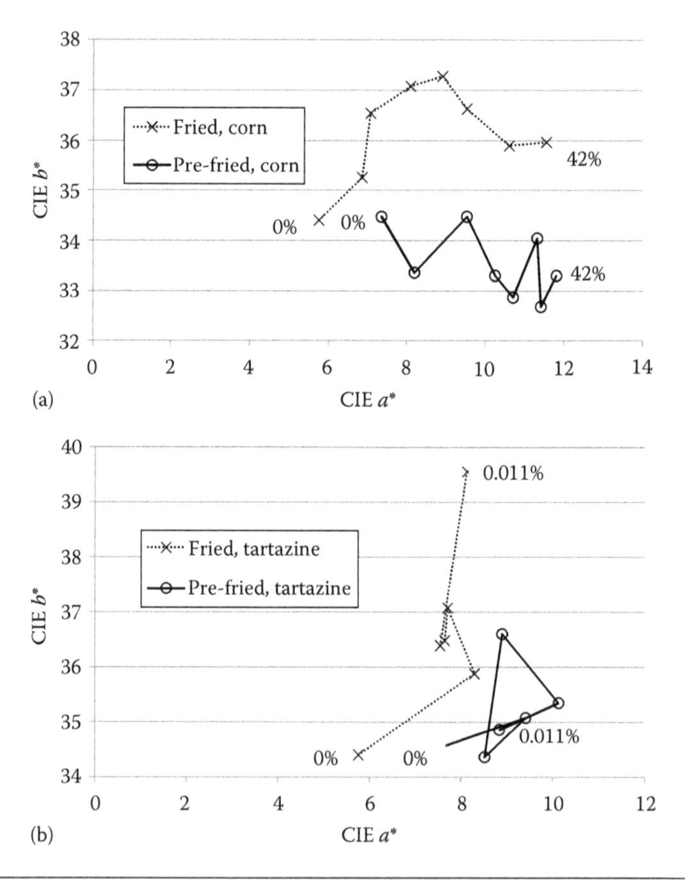

Figure 10.8 (a) CIE a^*-b^* diagram showing the effect of increasing corn flour concentration from 0% to 42% on the color of fried, battered squid rings. (b) CIE a^*-b^* diagram showing the effect of increasing tartrazine concentration from 0% to 0.011% on the color of fried, battered squid rings. (Based on data from Baixauli, R. et al., *Eur. Food. Res. Technol.*, 215, 457, 2002.)

The increase in colorant concentration also seems to increase the yellowness indices, but plotting the data on a CIELAB $a^* - b^*$ diagram shows a lack of clear tendency in the color development due to increased colorant concentration (Figure 10.8b).

10.4 Conclusions

Whiteness, yellowness, and browning are important characteristics of food products, either because consumers may prefer one or the other, depending on the products, or because yellowness or browning indicates change in the quality of the product due to processing or storage. In spite of their importance, these indices are very often used in an inappropriate manner (as illustrated by some of the case studies earlier).

The most widely used figures for whiteness are L^* or WI_{JUDD}, but these seldom have the sensitivity to describe changes, and sometimes show tendencies contrary to those found when more appropriate indices, for example, WI_{HUNTER}, are used. Except for some very special cases, the internationally standardized WI_{CIE} cannot be used because the color of most food products falls outside its limits of validity.

Very often the CIELAB b^* yellowness coordinate is used as a measure of yellowness (the higher the b^* value, the yellower the specimen), but this does not take the lightness dimension into consideration. Generally, YI_{E313} or YI_{FC} are better descriptors of yellowness. Depending on the sample set they may be very strongly correlated (even with b^*), but in other cases the correlation may be very low.

The BR_{BUERA} may be a good descriptor of the browning of food products, depending on the sample set it may or may not be strongly correlated with b^* or one of the yellowness indices.

Many authors use the total color difference (ΔE^*_{ab}) to describe changes in color, but it must be emphasized that ΔE^*_{ab} can show only the "amount" of color difference and not its direction. If it is important to show in which direction the color changes, it is better to use one of the whiteness, yellowness, or browning indices.

When choosing one or the other index, we should consider first of all whether we want to measure a *change in the appearance* of the product or simply *follow a chemical reaction* due to processing or storage. In the first case, the index has to be related to a perceptual quantity ("whiter," "yellower," or "browner") and in the second, it may be any physical or psychophysical quantity (such as reflectance, transmittance, or absorbance at a given wavelength or a combination of these at different wavelengths).

Finally, we have to draw attention to the importance of clearly describing measurement parameters in publications discussing color. In many of the reviewed articles, the authors do not specify the illuminant, the observer, or the measurement geometry, and it is often not clear if the L, a, b or L^*,

a^*, b^* coordinates given refer to the Hunter or the CIELAB values (which are most certainly not interchangeable). There seems to be a crying need for more color education in the proper application of color-measuring equipment in the food industry.

Acknowledgment

The author thanks Dr. Ana Salvador Alcaraz, Ms. Derya Arslan, Dr. Yi-Zhong Cai, Dr. María Vargas Colás, Dr. Jacek Jaczynski, Dr. Diarmuid Sheehan, Dr. Ivan Švec, and Dr. Mayoyes Alvarez Torres, corresponding authors of the publications on which the case studies have been based, and to which they have kindly provided measurement details and additional data.

References

Arslan, D. and M. M. Özcan. 2010. Study the effect of sun, oven and microwave drying on quality of onion slices. *LWT—Food Science and Technology* 43: 1121–1127.

ASTM (American Society for Testing and Materials). 2005. *E 313-05 Standard Practice for Calculating Yellowness and Whiteness Indices from Instrumentally Measured Color Coordinates*. West Conshohocken, PA: ASTM International.

Baixauli, R., A. Salvador, S. M. Fiszman, and C. Calvo. 2002. Effect of the addition of corn flour and colorants on the colour of fried, battered squid rings. *European Food Research and Technology* 215: 457–461.

Buera, M. P., R. D. Lozano, and C. Petriella. 1985. Definition of color in the non-enzymatic browning process. *Die Farbe* 32/33: 316–326.

CIE (Commission Internationale de l'Eclairage). 2004. *Colorimetry*, 3rd edn. Vienna: CIE Central Bureau, Publ. 15-2004.

Feillet, P., J. C. Autran, and C. Icard-Vernière. 2000. Pasta brownness: An assessment. *Journal of Cereal Science* 32: 215–233.

Fernández, C., M. D. Alvarez, and W. Canet. 2008. Steady shear and yield stress data of fresh and frozen /thawed mashed potatoes: Effect of biopolymers addition. *Food Hydrocolloids* 22: 1381–1395.

Francis, F. J. and F. M. Clydesdale. 1975. *Food Colorimetry: Theory and Applications*. Westport, CN: AVI Publishing.

Hunter, R. S. 1960. New reflectometer and its use for whiteness measurement. *Journal of the Optical Society of America* 50: 44–48.

Judd, D. B. and G. Wyszecki. 1963. *Color in Business, Science and Industry*. New York: John Wiley & Sons.

Kotwaliwale, N., P. Bakane, and A. Verma. 2007. Changes in textural and optical properties of oyster mushroom during hot air drying. *Journal of Food Engineering* 78: 1207–1211.

Sheehan, J. J., A. D. Patel, M. A. Drake, and P. L. H. McSweeney. 2009. Effect of partial or total substitution of bovine for caprine milk on the compositional, volatile, non-volatile and sensory characteristics of semi-hard cheeses. *International Dairy Journal* 19: 498–509.

Švec, I., M. Hrušková, M. Vítová, and H. Sekerová. 2008. Colour evaluation of different pasta samples. *Czech Journal of Food Sciences* 26: 421–427.

Taskaya, L., Y. Chen, and J. Jaczynski. 2010. Color improvement by titanium dioxide and its effect on gelation and texture of proteins recovered from whole fish using isoelectric solubilization/precipitation. *LWT—Food Science and Technology* 43: 401–408.

Vargas, M., C. Maite, A. Albors, C. Amparo, and C. Gonzáles-Martínez. 2008. Physicochemical and sensory characteristics of yoghurt produced from mixtures of cows' and goats' milk. *International Dairy Journal* 18: 1146–1152.

Xiong, G., W. Cheng, L. Ye, X. Du, M. Zhou, R. Lin, S. Geng, M. Chen, H. Corke, and Y. Cai. 2009. Effects of konjac glucomannan on physicochemical properties of myofibrillar protein and surimi gels from grass carp (*Ctenopharyngodon idella*). *Food Chemistry* 116: 413–418.

Zapotoczny, P., M. Markowski, K. Majewska, A. Ratajski, and H. Konopko. 2006. Effect of temperature on the physical, functional, and mechanical characteristics of hot-air-puffed amaranth seeds. *Journal of Food Engineering* 76: 469–476.

CHAPTER 11

Color Classification of Veal Carcasses

Past, Present, and Future

MARCEL LUCASSEN, JOHAN ALFERDINCK, and RON VAN MEGEN

Contents

11.1 Introduction

In compliance with EC Directives, in Dutch slaughterhouses veal carcasses are classified on the basis of three factors: conformation, fatness, and color. Of these factors, color has become a very important parameter in the pricing system, involving both the farmers who breed the animals and buyers who deliver the meat to the consumer market. In the past 20 years, major efforts have been put into the development of a reliable color classification system for veal carcasses. Such a system is necessary to guarantee uniform classification results among the different slaughterhouses in the Netherlands, but also to provide a sound basis for international trading since the majority of veal meat produced in the Netherlands is exported to other countries. In this field, the Netherlands has a leading global position.

This chapter presents the main issues involved in the development of the color classification method, discussing both the historical perspective and current state of the art. We also take a look at future possibilities.

11.2 Past: From Visual to Instrumental Color Classification

Initially, the color classification at the different slaughterhouses was performed visually by certified employees of the Central Office for Slaughter Livestock Services (BV CBS). At 45 min postmortem, and under prescribed lighting conditions, the color of the *musculus rectus abdominis* (muscle tissue) was visually matched to a 10-point scale ranging from light (1) to dark (10). Figure 11.1 shows a photograph of the color scales used for classification of white veal (image on the left-hand side) and pink veal (image on the right-hand side). In the remainder of this chapter, we will only discuss the classification of white veal. The exact color specifications of the 10-point scale were determined from an analysis of the gamut in CIELAB color space encompassed by representative variations in veal meat samples. The color scale was made by paint samples whose spectral reflectance was matched against the spectra derived from the 10 selected colors in CIELAB space. Due to inevitable illuminant metamerism associated with spectral matching, the developed color scale was optimized for a specific light source, and thus required controlled lighting conditions (involving both color and intensity) in the slaughterhouses. Fluorescent lamps with a daylight color (TL57, Philips) and a color rendering index Ra > 90 (CIE 1995b) were prescribed for the slaughterhouses. Although the visual method is simple and effective, it depends on subjective criteria and illumination. As a logical next step, the subjective visual color classification was replaced by objective instrumental color classification. A handheld tristimulus colorimeter (Konica-Minolta CR300), operated by certified personnel, measures *XYZ* tristimulus values of the same muscle tissue (*musculus rectus abdominis*), which are then converted into a color class on the 10-point scale. The algorithms underlying this conversion were derived from comparison of the measured $L^*a^*b^*$ values with the visually assigned color classes. Functions derived by discriminant analysis were applied to calculate the most likely color class belonging to the

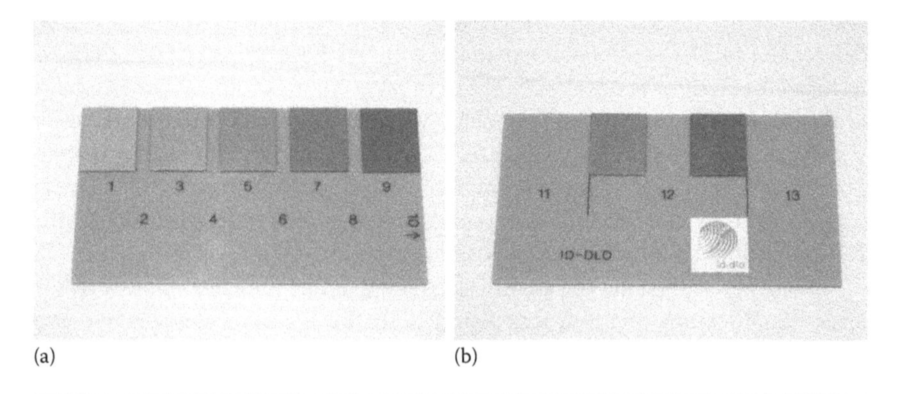

(a) (b)

Figure 11.1 Gray-scale pictures of the standards used for visual color classification of white veal (a) and pink veal meat (b).

measured $L^*a^*b^*$ values, as described in Hulsegge et al. (2001). The measured $L^*a^*b^*$ values result from application of the 2° CIE standard observer and standard daylight illuminant D65, whereas the illumination in the slaughterhouses was TL57. This difference in illumination gives rise to a structural but limited difference in colorimetric values which we assume to be modeled by the discriminant functions.

11.3 Present: Improvements in Hardware and Software

Today, the instrumental classification as sketched earlier is still in place, albeit in optimized format. To increase the quality and stability of the color measurements, the instruments have been replaced by newer versions (Konica-Minolta CR400). Before doing so, however, several options were considered. A switch from using a tristimulus meter to a spectrophotometer would allow illuminant metamerism to be considered (Berns 2000), but at the same time it would increase the risk of fouling the integrating sphere of the spectrophotometer. Also, a larger measurement aperture would minimize the effect of spatial inhomogeneities in the meat tissue being measured. However, in order not to lose the connection with historical databases containing both measurements and visual assessments, and knowing that sufficient accuracy could be reached using a proper calibration procedure, it was decided to stick to the same color measurement technique with a tristimulus colorimeter (Figure 11.2).

After introduction of the new color measurement instruments in the practice of the slaughterhouses, the instrument calibration procedure was improved. Up till then, the daily instrument calibration check involved verification of XYZ values measured on a white calibration tile. Whenever the difference between any of the measured X, Y, or Z values with the target values exceeded a given percentage, it was required to inspect the instrument

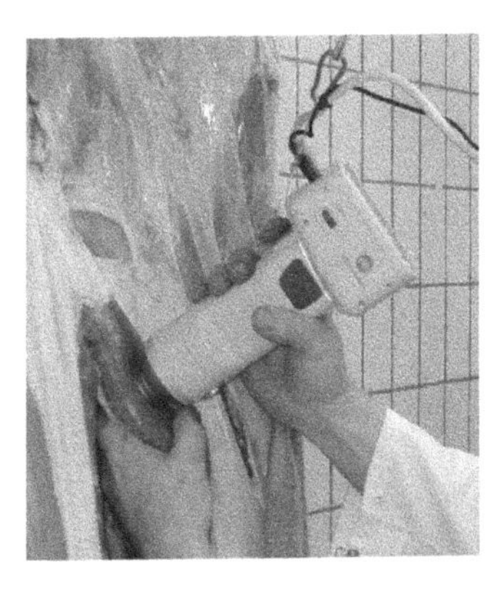

Figure 11.2 Instrumental color classification at the slaughterhouse. The operator uses a handheld colorimeter to measure XYZ tristimulus values from the *musculus rectus abdominis*.

for possible contamination, clean the instrument and calibration tile and remeasure, or even replace the instrument by a spare one, if necessary. In the new procedure, an additional calibration tile having a color representative for veal meat is also measured. This so-called *user calibration* is not only a direct verification of the proper functioning of the instrument in the target area of color space, but also beneficial for maintaining the *inter-instrument agreement* (IIA) between different instruments (of the same type) at an acceptable level. Differences between measured and target values are now expressed in the ΔE_{94} color difference metric (CIE 1995a), the value of which lies in one of three categories indicating the instrument's operational status. The latter is labeled either code green (instrument OK), code yellow (instrument still OK but may need attention), or code red (instrument not OK: clean and remeasure). After 1 year of testing in practice, the stability and IIA are considered as excellent. The choice for the relatively simple ΔE_{94} metric, and not the more recent and complex ΔE_{00} (CIE 2001), was motivated by the fact that the colors of white veal meat are restricted to a limited "reddish" area in CIELAB space (see Figure 11.3). Application

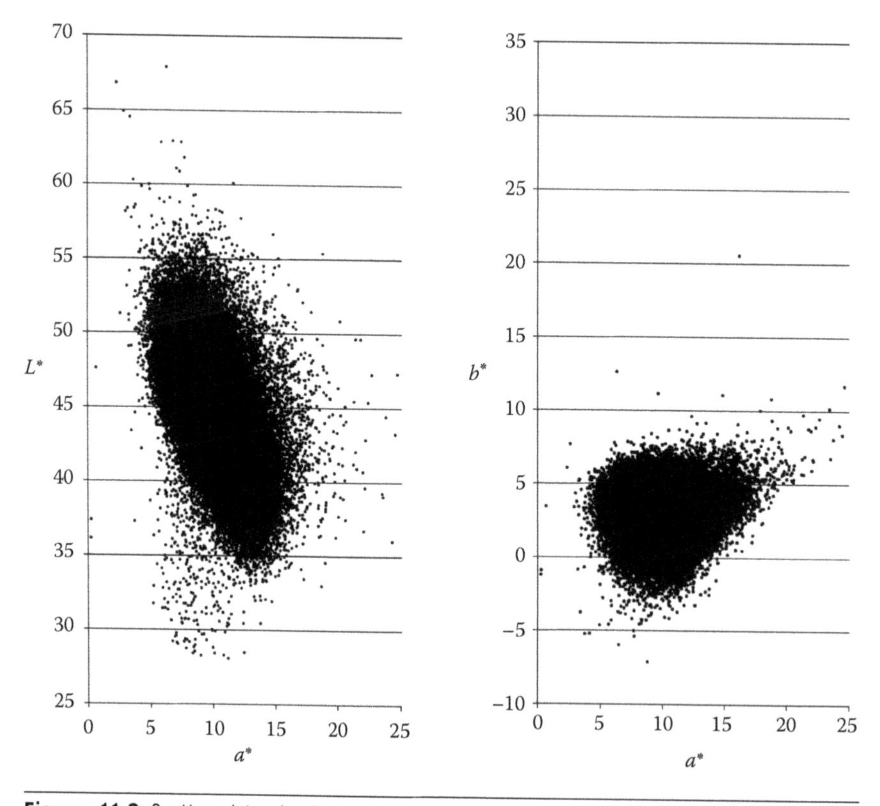

Figure 11.3 Scatter plots showing the gamut of measured white veal in CIELAB color space (n=113,556).

of the ΔE_{00} is said to be most helpful in the blue area of color space and for near-achromatic colors. Also, the equations underlying the computation of ΔE_{00} do not support an easy interpretation of visually perceived differences.

Finally, a new algorithm was developed to convert measured *XYZ* values into a color class along the 10-point scale. The "old" algorithm, using discriminant functions that disregarded the measured b^* value, did not seem to optimally cover the gamut of veal meat color anymore. The new algorithm incorporates the same metric as used for the instrument calibration check, ΔE_{94}, for which we now optimized the parametric factors k_L, k_C, and k_H. Ten center points were selected in CIELAB space, which represent the 10 color classes and have a structured pattern in their mutual spacing which was on average $\Delta E_{ab} = 2$. For each color measurement, we compute the color difference between the measured L^*, a^*, b^* values with each of the 10 central points. The class associated with the central point having the smallest ΔE_{94} is then assigned as the final color class. The optimization of the parameters k_L, k_C, and k_H and the selection of the 10 center points are done with respect to historical databases containing both the classifications of the previous algorithm and visual classifications. As Figure 11.4 shows, the distribution of assigned color classes of the new algorithm closely resembles that of the old algorithm. This is important from both technical and commercial points of view.

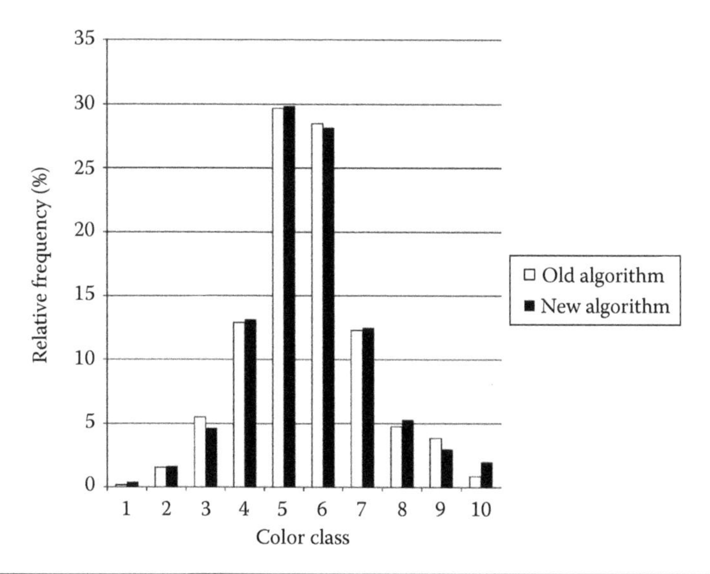

Figure 11.4 Relative frequency distribution of veal carcass color classes, determined with the old and the new algorithm. The latter is based on the minimum color difference with the centers of the 10-point scale in CIELAB space, being more stable and easier to comprehend.

11.4 Future Perspective

In the future, we may expect to further benefit from technological breakthroughs in color measurement. Developments in light-emitting diode (LED) technology already offer the possibility to use LEDs as the internal light source in color measurement systems, which are even more stable, energy efficient, and have a longer lifetime, and thus may require less calibration efforts.

Operational research can be conducted in the different slaughterhouses to investigate local factors that may cause differences between slaughterhouses. Differences in conditions such as temperature, humidity, but also transportation and animal stress, may lead to unwanted variations in the measured color and hence in the assigned color class.

With the upcoming possibilities of image processing as described by Du and Sun (2004), camera-based, noncontact color measurement would seem to be the next step. However, this implies that the color of veal meat should be measured on the outside of the carcass. It is questionable whether this correlates well enough with the color of the *musculus rectus abdominis*, which has been previously suggested by Denoyelle and Berny (1999) as the preferred indicator of veal meat color. In addition, it would require substantial attention to calibrated lighting conditions in the slaughterhouses. In particular, the glossiness of the moist carcasses as well as of the fat coverage is considered a difficult hurdle to pass. It is known that noncontact color measurement is more suitable to assess differences in color than to determine the absolute color. Perhaps, the need for exact (calibrated) lighting may be relaxed by using smart portions of the wavelength area (Aporta et al. 1996) or by smart calibrations to known color references (Connolly et al. 1996, Pointer 2000).

References

Aporta, J., B. Hernández, and C. Sañudo. 1996. Veal colour assessment with three wavelengths. *Meat Science* 44: 113–123.

Berns, R. S. 2000. *Billmeyer and Saltzman's Principles of Color Technology*. New York: John Wiley & Sons.

CIE (Commission Internationale de l'Eclairage). 1995a. *Industrial Colour-Difference Evaluation*. Vienna, Austria: CIE Central Bureau, Publ. 116.

CIE (Commission Internationale de l'Eclairage). 1995b. *Method of Measuring and Specifying Colour Rendering Properties of Light Sources*. Vienna, Austria: CIE Central Bureau, Publ. 13.3-1995.

CIE (Commission Internationale de l'Eclairage). 2001. *Improvement to Industrial Colour-Difference Evaluation*. Vienna, Austria: CIE Central Bureau, Publ. 142.

Connolly, C., T. W. W. Leung, and J. Nobbs. 1996. The use of video cameras for remote colour measurement. *Journal of the Society of Dyers and Colourists* 112: 40–43.

Denoyelle, C. and F. Berny. 1999. Objective measurement of veal color for classification purposes. *Meat Science* 53: 203–209.

Du, C. J. and D. W. Sun. 2004. Recent developments in the applications of image processing techniques for food quality evaluation. *Trends in Food Science & Technology* 15(5): 230–249.

Hulsegge, B., B. Engel, W. Buist, G. S. M. Merkus, and R. E. Klont. 2001. Instrumental colour classification of veal carcasses. *Meat Science* 57: 191–195.

Pointer, M. R. 2000. Digital cameras for colour measurement. *Colour Image Science 2000*. Derby, U.K.: University of Derby, pp. 71–83.

CHAPTER 12

Application of Image Analysis to the Color–Phenolic Composition Relationships of Grape Seeds

FRANCISCO JOSÉ RODRÍGUEZ-PULIDO, FRANCISCO J. HEREDIA, JUAN MANUEL ZALDÍVAR-CRUZ, and M. LOURDES GONZÁLEZ-MIRET

Contents

12.1 Introduction

The study of phenolic compounds in grape seeds of *Vitis vinifera* has achieved great interest in recent years. These compounds are responsible for structure and aroma of wine. Furthermore, these compounds are implicated in color of wine. Phenols such as catechin and its polymers are commonly present in grape seeds, which pass to the wine during fermentation process. Grape seeds are composed of water (25%–45%), polysaccharides (34%–36%), tannins (4%–10%), nitrogen compounds (4%–6.5%), minerals (2%–4%), and lipids (13%–20%). Although some of these phenolics are noncolored components, they influence the final color of red wine due to copigmentation reactions with anthocyanins (Boulton 2001). Copigmentation phenomenon consists of noncovalent molecular associations between anthocyanins and other organic compounds such as flavonoids, yielding to changes or increments of the color intensity.

The composition of the seeds changes along the maturation until the grapes reach the ripeness, affecting the sensory properties of wine. In insufficiently mature grapes, seeds give aggressive and herbaceous flavor, yielding astringent taste to wine (Ristic and Iland 2005).

In warm climates, the wine is exposed to dry springs and hot summers. Due to this, the period between veraison and industrial maturity decreases. This phenomenon causes the grape pulp quickly reach the suitable sugar level to be fermented, while the phenolic compounds of the seeds have not yet reached optimum maturity. This will cause loss of color due to lack of phenols to stabilize the anthocyanic color in the wine. Currently, the moment of harvesting is determined based on chemical properties of the must, and the phenolic maturation of the seeds is not usually considered. The chemical changes occurring during the phenolic maturation induce changes of appearance of the seeds, modifying their color (from pale green to dark brown) as well as their shape.

On the other hand, colorimetry is widely used for evaluating the quality and composition of foods. Digital image analysis appears as a successful complement since it can be used to determine not only color but also other characteristics such as shape, texture, and homogeneity (Savakar et al. 2009, Zheng et al. 2006). In previous studies, we have determined the usefulness of the digital image analysis to assess the phenolic maturation evolution of the seeds in grapes for vinification. This has achieved interest in the industry. The morphological differences between varieties have been studied using general discriminant analysis models; these can classify grape seeds with high accuracy (Rodríguez-Pulido et al. 2010). This study has been continued using a deeper knowledge of the phenolic composition of the seeds and determining their relationships with appearance using digitization techniques.

12.2 Sampling

A red grape variety (*Tempranillo* cv.) grown in vineyards located in Condado de Huelva (Southwest of Spain) was sampled for this study twice a week from July until the harvest. A gross sample of 2 kg was divided into two fractions. One of these fractions was used for physicochemical analysis of the must, which determined average weight and berry size, pH, sugars, and total acidity. In the other fraction, 100 seeds were separated, cleaned, and imaged. Once the images were taken, the seeds were kept frozen until later HPLC analysis of phenols.

12.3 Image Analysis

Image analysis is framed by computer vision, a subfield of artificial intelligence, and its aim is to "teach" a computer to understand a scene and the characteristics of an image. A computer vision system includes an illumination system, a charge-coupled device (CCD), a frame grabber that converts the analog image from the camera into a digitized one, and a computer with the suitable software for image processing and interpretation of results (Wang and Sun 2002). The digital camera receives images onto a CCD that registers them in gradations of three basic colors: red, green, and blue (RGB). This color space is widely used in electronic devices but it depends on the device where it is visualized. Thus, we cannot use this space for absolute measures of color. It is therefore necessary to convert between color spaces using a calibration. This requires control of illumination (CIE 2007, León et al. 2006). In this study, CIELAB space has been used. This was proposed by the International Commission on Illumination, CIE (1976), which defines any color by three attributes: lightness (L^*), chroma (C^*_{ab}), and hue (h_{ab}). Besides color calibration, image analysis such as segmentation is an important feature of the analysis process. Segmentation refers to the process of partitioning a digital image into multiple segments (sets of pixels). The goal of segmentation is to simplify and/or change the representation of an image into something more meaningful and easier to analyze.

The main advantage of measuring color by digitalization derives from the fact that we can identify the seeds in the images by combining image analysis and tristimulus colorimetry technologies, and then morphological and colorimetric characteristics can be extracted from each one. Moreover, thanks to advances in computer science, this technique is becoming a powerful tool to test quality in the food industry (Brosnan and Sun 2004). Choosing the appropriate segmentation criteria, it is possible to analyze color and measure morphological properties in an automated manner (Zheng and Sun 2008).

In this chapter, the DigiEye® imaging system was used (Luo et al. 2001) (Figure 12.1). This includes an illumination box specially designed by VeriVide Ltd. (Leicester, United Kingdom) to illuminate the samples consistently in D65 illuminant and a digital camera connected to a computer (Figure 12.2). The camera parameters were set for all images with an exposure time of 1/15 s, aperture *f*/6.3, ISO 200 sensitivity, and were stored in TIF format with a size of 3872 × 2592 pixels and a resolution of 96 PPI (pixels per inch).

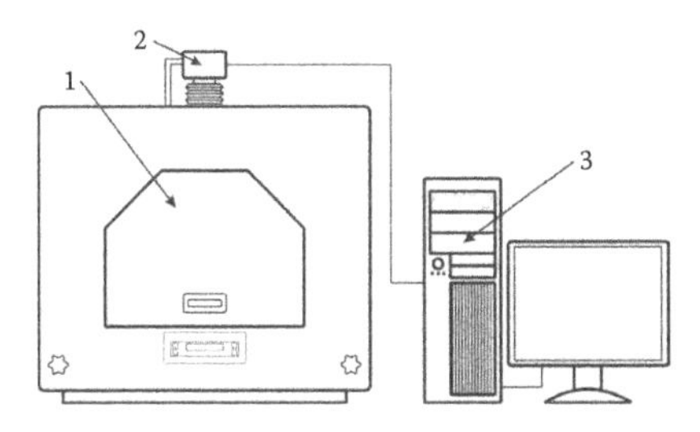

Figure 12.1 The DigiEye® system: (1) illumination box, (2) digital camera, and (3) computer.

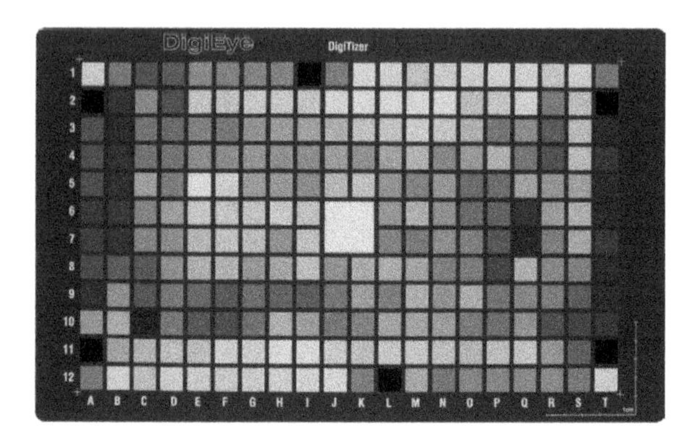

Figure 12.2 Chart of calibration of digital camera used by the DigiEye® system.

Using image analysis, morphological parameters and appearance were obtained as well as conversion from *RGB* values to CIELAB coordinates with original software DigiFood® (Heredia et al. 2006). Segmentation criteria were based on HSI (hue/saturation/intensity) color space and morphological features of seeds, such as area and size. HSI color space stems from *RGB* but its attributes have the same psychophysical meaning as that of the angular coordinates of CIELAB space. The colorimetric segmentation criteria were hues (H) between 0 and 255 (all hues), saturation (S) between 0 and 180, and intensity (I) between 0 and 140. Furthermore, these objects must have an area between 6 and 35 mm^2 and length not greater than 15 mm. These dimensions are consistent with the sizes of the seeds. For this purpose and in order to establish a correspondence between both magnitudes, the software performed a spatial calibration to convert pixels to millimeters.

From segmented images, the following parameters were measured:

- Length, width, and area
- Aspect: Ratio between major and minor axes of the ellipse equivalent to seed
- Roundness $= \dfrac{\text{Perimeter}^2}{4 \times \pi \times \text{area}}$
- Heterogeneity: Fraction of pixels that deviate >10% from the average intensity
- Browning level: Fraction of pixels having L^* value is lower than 50 units
- Mean color difference from the mean (MCDM) (Berns 2000):

$$\text{MCDM} = \frac{\sum_{i=1}^{N} \left[\left(L_i^* - \bar{L}^* \right)^2 + \left(a_i^* - \bar{a}^* \right)^2 + \left(b_i^* - \bar{b}^* \right)^2 \right]^{1/2}}{N}.$$

- RGB color

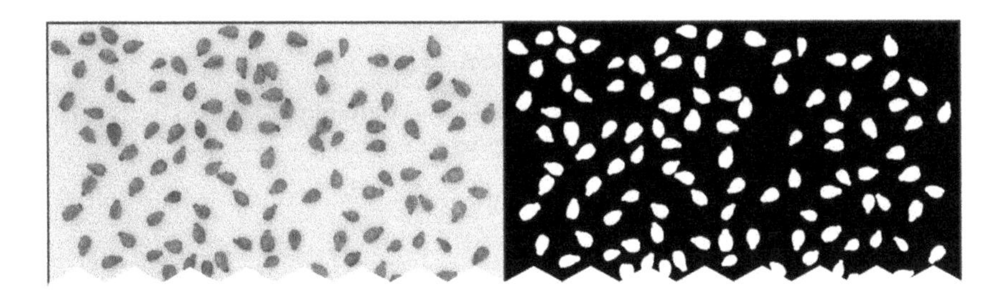

Figure 12.3 Image of grape seeds and resulting mask after apply segmentation process.

Table 12.1 Colorimetric Values for Grape Seeds during Maturation

Sample	Date	L^*	a^*	b^*	C^*_{ab}	h_{ab}
1	Jul 20	57.97	10.08	29.10	31.09	69.65
2	Jul 23	54.97	10.36	27.33	29.60	67.70
3	Jul 27	55.97	11.24	27.55	29.98	67.25
5	Aug 3	51.49	11.17	23.92	26.60	64.24
6	Aug 6	51.01	11.72	24.05	26.95	62.89
7	Aug 10	50.92	12.02	23.74	26.79	62.26
8	Aug 13	50.15	11.11	21.41	24.31	61.58

Once the images were acquired and the criteria of segmentation were applied, all seeds were recognized by the software (Figure 12.3).

Table 12.1 shows the CIELAB colorimetric parameters at each stage of maturation. Lightness, chroma, and hue all decreased with time due to the browning of the seed.

12.4 Chemical Analysis

For the analysis of phenols, a modification of the method proposed by Sandhu and Gu (2010) was applied. 0.5 g of freeze-dried seeds were milled and extracted with 10 mL of acetone/water/acetic acid (70:29.3:0.3 v/v) solvent. After this, acetone from extraction was removed with the help of a vacuum concentrator. The remaining fraction was redissolved in an aqueous solution of formic acid. This solution was filtered in nylon and was encapsulated for the injection. Chromatographic analysis was performed on an Agilent 1200 series HPLC system equipped with a UV–visible diode array detector.

The phenolic profile of seeds was analyzed. For identification of these compounds, retention times and spectra were compared. Although other compounds were identified, Table 12.2 shows only those compounds whose concentration changed.

For the study of relationship between chemical composition and color in seeds, linear and multiple regression models were applied. As shown in

Table 12.2 Content of Phenolic Compounds in Grape Seeds

Date	Protocatechuic Acid (mg/kg)	Catechin (mg/kg)
Jul 20	8	867
Jul 23	8	985
Jul 27	12	674
Aug 3	26	504
Aug 6	20	604
Aug 10	38	443
Aug 13	24	491

Table 12.3 Correlation Matrix between Colorimetric and Chemical Values

	L^*	a^*	b^*	C^*_{ab}	h_{ab}
Protocatechuic acid (mg/kg)	**−0.82**	**0.81**	**−0.78**	−0.75	**−0.84**
Catechin (mg/kg)	**0.80**	**−0.83**	**0.82**	**0.80**	**0.86**

Note: Results in bold are significant at $p < 0.05$.

Table 12.3, the quantitative colorimetric variables correlated with not only some phenolic compounds, but also the hue (h_{ab}), demonstrating their involvement in browning phenomenon from a qualitative point of view.

12.5 Conclusions

Features in images such as color, size, and shape can be extensively applied to seed maturity evaluation. The results suggest that image analysis has good potential to be used for color and appearance measurements in grape seeds. Moreover, as has been demonstrated in other food products, there is a high correlation between the chemical and food color. For this reason, this technique provides a fast, nondestructive, and low-cost alternative in the winery industry to conventional chemical analysis. The digital imaging method provides measurements and analyses of the color of food surfaces that are adequate for food science research. While it is not yet a replacement for sophisticated color measurement instruments, it is an attractive alternative due to its simplicity, versatility, and low cost.

Acknowledgment

This work was supported by funding from the Ministerio de Ciencia e Innovación, Gobierno de España by the project AGL2008-05569-C02-02.

References

Berns, R. 2000. *Billmeyer and Saltzman's Principles of Color Technology*. New York: John Wiley & Sons.

Boulton, R. 2001. The copigmentation of anthocyanins and its role in the color of red wine: A critical review. *American Journal of Enology and Viticulture* 52(2): 67–87.

Brosnan, T. and D. W. Sun. 2004. Improving quality inspection of food products by computer vision—A review. *Journal of Food Engineering* 61(1): 3–16.

CIE (Commission Internationale de l'Éclairage). 1976. *Recommendations on Uniform Color Spaces, Color-Difference Equations, Psychometric Color Terms*. Vienna, Austria: CIE Central Bureau.

CIE (Commission Internationale de l'Éclairage). 2007. *Standard Illuminants for Colorimetry*. Vienna: CIE Central Bureau. ISO 11664-2:2007.

Heredia, F. J., M. L. González-Miret, C. Álvarez, and A. Ramírez. 2006. DigiFood. Registration No. SE-01298.

León, K., D. Mery, F. Pedreschi, and J. León. 2006. Color measurement in $L^*a^*b^*$ units from RGB digital images. *Food Research International* 39(10): 1084–1091.

Luo, M. R., G. H. Cui, and C. Li. 2001. Apparatus and method for measuring colour (DigiEye System), Derby University Enterprises Limited. British Patent 0124683.4.

Ristic, R. and P. G. Iland. 2005. Relationships between seed and berry development of *Vitis vinifera* L. cv. Shiraz: Developmental changes in seed morphology and phenolic composition. *Australian Journal of Grape and Wine Research* 11(1): 43–58.

Rodríguez-Pulido, F. J., L. Gómez-Robledo, M. L. González-Miret, and F. J. Heredia. 2010. Seguimiento de la maduración de variedades de uva mediante colorimetría y análisis de imagen de semillas. In *IX Congreso Nacional del Color*. Alicante, Spain.

Sandhu, A. K. and L. Gu. 2010. Antioxidant capacity, phenolic content, and profiling of phenolic compounds in the seeds, skin, and pulp of *Vitis rotundifolia* (Muscadine Grapes) as determined by HPLC-DAD-ESI-MSn. *Journal of Agricultural and Food Chemistry* 58(8): 4681–4692.

Savakar, D., G. Anami, and S. Basavaraj. 2009. Classification of food grains, fruits and flowers using machine vision. *International Journal of Food Engineering* 5(4): art. 14.

Wang, H. H. and D. W. Sun. 2002. Correlation between cheese meltability determined with a computer vision method and with Arnott and Schreiber tests. *Journal of Food Science* 67(2): 745–749.

Zheng, C. and D. W. Sun. 2008. Object measurement methods. In *Computer Vision Technology for Food Quality Evaluation*. Amsterdam, the Netherlands: Academic Press.

Zheng, C., D. W. Sun, and L. Zheng. 2006. Recent developments and applications of image features for food quality evaluation and inspection—A review. *Trends in Food Science & Technology* 17(12): 642–655.

CHAPTER **13**

Sensing Chlorophyll, Carotenoids, and Anthocyanins Concentrations in Leaves with Spatial Resolution from Digital Image

GABRIELA CORDON and MARÍA GABRIELA LAGORIO

Contents

13.1 Introduction

Prediction of nutrients shortage in plants from the observation of visual symptoms is a current practice in agronomy. In fact, chlorosis of older leaves is a sign for nitrogen deficiency while redness in mature leaves, caused by the increase in anthocyanins content, can be a symptom for phosphorus deficiency (Bonilla 2000).

As the color of plant leaves is an indicator of their pigments' quality and content, it makes sense to think in the possible application of color images of leaves to the nondestructive assessment of plant health. Digital images and color analysis of leaves have been studied in literature, in connection with the evaluation of the effect of environmental stress on plants. These works included stress produced by water shortage and nitrogen deficiency (Ahmad and Reid 1996), low temperature (Bacci et al. 1998), senescence (Schaberg et al. 2008), and micropropagation of plants (Yadav and Ibaraki 2010).

The present study shows a method to infer the concentration of pigments in a leaf from images obtained with a commercial scanner. The methodology presented here displays a great potential due to the low cost of implementation, as it only requires a commercial scanner and software for the digital processing of the images. Additionally, this method has practical advantages as an alternative procedure to the analysis of pigments by traditional chemical methods. Chemical methods are destructive and time-consuming whereas imaging methods may be performed quickly, on intact plant material, allowing additionally the spatial resolution of pigment concentration in leaves.

13.2 Materials and Methods

13.2.1 Correlations between R, G, and B Coordinates and Pigment Concentration

Uniformly colored leaves of different species were selected to get a correlation between color coordinates and the pigment concentration. Reflectance spectra of these leaves were then obtained from 300 to 700 nm, using a spectrophotometer (UV3101PC, Shimadzu, Tokyo, Japan) equipped with an integrating sphere (ISR-3100, Shimadzu, Tokyo, Japan). Barium sulfate was used as 100% reflectance (white reference). Seven reflectance spectra were averaged for each species and the resultant spectrum was used for further derivation of color coordinates (*R*, *G*, and *B*). Spectral measurements were performed at room temperature on leaves detached from the plant a few minutes before the determinations were carried out. The color uniformity was proved by measuring reflectance spectra of different areas in the leaves: differences lower than 2% in reflectance values for the same set of leaves were considered acceptable. The selected species were *Hedera helix*, *Liquidambar styraciflua*, *Populus alba*, *Rosa* sp., *Gardenia jasminoides*, *Schefflera arborícola*, *Aloysia triphylla*, and *Ficus benjamina*. On these leaves, chlorophyll a (Chl_a), chlorophyll b (Chl_b), carotenoids (Car), and anthocyanins (Anth) were determined by extraction and posterior spectrophotometric measurement (Sims and Gamon 2002). The foliar area was also determined and the pigment concentration was finally expressed as nmol/cm^2 of leaf.

It is important to note that the variety of plant species used are not related at all. This fact guarantees that the correlations arising from these data can be generalized to other types of species. Not only green leaves but also yellow and red leaves were selected for the present study to avoid too narrow margins of pigment concentration.

From the reflectance spectrum defined as $S(\lambda)$ of uniformly colored leaves, it was possible to calculate the tristimulus values XYZ using Equations 13.1 through 13.3. In these equations, $\bar{x}$, $\bar{y}$, and $\bar{z}$ are the functions defined by the CIE 1931 standard observer and $I(\lambda)$ is the spectral distribution of a known reference illuminant (in this case, D65). The color space used in this case was sRGB. The normalization factor N was calculated by Equation 13.4:

$$X = \frac{1}{N}\int_{\lambda} \bar{x}(\lambda)S(\lambda)I(\lambda)d\lambda \tag{13.1}$$

$$Y = \frac{1}{N}\int_{\lambda} \bar{y}(\lambda)S(\lambda)I(\lambda)d\lambda \tag{13.2}$$

$$Z = \frac{1}{N}\int_{\lambda} \bar{z}(\lambda)S(\lambda)I(\lambda)d\lambda \tag{13.3}$$

$$N = \int_{\lambda} \bar{y}(\lambda)I(\lambda)d\lambda \tag{13.4}$$

Given an XYZ color whose components are in the nominal range (0, 1), the transformation from XYZ to the RGB values was performed according to the following equations:

$$\begin{bmatrix} r \\ g \\ b \end{bmatrix} = [M]^{-1} \begin{bmatrix} X \\ Y \\ Z \end{bmatrix} \tag{13.5}$$

where

$$[M]^{-1} = \begin{bmatrix} 3.2404542 & -1.5371385 & -0.4985314 \\ -0.9692660 & 1.8760108 & 0.0415560 \\ 0.0556434 & -0.2040259 & 1.0572252 \end{bmatrix} \tag{13.6}$$

and

$$R = \begin{cases} 12.92r & r \leq 0.0031308 \\ 1.055r^{1.0/2.4} - 0.055 & r > 0.0031308 \end{cases} \tag{13.7}$$

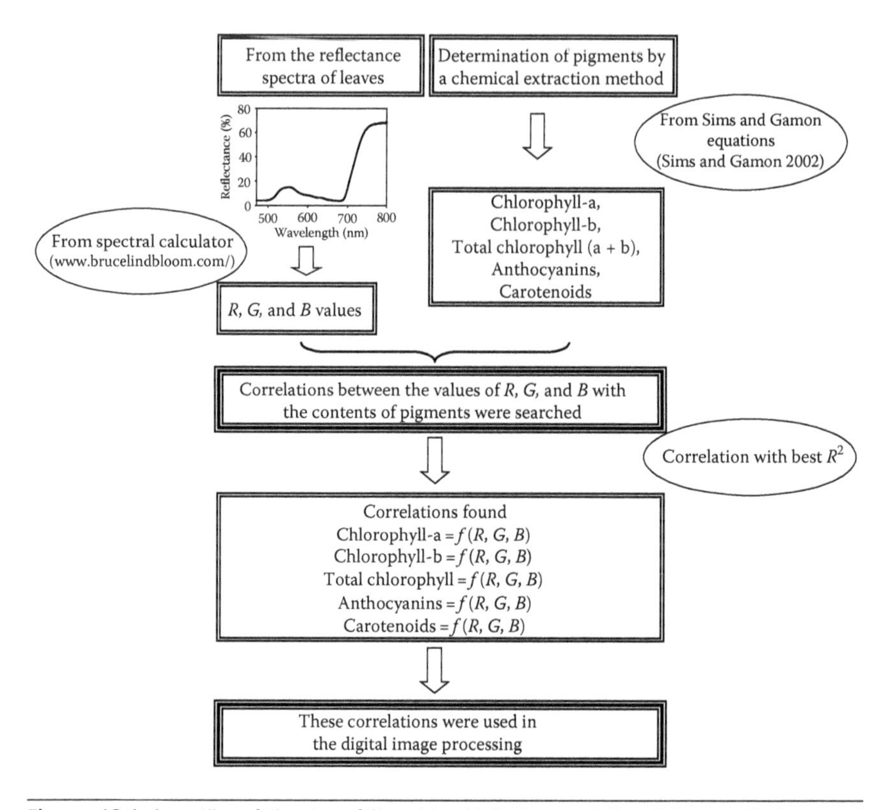

Figure 13.1 An outline of the steps followed to obtain the correlations between reflectance and pigment content.

$$G = \begin{cases} 12.92g & g \le 0.0031308 \\ 1.055g^{1.0/2.4} - 0.055 & g > 0.0031308 \end{cases} \tag{13.8}$$

$$B = \begin{cases} 12.92b & b \le 0.0031308 \\ 1.055b^{1.0/2.4} - 0.055 & b > 0.0031308 \end{cases} \tag{13.9}$$

Finally, to express the values for *R*, *G*, and *B* in the range (0, 255), each component must be multiplied by 255 (Lindbloom 2001, Pascale 2003).

Then, correlations between the color coordinates and the pigment concentration in nmol/cm^2 were obtained. In Figure 13.1, an outline of the general methodology followed to find these correlations is presented.

13.2.2 Determination of Pigment Concentration from Images

Images from another set of colored leaves (*Paspalum dilatatum*, *Ipomoea indica*, *Ricinus communis* L., *Setaria palmifolia*, *Sorbus foliolosa*, and *Rosa* spp.)

were captured using a commercial scanner (HP-Deskjet F380 from Hewlett-Packard). The scanned images in TIFF format with a resolution of 300 ppi and a depth of 24 bits were digitally processed using the Erdas Imagine 8.4 program. Using this software, the TIFF images were imported to IMG format and they were subsequently separated in *R*, *G*, and *B* bands.

Using the correlations between the color coordinates (*R*, *G*, and *B*) and pigment concentration, inferred for homogeneous leaves, the concentration of pigments per pixel in the new set of leaves was estimated (see Figure 13.2).

From the concentration values for each pixel, Erdas Imagine 8.4 program displays a concentration gray-scale image. To improve the visual interpretation, these gray-scale images were further processed. For this purpose, a classification of pixels with the same value of pigment concentration was made. After performing this classification, the pixels with similar values were stained with the same pseudo color. Following this procedure, concentration maps for chlorophyll, anthocyanins, and carotenoids were obtained.

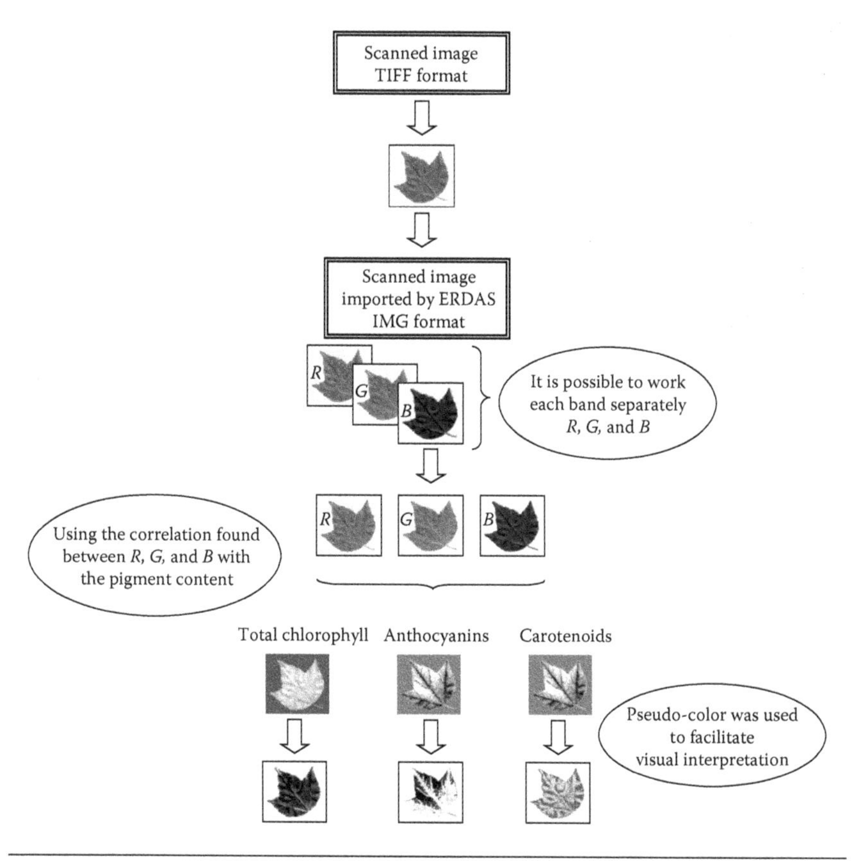

Figure 13.2 General outline summarizing the digital image processing for concentration images of leaf pigments.

13.2.3 Validation of the Proposed Method

For leaves of *Paspalum dilatatum*, *Ipomoea indica*, *Ricinus communis* L., *Setaria palmifolia*, *Sorbus foliolosa*, and *Rosa* spp., whose images had been previously obtained, the actual content of pigments was measured by the standard analytical extraction method (Sims and Gamon 2002), and the resultant values were compared with the average pigment concentration estimated from the optical nondestructive approach.

13.3 Results and Discussion

13.3.1 Correlations between R, G, and B Coordinates and Pigment Concentration

For the different studied species, the total chlorophyll concentration ($Chl_a + Chl_b$ concentration) was plotted as a function of *R*, *G*, and *B* coordinates, respectively. Values for the *R* component increased nonlinearly as chlorophyll concentration declined. Values for the *G* component decreased linearly with increasing chlorophyll content. Values for the *B* component, on the other hand, resulted insensitive to chlorophyll concentration (see Figure 13.3). It should be noted that each point in Figure 13.3 corresponds to unrelated plant species with highly different pigment concentrations, thus covering application for a wide range of leaves varieties.

Intuitively, one would expect that the values of the *G* component should increase when chlorophyll concentration rises. Surprisingly, both values of *G* and *R* increased when the total chlorophyll concentration decreased. This fact may be understood if one takes into account that leaves containing higher chlorophyll content also absorb more radiation in the red and green region. As a consequence, leaf reflectance in these regions is lower and so are *R* and *G* values. On the other hand, the leaf absorption in the blue region is so high that the reflectance is low and insensitive to changes in pigment concentration. As a result, the component *B* is saturated and does not respond to variations in the chlorophyll content of leaves. This result is expected because the absorption spectrum of chlorophyll has a maximum in the blue band of the electromagnetic spectrum, around 430 nm.

Excellent correlations were finally found for the sum of *R* and *G* with the total chlorophyll, chlorophyll a, and chlorophyll b contents (see Figure 13.4). Other good correlations were found between the ratio *R*/*G* and the proportion of anthocyanins/total chlorophyll and the ratio *B*/*G* and the carotenoids content (see Figures 13.5 and 13.6, respectively).

13.3.2 Determination of Pigment Concentration on Images

By applying the correlations shown in Figures 13.4 through 13.6 on leaf images obtained with a scanner, it was possible to obtain images of contents of chlorophyll, anthocyanins, and carotenoids. Figure 13.7 shows the scanned

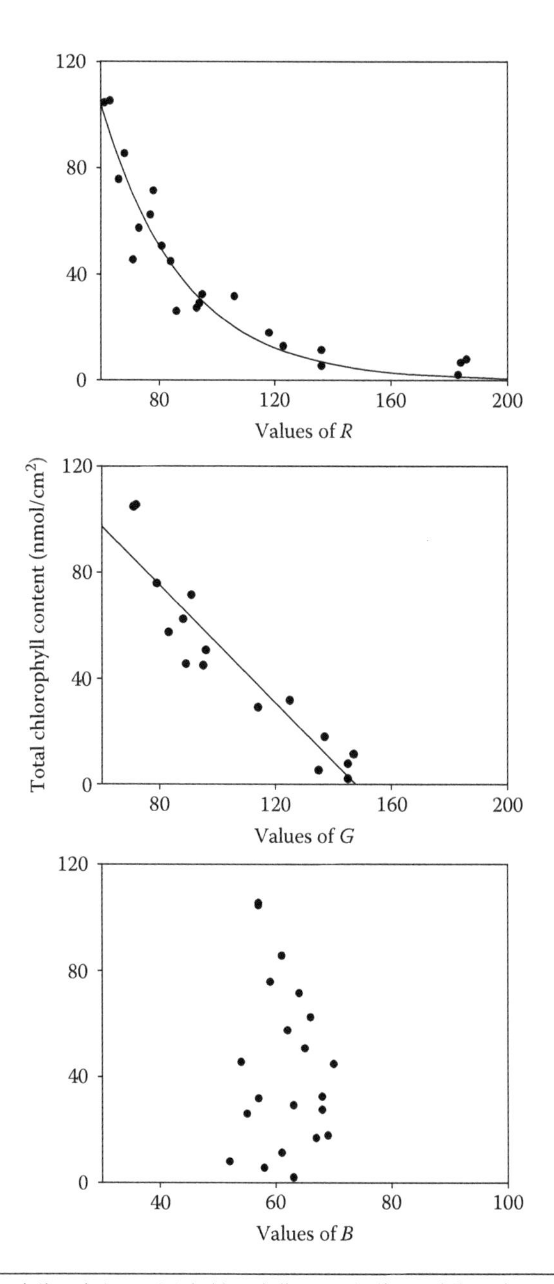

Figure 13.3 Correlations between total chlorophyll concentration and the values of the coordinates *R*, *G*, and *B*.

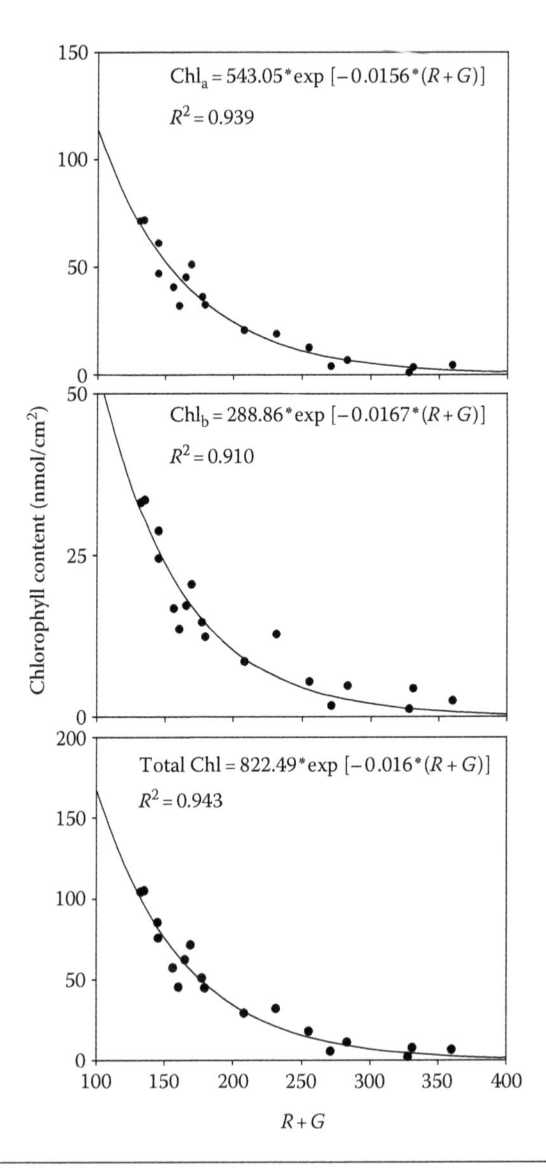

Figure 13.4 Correlations between the sum of the components R and G and the chlorophyll content of leaves.

image of a leaf of *Ipomoea indica* (a), the image of total chlorophyll content from the same leaf (b), and the image of carotenoids content (c). Image of anthocyanins is not shown in this figure because its content was too low to be detected.

In Figure 13.8, the scanned image of a leaf of *Sorbus foliolosa* (a), the image of total chlorophyll content (b), the image of carotenoids content (c), and the image of anthocyanins content from the same leaf (d) are shown.

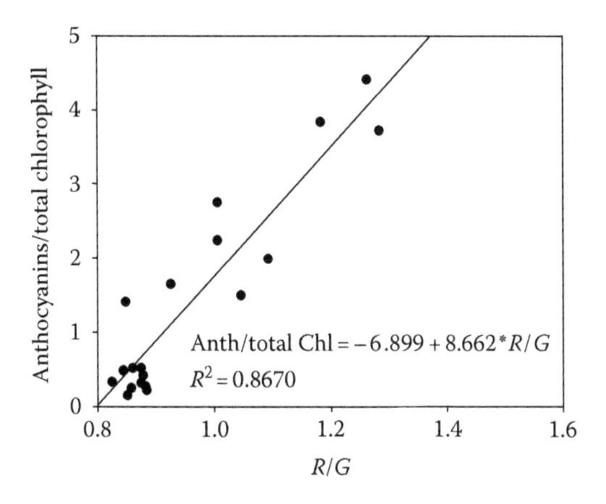

Figure 13.5 Correlations between the ratio *R/G* and the proportion anthocyanins/total chlorophyll of leaves.

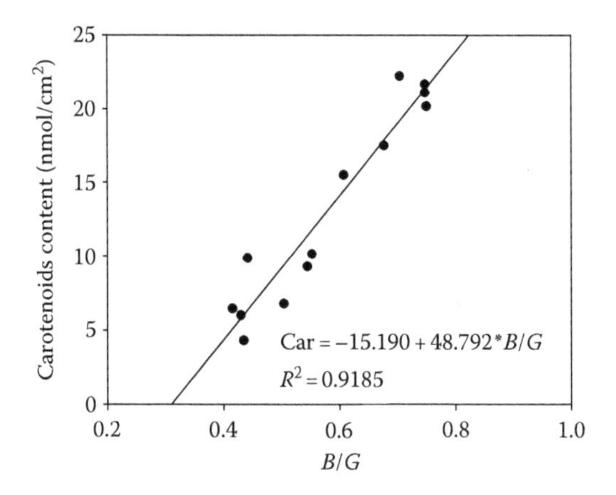

Figure 13.6 Correlations between the ratio *B/G* and the carotenoids content of leaves.

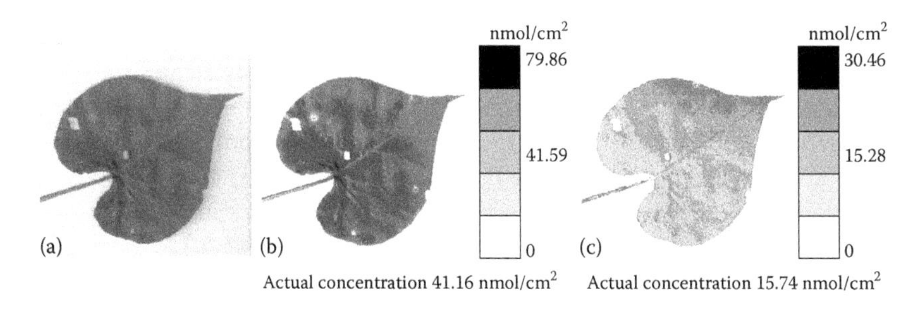

Figure 13.7 (a) Scanned image of a leaf of *Ipomoea indica*, (b) image of total chlorophyll content, and (c) image of carotenoids content from the same leaf. The actual concentration reported corresponds to the value obtained from the analytical classical method.

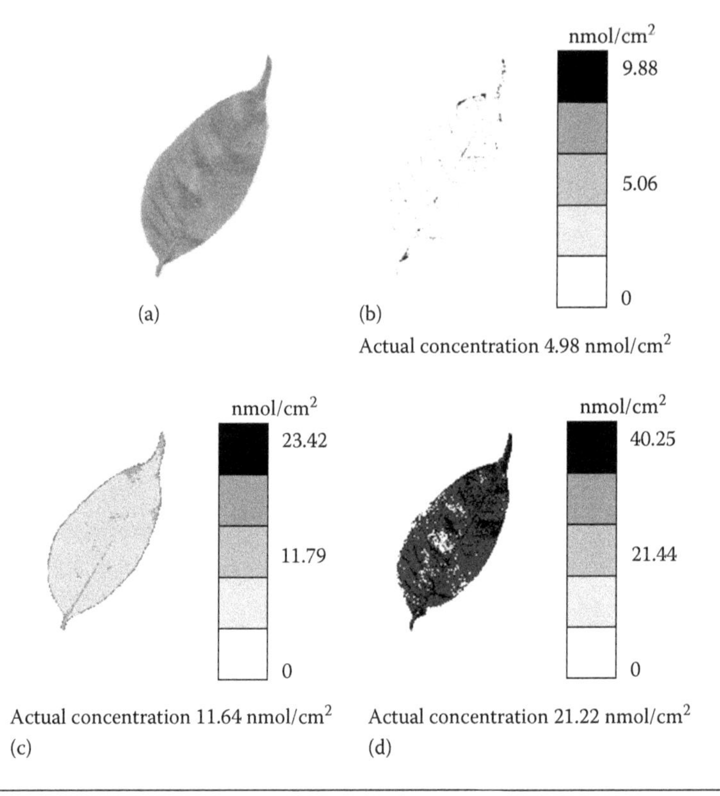

Figure 13.8 (a) Scanned image of a leaf of *Sorbus foliolosa*, (b) image of total chlorophyll content, (c) image of carotenoids content, and (d) image of anthocyanins content from the same leaf. The actual concentration reported corresponds to the value obtained from the analytical classical method.

13.3.3 Validation of the Proposed Method

A very good agreement between nondestructive estimation of pigment concentration and the values determined by the traditional analytical method are reflected in high values of coefficients of determination. $R^2 = 0.9976$ for total chlorophyll concentration (see Figure 13.9) and $R^2 = 0.9736$ for carotenoids (see Figure 13.10).

In the case of anthocyanins, the results were more complex and a general validation for a wide range of species could not be obtained. In fact, good correlations between the anthocyanins concentration found by wet treatment and the averaged value predicted by imaging techniques could only be obtained for slightly reddish yellow leaves, where the ratio Anth/Chl was high enough (Anth/Chl $\approx$ 4). For low Anth/Chl values (Anth/Chl $<$ 1), no correlation was obtained. So, the concentration imaging of anthocyanins proposed in this work should be considered restricted to the first cases. Further detailed work should be performed in future on this point to achieve a wider application.

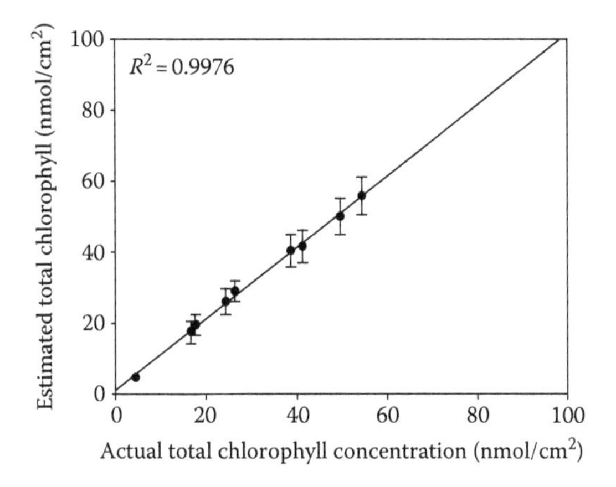

Figure 13.9 Agreement between predicted total chlorophyll concentration by the method proposed and values of total chlorophyll determined by the traditional analytical method.

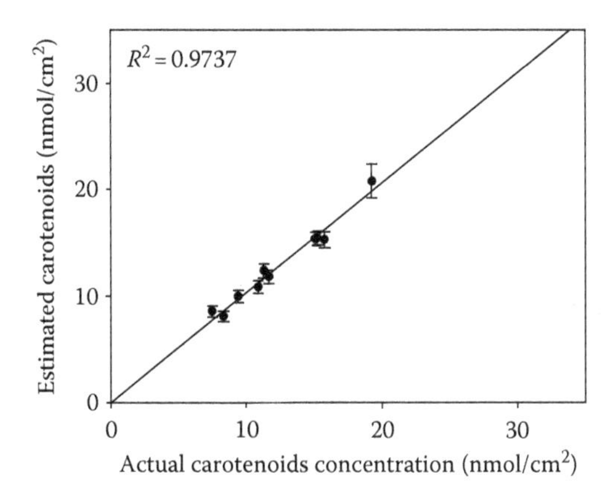

Figure 13.10 Agreement between predicted carotenoids concentration by the method proposed and values of carotenoids determined by the traditional analytical method.

13.4 Conclusions

In this work, it was possible to successfully develop a method for monitoring nondestructively pigment content in leaves. This procedure allows knowing the contents of chlorophylls and carotenoids in a wide range of leaf species using color images obtained with a commercial scanner. Anthocyanins concentration map can also be inferred, but in this case, it is restricted to leaves with high ratios for the anthocyanins to chlorophyll contents. The application of this methodology to the diagnosis of plant health inferred from leaf pigment concentration is highly effective and easily implemented due to the low cost of operation.

Acknowledgments

The authors are grateful to the University of Buenos Aires (Project UBACyT X114 2008–2010) and to the Agencia Nacional de Promoción Científica y Tecnológica (BID 1201/OC-AR PICT 938) for the financial support. Gabriela Cordon is supported by a fellowship from Conicet.

References

Ahmad, I. and J. Reid. 1996. Evaluation of colour representations for maize images. *Journal of Agricultural Engineering Research* 63: 185–196.

Bacci, L., B. De Vincenzi, and B. Rapi. 1998. Two methods for the analysis of colorimetric components applied to plant stress monitoring. *Computers and Electronics in Agriculture* 19: 167–186.

Bonilla, I. 2000. Chapter 6: Introduction to the mineral nutrition of plants. Mineral elements. In *Fundamentals of Plant Physiology*, ed. Azcón-Bieto y Talón. Barcelona, Spain: McGraw-Hill/Interamericana de España.

Lindbloom, B. J. 2001. http://www.brucelindbloom.com. Accessed May 12, 2011.

Pascale, D. 2003. A review of RGB color spaces. Technical report, Montreal, Canada: The BabelColor Company.

Schaberg, P., P. Murakami, M. Turner, H. Heitz, and G. Hawley. 2008. Association of the red coloration with senescence of sugar maple leaves in autumn. *Trees* 22: 573–578.

Sims, D. and J. Gamon. 2002. Relationships between leaf pigment content and spectral reflectance across a wide range of species, leaf structures and developmental stages. *Remote Sensing of Environment* 81: 337–354.

Yadav, S. and Y. Ibaraki. 2010. Estimation of the chlorophyll content of micropropagated potato plants using RGB based image analysis. *Plant Cell Tissue and Organ Culture* 100: 183–188.

CHAPTER 14

Color of Honeys from the Southwestern Pampas Region

Relationship between the Pfund Color Scale and CIELAB Coordinates

LILIANA M. GALLEZ, ALFREDO MARCONI, ELIAN TOURN, M. LOURDES GONZÁLEZ-MIRET, and FRANCISCO J. HEREDIA

Contents

14.1 Introduction

Color is the most important physical trait used to characterize honey. It relates to the botanical and geographical origin (Mateo Castro et al. 1992, Morse and Hooper 1992, Terrab et al. 2003a–c); however, heating treatments darken honeys, modifying the initial color. Minerals are among the components related to honey color (Crane 1990, Balbarrey et al. 2010), and phenolic compounds, associated to its antioxidant activity, were also found to be positively correlated to color (Frankel et al. 1998, Beretta et al. 2005, Bertoncelj et al. 2007, Vit et al. 2008).

A serious problem posed to researchers is that honey colors lack a satisfactory vocabulary that is accepted in all countries (Crane 1990). Data sets involving honey color cannot be compared, because this variable is recorded using very different systems. Some papers report color in millimeter (in terms of the Pfund scale) (Piazza et al. 1991, Ciappini et al. 2008, Vit et al. 2008, Vanhanen et al. 2011) whereas others establish color coordinates in the CIELAB system (L^*, a^*, b^*), chroma (C^*_{ab}), and hue angle (h_{ab}) (Terrab et al. 2002, González-Miret et al. 2005, 2007).

The international market usually classifies honey according to the Pfund color scale, light honey being generally more valued than dark honey (Bogdanov et al. 2004). The Pfund grader is an optical comparator device manufactured by Köeler Inc., adopted for color grading in the United States at the beginning of the twentieth century (Aubert and Gonnet 1983). It gives a measure of the point along a calibrated amber glass prism where the liquid honey sample placed in a trough-shaped glass cuvette matches the prism. The Pfund scale ranges from 0 to 140 mm, beginning with very light-colored honey and increasing up to the darkest honey. The scale shows seven ranges with different color designations named water white, extra white, white, extra light amber, light amber, amber, and dark. This classification was officially admitted in the *United States standards for grades of extracted honey* (effective date May 23, 1985).

Even when other comparators have been adopted, results are expressed in Pfund units. Two of them are described as official methods of the AOAC International: the USDA color comparator and the Lovibond 2000 (AOAC 2000).

The Pfund grader as well as the other color comparators does not allow registering greenish or reddish hues distinguished by consumers. These nuances could be related to botanical origins, for example, greenish from lime (*Tilia* sp.), grayish yellow from borage (*Borago officinalis*), and greenish brown from tree of heaven (*Ailanthus altissima*) and maple (*Acer* sp.) (Crane 1990).

The tristimulus methods (e.g., CIELAB space) provide more information on honey color than the Pfund grader (Aubert and Gonnet 1983, Terrab et al. 2003a–c, Bogdanov et al. 2004, Bertoncelj et al. 2007, González-Miret et al. 2007). The first attempt to use the CIE colorimetric coordinates to assess honey color was that of Aubert and Gonnet (1983). They conclude that this method should be a standard research system for a more accurate floral classification of honeys; however, the Pfund index might remain, due to its easiness in practical use, as the commercial reference.

Bogdanov et al. (2004) state in a review dealing with physicochemical honey analysis that in order to apply colorimetric (CIELAB) methods to routine honey analysis, they have to be validated by a comparison with the classical color grading methods. The aim of this work is to find a relationship between data recorded with the Pfund and the CIELAB methods.

14.2 Materials and Methods

14.2.1 Honey Samples

A set of 50 samples were selected as representative of the color range of honeys from the southwestern Pampas region (Argentina). Samples, provided by professional beekeepers, were obtained by centrifugation and stored at 5°C in the dark until analysis.

14.2.2 Pollen Analysis

Pollen analysis was performed after counting a minimum of 1000 pollen grains on three slides from each sample. The frequency classes of pollen grains were given as dominant pollen (>45%), secondary pollen (16%–45%), important minor pollen (3%–15%), and traces (<3%), according to Louveaux et al. (1978). The pollen types were identified to species whenever possible, otherwise to genus, tribe, or family taxa.

14.2.3 Color Measurement

The color of honey samples was assessed by a Pfund color grader (Koehler Instrument Company Inc., New York) and by a HunterLab colorimeter. Since both methods require liquid honey, samples were slightly heated to 40°C and centrifuged for 5 min at 3000 rpm in order to eliminate bubbles. Other authors heated samples to 50°C without altering their color (Bertoncelj et al. 2007, Vanhanen et al. 2011).

With the Pfund grader, six readings were taken for each honey sample, three from each side of the glass wedge (Fell 1978).

In order to record color with the HunterLab Ultra Scan XE colorimeter in transmission mode, honey was poured into an optical glass cuvette (10 × 50 × 50 mm). D65 standard illuminant and CIE 1964 standard observer (10° visual field) were used in the calculus.

14.2.4 Conductivity Measurement

Electric conductivity of the honey solutions at 20% dry weight basis in distilled water, free from CO_2, was measured with a conductivity meter 0–20 mS/EC with temperature compensation (EC-Controller ECCO), according to Louveaux et al. (1973).

14.2.5 Statistical Analysis

Different mathematical algorithms were tested in order to find the possibility of predicting the Pfund color of honeys based on tristimulus colorimetric variables. Simple and multiple correlation models were applied, including scalar (L^* a^* b^*) or angular (L^* C^*_{ab} h_{ab}) CIELAB variables.

14.3 Results and Discussion

14.3.1 Honey Samples Characterization

Average values of the color and conductivity of honey samples are shown in Table 14.1. When compared with other reports (Terrab and Heredia 2004, Terrab et al. 2004a,b), all samples showed high L^* values, possibly due to the local melliferous flora. These values were similar to those of rosemary and orange honeys measured by Mateo Castro et al. (1992).

Pollen analysis revealed that 56 pollen types were present in the honey samples. The families Brassicaceae (*Diplotaxis tenuifolia* and other Brassicaceae), Myrtaceae (represented by genus *Eucalyptus*), Asteraceae (represented by several pollen types), and Fabaceae (represented by genera *Prosopis*, *Lotus*, *Melilotus*, *Prosopidastrum*, *Trifolium*, and *Medicago*) were the most represented (Figure 14.1). The combination of three pollen types, *D. tenuifolia*, *Eucalyptus* sp., and *Centaurea solstitialis*, points out the geographical origin of the samples (Valle et al. 2007). The pollen content of *D. tenuifolia*, *Centaurea* sp., and *Prosopis* sp. was positively correlated to h_{ab} value ($R = 0.39$; $p < 0.05$), that is, to slightly greenish hue, while another group (*Eucalyptus* sp., *Baccharis* sp., *Schinus* sp., *Condalia microphyla*, and *Larrea* sp.) showed a relationship to reddish hue, showing lower h_{ab} values than the former ($R = -0.33$; $p < 0.05$).

14.3.2 Statistical Relationship between Pfund Scale and CIELAB Parameters

Multivariate analysis showed a direct relationship between Pfund values, conductivity, and CIELAB variables (Table 14.2). Light honeys (on the Pfund scale) presented slight greenish hue (Figure 14.2), high lightness, low chroma, and low conductivity (Figure 14.3), whereas the darker ones behave in the opposite way (Figures 14.2 and 14.3).

It has been found that color according to the Pfund scale can be estimated accurately as a function of CIELAB parameters by means of multiple linear regression analysis, based on the regression equation 14.1 ($R^2 = 0.91$; $F_{3.46} = 152.55$, $p < 0.0001$):

$$\text{mm Pfund} = -0.631L^* + 0.840C^*_{ab} - 1.026h_{ab} + 155.89 \tag{14.1}$$

Table 14.1 Descriptive Statistical Data for Color Parameters and Electrical Conductivity Measured in Honey Samples

	L^*	a^*	b^*	C^*_{ab}	h_{ab}	Pfund (mm)	Conductivity (mS/cm)
Average	83.83	0.32	43.74	43.88	91.67	45.75	0.26
SD	5.00	3.51	20.02	20.02	5.00	25.71	0.09
Minimum	73.59	−3.75	14.09	14.32	82.95	3.93	0.14
Maximum	91.17	9.94	81.58	82.18	101.07	85.23	0.42

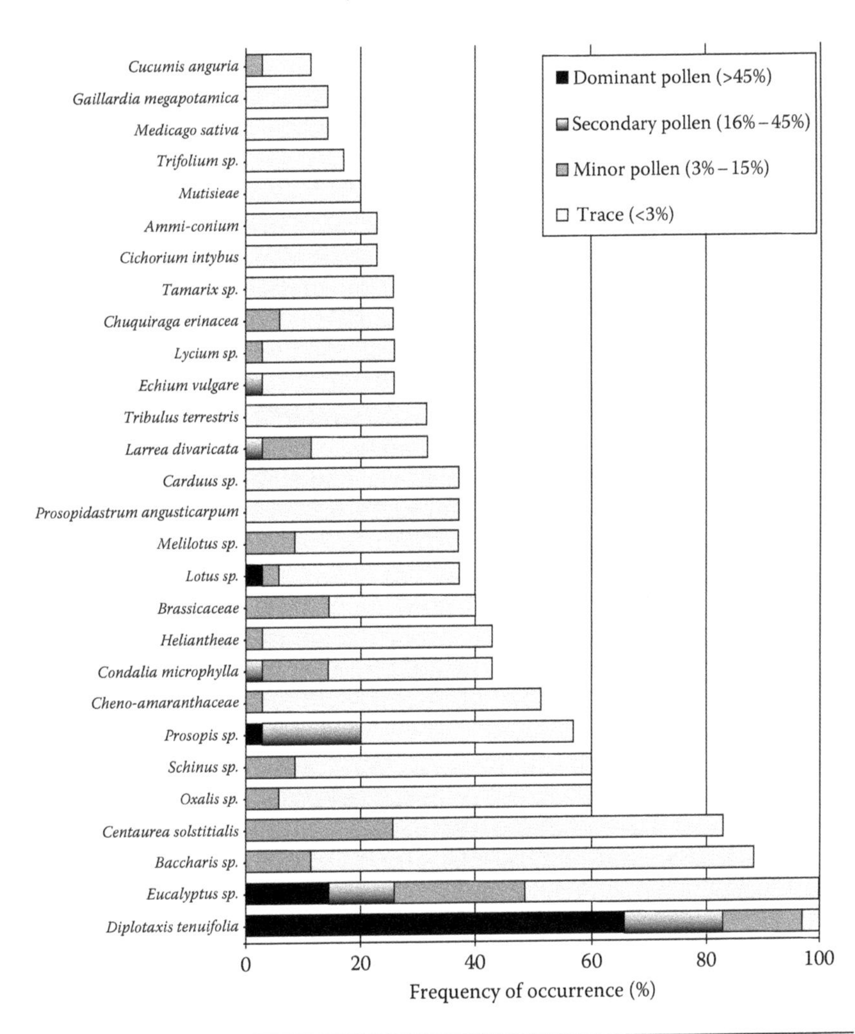

Figure 14.1 Pollen spectrum of the 50 honey samples showing frequency of occurrence and frequency classes. D, dominant pollen (>45%); S, secondary pollen (16%–45%); M, minor important pollen (3%–15%); and T, trace pollen (<3%).

Table 14.2 Correlation Matrix Indicating Association among Color Parameters and Electrical Conductivity

	L^*	C^*_{ab}	h_{ab}	mm Pfund
C^*_{ab}	−0.86**			
h^*_{ab}	0.96**	−0.94**		
mm Pfund	−0.87**	0.95**	−0.93**	
Conductivity	−0.64**	0.75**	−0.69**	0.72**

** Statistically highly significant ($p < 0.01$).

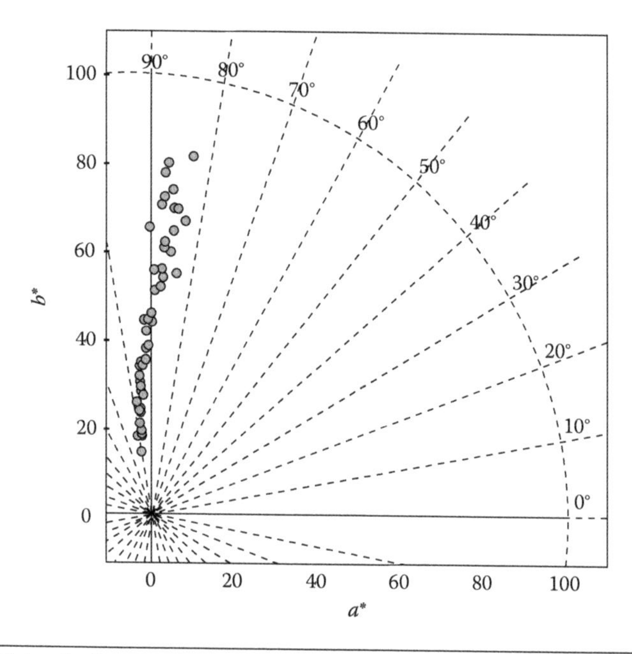

Figure 14.2 Localization area of samples on a^* and b^* coordinates diagram.

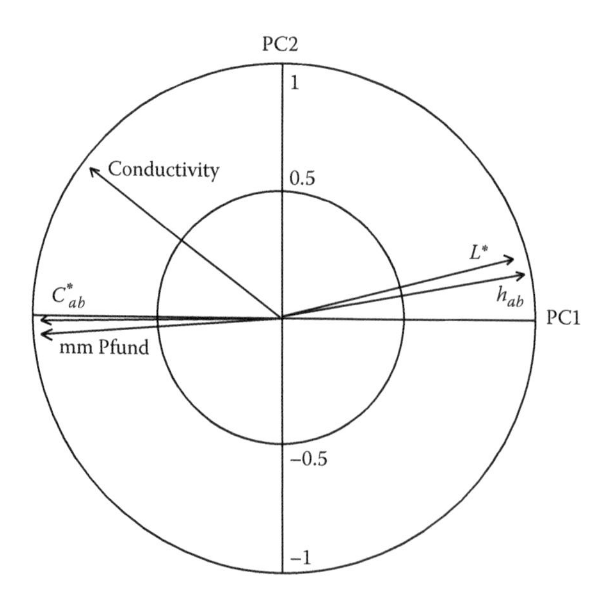

Figure 14.3 Principal component analysis of honey samples considering color (L^*, C^*_{ab}, h_{ab}, and mm Pfund) and electrical conductivity as variables. PC1 (principal component 1) and PC2 (principal component 2).

Observers usually interpret the Pfund color index as the "total color" of honeys by comparison with a standard colored glass prism. In order to find an equation that relates Pfund and the color of honeys, we have defined the total color (E) as the CIELAB color difference (ΔE^*_{ab}) between the honey and the white reference. Taking this parameter into account, a slight improvement in the linear correlation coefficients has been obtained. Moreover, the correlation reached higher values ($R^2 = 0.96$) when a nonlinear quadratic polynomial regression is used:

$$\text{mm Pfund} = -32.82 + 2.48E - 0.013E^2 \tag{14.2}$$

These results allow calculating the Pfund value by means of instrumental colorimetric measurements, without involving visual comparisons. Considering that CIELAB is the method currently used in color measurement and that the Pfund scale is widely used in honey trade, relating both methods mathematically by means of a single equation can be of great interest for the industry.

Acknowledgments

The authors thank Dr. Ana Andrada for her kind advice in pollen analysis, Dr. Diana Constenla for valuable technical advice, and Dr. Cecilia Pellegrini for her generous assistance. Elian Tourn was granted a CIC (Comisión de Investigaciones Científicas de la Provincia de Buenos Aires) scholarship during this research.

References

AOAC (Association of Official Analytical Chemists). 2000. *Official Methods of Analysis*, 17th edn. Gaithersburg, MD: AOAC International.

Aubert, S. and M. Gonnet. 1983. Mesure de la couleur des miels. *Apidologie* 14(2): 105–118.

Balbarrey, G., A. Andrada, J. Echazarreta, D. Iaconis, and L. Gallez. 2010. Relationship between mineral content and color in honeys from two ecological regions in Argentina. In *AIC 2010 Color and Food, Proceedings*, eds. J. Caivano and M. López. Buenos Aires, Argentina: Grupo Argentino del Color, pp. 552–555.

Beretta, G., P. Granata, M. Ferrero, M. Orioli, and R. Maffei Facino. 2005. Standardization of antioxidant properties of honey by a combination of spectrophotometric/fluorimetric assays and chemo-metrics. *Analytica Chimica Acta* 533: 185–191.

Bertoncelj, J., U. Doberšek, M. Jamnik, and T. Golob. 2007. Evaluation of the phenolic content, antioxidant activity and colour of Slovenian honey. *Food Chemistry* 105(2): 822–828.

Bogdanov, S., K. Ruoff, and L. Persano Oddo. 2004. Physico-chemical methods for characterization of unifloral honeys: A review. *Apidologie* 35: S4–S17.

Ciappini, M. C., M. B. Gatti, M. V. Di Vito, S. Gattuso, and M. Gattuso. 2008. Characterization of different floral origins honey samples from Santa Fe (Argentine) by palynological, physicochemical and sensory data. *Apiacta* 43: 25–36.

Crane, E. 1990. *Bees and Beekeeping. Science, Practice and World Resources*. London, U.K.: Heinemann Newnes.

Fell, R. D. 1978. The color grading of honey. *American Bee Journal* 118(12): 782–789.

Frankel, S., G. E. Robinson, and M. R. Berenbaum. 1998. Antioxidant capacity and correlated characteristics of 14 unifloral honeys. *Journal of Apicultural Research* 37: 27–31.

González-Miret, M. L., F. Ayala, A. Terrab, J. F. Echávarri, A. I. Negueruela, and F. J. Heredia. 2007. Simplified method for calculating colour of honey by application of the characteristic vector method. *Food Research International* 40: 1080–1086.

González-Miret, M. L., A. Terrab, D. Hernanz, M. A. Fernandez-Recamales, and F. J. Heredia. 2005. Multivariate correlation between color and mineral composition of honeys and by their botanical origin. *Journal of Agricultural and Food Chemistry* 53(7): 2574–2580.

Louveaux, J., A. Maurizio, and G. Vorwhol. 1978. Methods of melissopalynology by International Commission for bee Botany or IUBS. *Bee World* 59: 139–157.

Louveaux, J., M. Pourtallier, and G. Vorwohl. 1973. Méthodes de'analyses des miels. Conductivité. *Bulletin Technique Apicole* 16: 1–7.

Mateo Castro, R., M. Jiménez Escamilla, and F. Bosch-Reig. 1992. Evaluation of the color of some unifloral honey types as a characterization parameter. *Journal of AOAC International* 75: 537–542.

Morse, R. A. and T. Hooper. 1992. *Enciclopedia ilustrada de apicultura*. Buenos Aires, Argentina: El Ateneo.

Piazza, M., M. Accorti, and L. Persano Oddo. 1991. Electrical conductivity, ash, colour and specific rotatory power in Italian unifloral honeys. *Apicoltura* 7: 51–63.

Terrab, A., M. J. Diez, and F. J. Heredia. 2002. Chromatic characterisation of Moroccan honeys by diffuse reflectance and tristimulus colorimetry—Non-uniform and uniform colour spaces. *Food Science and Technology International* 8(4): 189–195.

Terrab, A., M. J. Diez, and F. J. Heredia. 2003a. Palynological, physico-chemical and colour characterization of Moroccan honeys: I. River red gum (*Eucalyptus camaldulensis* Dehnh) honey. *International Journal of Food Science and Technology* 38(4): 379–386.

Terrab, A., M. J. Diez, and F. J. Heredia. 2003b. Palynological, physico-chemical and colour characterization of Moroccan honeys. II. Orange (*Citrus* sp.) honey. *International Journal of Food Science and Technology* 38(4): 387–394.

Terrab, A., M. J. Diez, and F. J. Heredia. 2003c. Palynological, physico-chemical and colour characterization of Moroccan honeys: III. Other unifloral honey types. *International Journal of Food Science and Technology* 38(4): 395–402.

Terrab, A., M. L. Escudero, M. L. Gonzalez-Miret, and F. J. Heredia. 2004a. Colour characteristics of honeys as influenced by pollen grain content: A multivariate study. *Journal of the Science of Food and Agriculture* 84(4): 380–386.

Terrab, A., M. L. Gonzalez-Miret, and F. J. Heredia. 2004b. Colour characterisation of thyme and avocado honeys by diffuse reflectance spectrophotometry and spectroradiometry. *European Food Research and Technology* 218(5): 488–492.

Terrab, A. and F. J. Heredia. 2004. Characterization of avocado (*Pleirsea americana* Mill) honeys by their physicochemical characteristics. *Journal of the Science of Food and Agriculture* 84(13): 1801–1805.

Valle, A., A. Andrada, E. Aramayo, M. Gil, and S. Lamberto. 2007. A melissopalynological map of the south and southwest of the Buenos Aires Province, Argentina. *Spanish Journal of Agricultural Research* 5(2): 172–180.
Vanhanen L. P., A. Emmertz, and G. P. Savage. 2011. Mineral analysis of mono-floral New Zealand honey. *Food Chemistry* 128(1): 236–240.
Vit, P., M. G. Gutiérrez, D. Titěra, M. Bednař, and A. J. Rodríguez-Malaver. 2008. Mieles checas categorizadas según su actividad antioxidante. *Acta Bioquímica Clínica Latinoamericana* 42(2): 237–244.

CHAPTER 15

Influence of Different Backgrounds on the Instrumental Color Specification of Orange Juices

CARLA M. STINCO, ROCÍO FERNÁNDEZ-VÁZQUEZ, ANTONIO J. MELÉNDEZ-MARTÍNEZ, FRANCISCO J. HEREDIA, EMILIA BEJINES MEJÍAS, and ISABEL M. VICARIO

Contents

15.1 Introduction

Color can be defined as a mental response to the stimulus that a visible radiation produces on the retina, which is transmitted to the brain by the optical nerve. The perception of color is a psychophysical phenomenon, linked to the psychology of the observer, the physiology of vision, and the spectral radiant energy of a source of light (Wyszecki and Stiles 1982).

Color measurement is used in the food industry both in routine quality assessment and in research and development. Color is an attribute directly related to the consumer preferences in citrus juices, because generally the consumer associates an attractive color with a pleasant flavor (Huggart et al. 1977, 1979, Hutchings 1994, Tepper 1993).

Color evaluation can be approached from two perspectives: methods based on sensory analysis and methods based on instrumental measurement. Within the first group, the color is assessed visually; these methods are what we commonly call sensory analysis, that is, the evaluation of their characteristics by means of the senses. It includes preference tests, difference tests, acceptance tests (Luckow and Delahunty 2004), and descriptive analysis methods (Spoto et al. 1997, Stone et al. 2004). Besides, it is advisable to take into account the observing conditions such as illumination (source and angle), sample presentation, the background or surrounding on which the sample is presented (Meléndez-Martínez et al. 2005a). Furthermore, it also depends on the observer and the physical conditions of the experimental device (Hita and Romero 1981, MacDougall 2002). Also, color scales or atlases (Munsell system, DIN system, Natural Color System (NCS), Ostwald system, etc.) can be used. The United States Department of Agriculture (USDA) established a standard for the grading of the colors of orange juice (OJ) using 6 plastic color tubes, with visual scores ranging from 34 to 40 (USDA 1982).

In contrast to this, it is possible to define color measurement objectively by using appropriate instruments and following the recommendations established by the CIE (1991a,b). These measurement types have the advantage of being more straightforward, precise, and versatile; furthermore, they eliminate subjectivity; besides, they are nondestructive and allow automatization. Due to these facts, they are very fitting for quality control in the industry in those aspects in which color can play a key role, like those related with nutritional value, color standards, or conformity with regulations.

The color is expressed by means of color coordinates in the *Yxy* color space (based on the *XYZ* tristimulus values) or in the $L^*a^*b^*$ color space (CIELAB), both developed by the CIE or alternatively in the Hunter Lab color scale. Three types of color measurement instruments can be used for food color measurement: colorimeters, spectrophotometers, and spectroradiometers. The colorimeter directly measures the sample color by illuminating it with a given light source and measuring the light reflected after it passes through red, blue, and green standard filters. The spectrophotometer is designed to measure the spectral distribution of transmittance or reflectance of an object from which one can calculate the color under different conditions. Unlike color, both are intrinsic properties of the object that does not change with the illumination received or the observer. The spectroradiometer is designed to measure radiometric quantities as a function of wavelength. It can also be used for transmittance or reflectance measurements of any object (Meléndez-Martínez et al. 2005b).

Several methods for the instrumental measurement of OJ color have been reported in the literature. Eagerman (1978) employed a number of possible sample presentation methods on 5 different instruments using a variety of optical designs for the measurement of color in 150 samples, demonstrating that any general purpose tristimulus colorimeter can be used for OJ color specification.

Independent of the instrument used, it is very important to define the correct conditions of the color measurements. The CIE recommends a series of standard illuminants (CIE 1991b), observers (CIE 1991a), and so on, and color spaces (CIE *XYZ*, CIELUV, CIELAB) (CIE 1978) aimed at standardizing the objective measurement and the color definition. Also, the arrangement of the sample, related to the geometry of the system, the light source intensity, the thickness for sample, and the blank measurement have to be considered as they influence the instrumental measurement (Meléndez-Martínez et al. 2006). Other factors that can influence this measurement are the background and the surrounding area. In relation to this, the objectives of this study were to evaluate the effect of these two factors and that of the pulp content on the color specifications of OJs.

15.2 Material and Methods

To study the influence of the surroundings and the background on the color measurements, three backgrounds and surroundings (white, gray, and black) were used. Figure 15.1 shows the nine combinations of backgrounds and surroundings used, and Figure 15.2 shows the geometry employed in our spectroradiometric measurements.

For color specifications, we used the CIELAB color space, recommended by the CIE (2004), and color differences were calculated as the Euclidean distances between pairs of points in the three-dimensional space defined by L^*, a^*, and b^* according to the following formula:

$$\Delta E^*_{ab} = \sqrt{\left(\Delta L^*\right)^2 + \left(\Delta a^*\right)^2 + \left(\Delta b^*\right)^2} \tag{15.1}$$

Other CIELAB-based color difference formulas, such as CIE94 (CIE 1995) or CIEDE2000 (CIE 2001), improves CIELAB predictions of visual-perceived color differences (Melgosa 2000), but considering that this is a methodological work, we decided to use the most usual CIELAB formula.

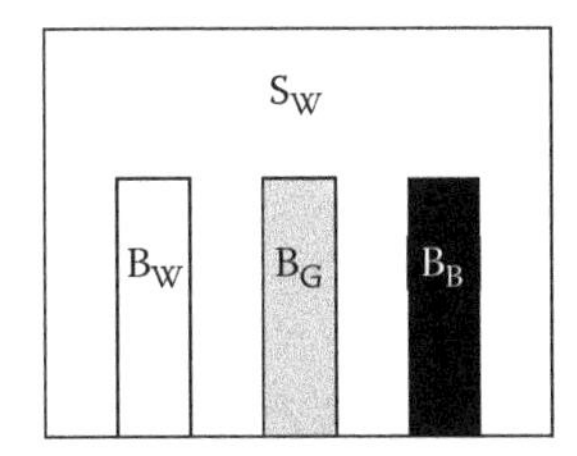

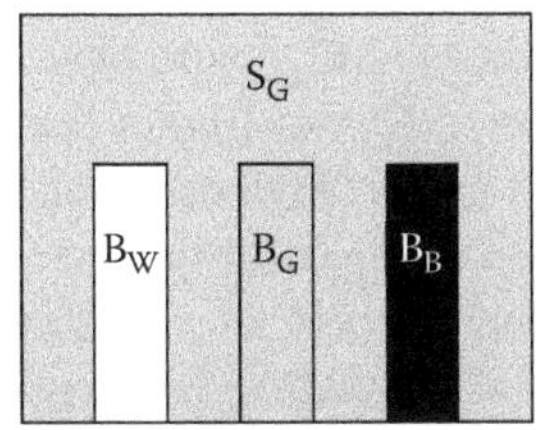

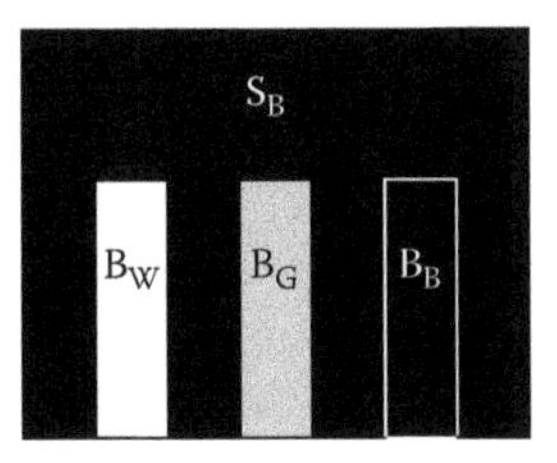

Figure 15.1 Combinations of backgrounds and surroundings.

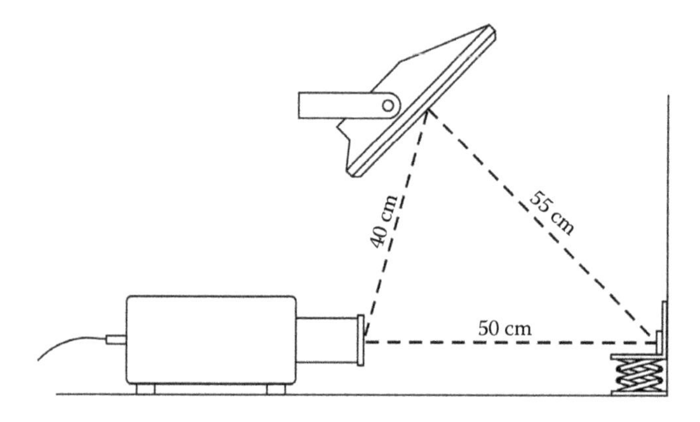

Figure 15.2 Outline of the geometry used in the spectroradiometric measurements: cross section.

Four OJs bought in local supermarkets were diluted with their own serum (obtained upon centrifugation) to give five levels of pulp concentration (100%, 75%, 50%, 25%, and 0%). The total number of samples analyzed was therefore 20.

In this work, the reflectance spectra were obtained by means of a CAS 140 B spectroradiometer (Instrument Systems, Germany) fitted with a Top 100 telescope optical probe, a Tamron zoom mod. SP 23A (Tamron USA, Inc., Commack, NY), and an external light source a white light 150 W metal-halide lamp Phillips MHN-TD Pro (12,900 lu, 4,200 K color temperature, and "cold white" chromaticity) as source of illumination. Blank measurements were made with distilled water against a white background. A plastic cuvette (10 × 10 × 45 mm) was used for the measurements.

15.3 Results and Discussion

15.3.1 *Influence of the Surroundings on Color Measurements*

To the best of our knowledge, no specific studies on the effects of the surroundings on the instrumental color specifications of OJ have been reported. The influence of the surroundings was evaluated by comparing the color differences for white/black (S_W/S_B), white/gray (S_W/S_G), and black/gray (S_B/S_G) surroundings in the different OJ samples. The results are shown in Figure 15.3.

The highest color differences (ΔE^*_{ab}) were observed for the pair S_W/S_B, followed by S_W/S_G. In both cases, ΔE^*_{ab} was higher than the visual discrimination threshold ($\Delta E^*_{ab} > 3$ CIELAB units) (Melgosa et al. 2001). The lowest color differences were obtained when comparing the surroundings S_B/S_G. According to these results and considering the CIE recommendations in the technical report (CIE 2004), the gray surrounding (uniform neutral gray with $L^* = 50$) was selected for the following measurements. It was also observed that, in general, the color differences decreased as the pulp content increased.

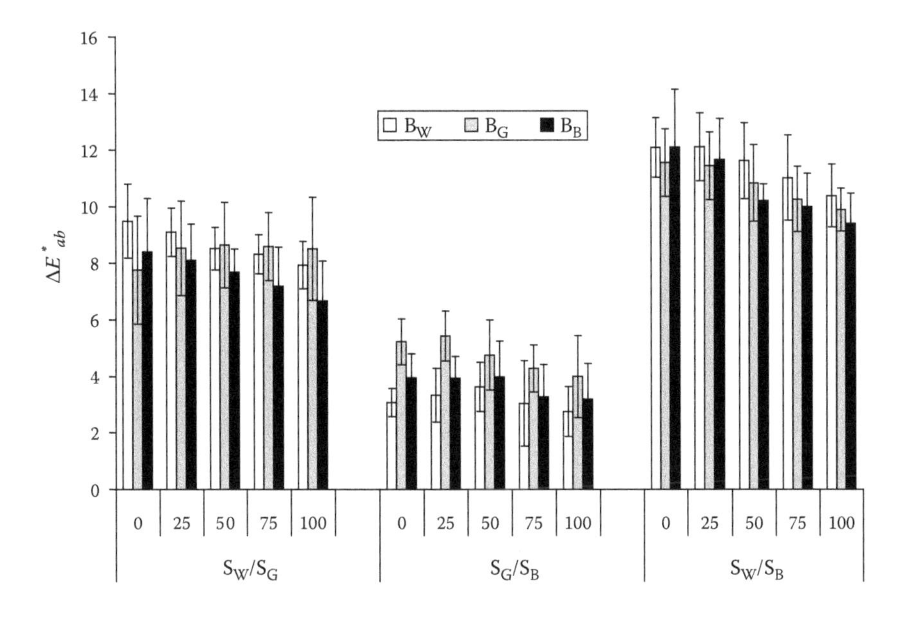

Figure 15.3 Means of color differences between surroundings with a white, gray, and black background for the OJs.

15.3.2 Influence of the Background on Color Measurements

Many studies have reported the influence of different backgrounds in the instrumental measurement of color (Eagerman 1978, Gullett et al. 1972, Meléndez-Martínez et al. 2005a, 2006).

In this work, we considered three possibilities to evaluate the effect of the background: white background (B_W), gray background (B_G), and black background (B_B). Figure 15.4a shows the location of the samples in the a^*b^* plane as a function of the background used. Two groups could be somewhat distinguished: one group including the samples measured with gray background (B_G) and black background (B_B) and another corresponding to the samples measured in white background (B_W). Obviously, the lowest values of L^* were obtained when the black background (B_B) was used while the highest values of L^* corresponded to the samples measured with white background (B_W) (Figure 15.4b).

It can be observed that the samples were localized in different areas of the CIELAB space, thus demonstrating the influence of the background on the color measurement in OJs with different pulp contents. As expected, the lowest color differences were obtained when comparing the pair B_B/B_G. Considering this pair, the differences ranged from 2 to 5 CIELAB units, while the highest differences (ranging from 6 to 18 CIELAB units) were obtained when comparing the pair B_W/B_B. In all cases, ΔE^*_{ab} appeared to have an inverse relationship with the pulp content, that is, the higher the pulp content the lower the effect of background on the color measurements (Figure 15.5).

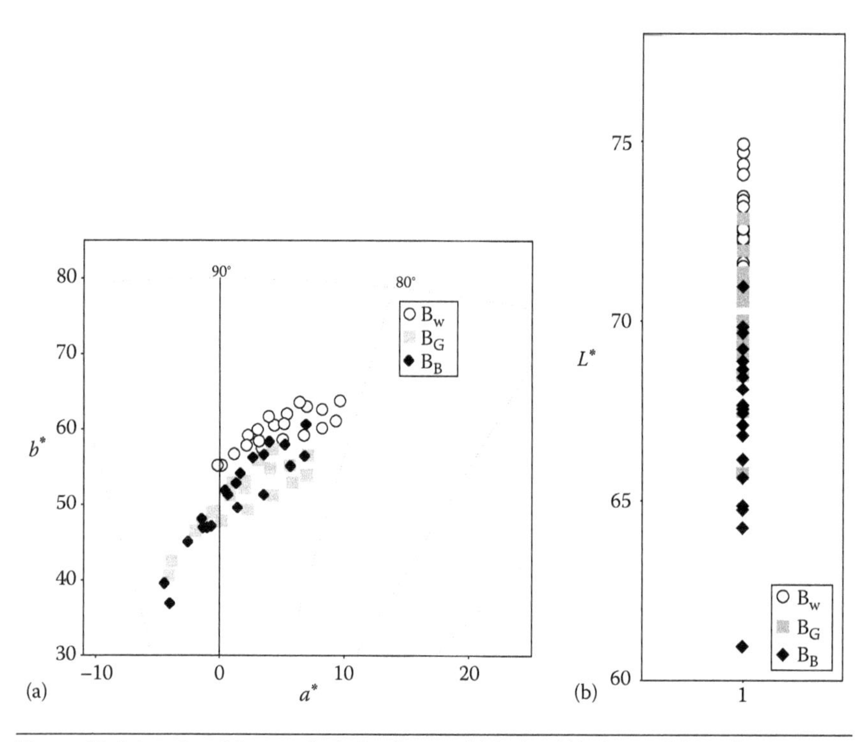

Figure 15.4 Location of the samples in the diagram: (a) in the a^*b^* plane and (b) in L^* values.

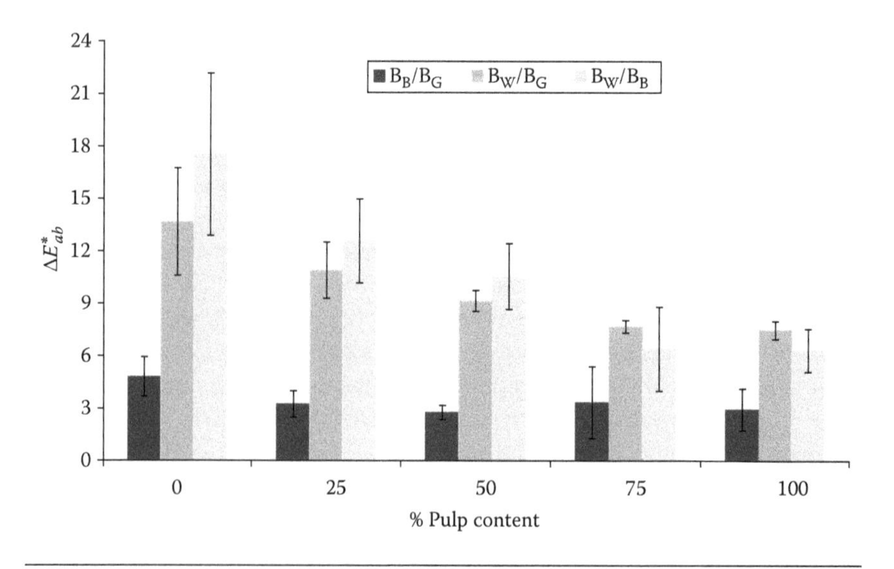

Figure 15.5 Color differences between the OJs.

15.3.3 *Influence of the Pulp Content on Color Measurements*

The influence of the pulp content on the objective color measurement of OJs color has been studied by several authors. Gullett et al. (1972) prepared a series of OJs with different pulp contents and evaluated their color both instrumentally and visually. Similarly, Arena et al. (2000) observed the differences in the structure of pulps by optical microscope and by image analysis. They reported that the modification of pulp structure and the different distribution of carotenoids between serum and pulps appear to be the more important factors on the color change in the reconstituted juices from concentrates.

Figure 15.6 shows that the background has a clear influence on lightness L^* (Figure 15.6a), C^*_{ab} (Figure 15.6b), and hue (Figure 15.6c). It can be observed that the lower the percentage of pulp the higher the values of L^* when using the white background, although the inverse trend was observed with the other two backgrounds.

This observation agrees with those of Meléndez Martínez et al. (2005a) who reported that the behavior of the color parameters as a function of concentration was similar independently of the background, except in the case of L^*. As for C^*_{ab}, its values increased as the pulp content did. The contrary was observed for hue, which changed clockwise, thus toward more orange hues.

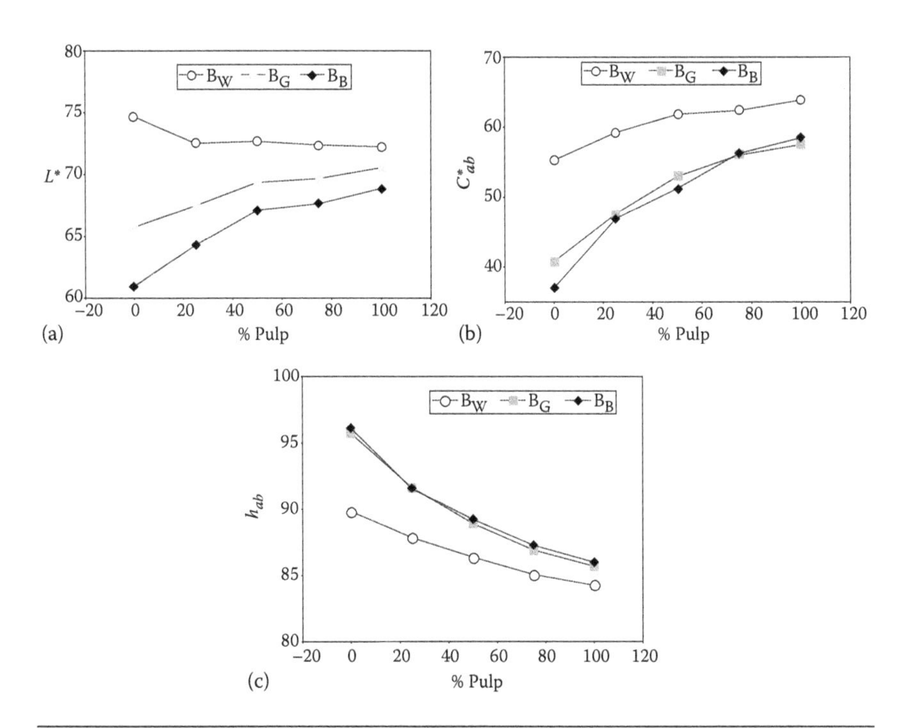

Figure 15.6 Representation of (a) L^*, (b) C^*_{ab}, and (c) h_{ab} vs. percentage of pulp as a function of the backgrounds (sample 2).

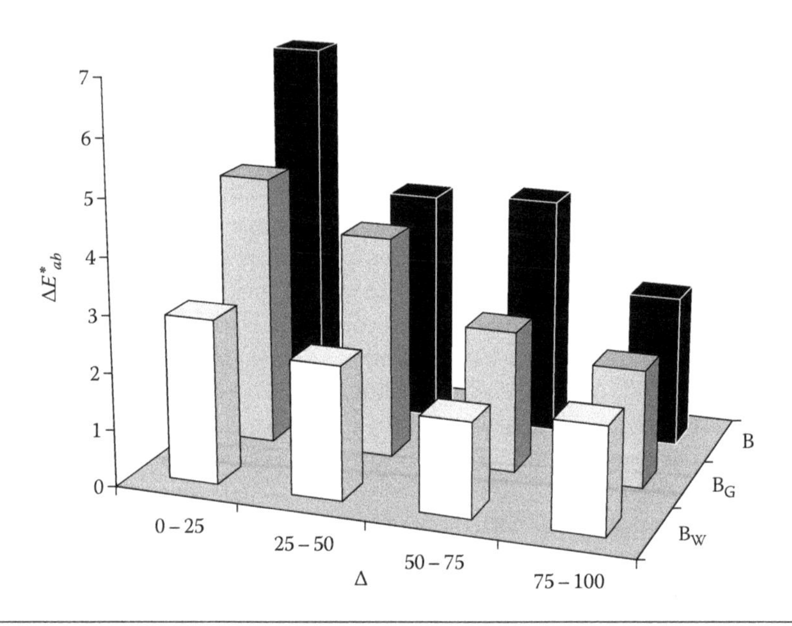

Figure 15.7 Color differences between the OJs.

Considering the backgrounds individually, it was observed that the worst color discrimination between samples with different pulp contents was achieved using $B_{W,}$ since ΔE^*_{ab} were lower than the visual discrimination threshold. In contrast, the best discrimination was obtained when the B_B was used ($\Delta E^*_{ab} > 3$) (Figure 15.7). These data are in agreement with a previous study (Meléndez-Martínez et al. 2005a), which concluded that the use of black background led to a better arrangement of OJ solutions as a function of their color intensity.

Acknowledgments

This work was partially supported by the project P08-AGR03784 (Consejería de Innovación Ciencia y Empresa, Junta de Andalucía). Carla M. Stinco holds a predoctoral research grant from the Universidad de Sevilla.

References

Arena, E., B. Fallico, and E. Maccarone. 2000. Influence of carotenoids and pulps on the color modification of blood orange juice. *Journal of Food Science* 65(3): 458–460.

CIE (Commission Internationale de l'Éclairage). 1978. Recommendations on uniform color spaces, color-difference equations, psychometric color terms. Vienna, Austria: CIE Central Bureau, Publ. 15 (E-1.3.1) 1971, Supplement 2.

CIE (Commission Internationale de l'Éclairage). 1991a. CIE standard colorimetric observers. Vienna, Austria: CIE Central Bureau, ISO/CIE 10527: 1991 (E).
CIE (Commission Internationale de l'Éclairage). 1991b. CIE standards colorimetric illuminants. Vienna, Austria: CIE Central Bureau, ISO/CIE 10526: 1991 (E).
CIE (Commission Internationale de l'Éclairage). 1995. Industrial color-difference evaluation. Technical report. Vienna, Austria: CIE Central Bureau, Publ. 116:1995.
CIE (Commission Internationale de l'Éclairage). 2001. Improvement to industrial colour-difference evaluation. Technical report. Vienna, Austria: CIE Central Bureau, Publ. 142:2001.
CIE (Commission Internationale de l'Éclairage). 2004. *Colorimetry*, 3rd edn. Vienna, Austria: CIE Central Bureau, Publ. 15:2004.
Eagerman, B. A. 1978. Orange juice color measurement using general purpose tristimulus colorimeters. *Journal of Food Science* 43(2): 428–430.
Gullett, E. A., F. J. Francis, and F. M. Clydesdale. 1972. Colorimetry of foods: Orange juice. *Journal of Food Science* 37: 389–393.
Hita, E. and J. Romero. 1981. Análisis de la influencia de las condiciones de observación en los procesos de discriminación en color. *Óptica Pura y Aplicada* 14: 11–17.
Huggart, R. L., P. J. Fellers, G. De Jager, and J. Brady. 1979. The influence of color on consumer preferences for Florida frozen concentrated grapefruit juices. *Proceedings of the Florida State Horticultural Society* 92: 148–151.
Huggart, R. L., D. R. Petrus, and B. S. Buslig. 1977. Color aspects of Florida commercial grapefruit juices, 1976–1977. *Proceedings of the Florida State Horticultural Society* 90: 173–175.
Hutchings, J. B. 1994. *Food Colour and Appearance*. Glasgow, U.K.: Blackie Academic & Professional.
Luckow, T. and C. Delahunty. 2004. Consumer acceptance of orange juice containing functional ingredients. *Food Research International* 37(8): 805–814.
MacDougall, D. B. 2002. *Colour in Food, Improving Quality*. Cambridge, U.K.: Woodhead Publishing.
Meléndez-Martínez, A. J., I. M. Vicario, and F. J. Heredia. 2005a. Correlation between visual and instrumental colour measurements of orange juice dilutions: Effect of the background. *Food Quality and Preference* 16(5): 471–478.
Meléndez-Martínez, A. J., I. M. Vicario, and F. J. Heredia. 2005b. Instrumental measurement of orange juice colour: A review. *Journal of the Science of Food and Agriculture* 85(6): 894–901.
Meléndez-Martínez, A. J., I. M. Vicario, and F. J. Heredia. 2006. Influence of white reference measurement and background on the colour specification of orange juices by means of diffuse reflectance spectrophotometry. *Journal of AOAC International* 89(2): 452–457.
Melgosa, M. 2000. Testing CIELAB-based color-difference formulas. *Color Research and Application* 25: 49–55.
Melgosa, M., M. M. Pérez, A. Yebra, R. Huertas, and E. Hita. 2001. Algunas reflexiones y recientes recomendaciones internacionales sobre evaluación de diferencias de color. *Óptica Pura y Aplicada* 34: 1–10.
Spoto, M., R. Domarco, J. Walder, I. Scarminio, and R. Bruns. 1997. Sensory evaluation of orange juice concentrate as affected by irradiation and storage. *Journal of Food Processing and Preservation* 21(3): 179–191.

Stone, H., J. Sidel, S. Oliver, A. Woolsey, and R. Singleton. 2004. *Sensory Evaluation by Quantitative Descriptive Analysis*. Trumbull, CT: Food & Nutrition Press.
Tepper, B. J. 1993. Effects of a slight color variation on consumer acceptance of orange juice. *Journal of Sensory Studies* 8: 145–154.
USDA (United States Department of Agriculture). 1982. United States standards for grades of orange juice.
Wyszecki, G. and W. S. Stiles. 1982. *Color Science. Concepts and Methods. Quantitative Data and Formulae*. New York: John Wiley & Sons.

PART III
Color Change as Quality Index of Food

CHAPTER 16

Color Determination in Dehydrated Fruits

Image Analysis and Photocolorimetry

LINA MARCELA AGUDELO LAVERDE, NURIA ACEVEDO,
CAROLINA SCHEBOR, and MARÍA DEL PILAR BUERA

Contents

16.1 Introduction

The appearance of food is the first quality aspect evaluated by consumers and is critical in the acceptance of the products. The color of the food surface is the first sensation that the consumer perceives and uses as a tool to accept or reject the product (León et al. 2006). Foods containing reducing sugars and proteins are particularly sensitive to browning reactions. Solubility decrease, color development, and the loss of nutritional value are the most important causes of deterioration associated with browning in foods (Buera et al. 1987). Dehydrated fruits are considered to be highly stable; however, they are prone to suffer discoloration during storage. Many natural pigments are unstable in dried media, and brown pigments can be formed. These color changes cause deleterious changes in food appearance and organoleptic quality and may be an indication of the decreased nutritional and functional properties of foods. Discoloration can occur homogeneously, but most of the times heterogeneous distribution of color is observed.

In the last few years, emphasis has been placed on techniques applying computer vision systems (CVS) and image analysis for assessing the browning degree and other properties related to food quality (Briones and Aguilera 2005). CVS use devices for image acquisition, hardware and software for processing the images to simulate the role of the eyes and brain. Computer vision offers a spatial resolution as it covers the whole object of interest, which is not possible with tristimulus colorimeters that require measurements point to point of the object and provide, therefore, average values of color from very small areas. Digital cameras capture the colors while the computers store them. The image acquisition by a camera and the processing using appropriate programs represent an interesting alternative for heterogeneous systems, because they provide a large quantity of information in a practical method (Jayas et al. 2000). Recent studies have shown that CVS are reliable to determine different color and appearance aspects (Briones and Aguilera 2005, Mendoza et al. 2006, Venir et al. 2007) and browning development (Acevedo et al. 2008) in food. The objective of this work was to evaluate the kinetics of color changes in different dehydrated fruits as a function of relative humidity (RH) at different storage times after heat treatment at 45°C.

16.2 Materials and Methods

16.2.1 Fruits

Fresh pear, melon, and strawberry were obtained from the local market in optimal ripeness degree for consumption and stored at 4°C until the moment of the experiment. The fruits were washed and were cut into disks (2.0 cm diameter and 0.5 cm thickness). The cut material was immediately frozen with liquid nitrogen and stored at −20°C.

16.2.2 Materials Preparation

Fruit disks were covered with liquid nitrogen before freeze-drying. An ALPHA 1-4 LD2 freeze-dryer (Martin Christ Gefriertrocknungsanlagen GMB, Germany) was used. The freeze-dryer was operated at −55°C, at a chamber pressure of 4 Pa, and the process lasted 48 h. Disks were equilibrated in a range of 11%–93% RH for 14 days at 20°C (Greenspan 1977).

16.2.3 Photocolorimeter Determinations

Color changes for pear and melon were determined using a photocolorimeter Minolta CM-508-d (Minolta Corp., Ramsey, NJ). The measurements were performed at 2° observer and D65 illuminant. L^*, a^*, and b^* functions (CIELAB color space) were obtained as an average of three determinations.

16.2.4 Computer Vision Systems

Strawberry color changes were determined by image analysis. The CVS consisted of three elements: a lighting system, a digital camera, and a personal computer. The lighting system included a D65 lamp inside a black chamber. The angle between the camera axis and the sample plane was 90° and the angle between the light source and the sample plane was 45°, achieving appropriate condition to measurement of diffuse reflection. Images were taken in a standardized black box using a D65 illuminant and a digital camera, a Power Shot A70 (Canon Inc., Tokyo, Japan) was used. The CIELAB coordinates L^*, a^*, and b^* were obtained from color images using a Photoshop 7.0 PC program and mathematical formulas described previously by Papadakis et al. (2000). Calibration was performed employing 10 colored standards.

The images of the strawberry slices were segmented in four different zones, according to their a^* values. The selected intervals of a^* values were from −3.8 to 6.8; from 6.9 to 17.4; from 17.4 to 28; and from 28 to 38.6. The number of pixels of each zone of the strawberry slices was quantified and their corresponding percentage was calculated.

16.2.5 Evaluation of Browning Degree

Browning was evaluated in all the samples upon heat treatment at 45°C. After equilibration, the fruit disks were located between two glass plates separated by rubber o-rings, which allowed hermetic seal (to avoid water loss), and placed in an oven at 45 ± 1°C. This storage temperature was selected because it is a moderate value, higher but not too far from room temperature, and it is usually employed in accelerated storage assays (Acevedo et al. 2006). Each glass plate containing six fruit slices was removed every certain time to perform color analysis and placed back in the oven. The color functions selected to follow browning changes in fruit disks were the CIEDE 2000 global color changes function, ΔE_{00} (Sharma et al. 2005), and L^*, a^*, and b^* coordinates.

16.3 Results and Discussion

Color changes occurring in fruit disks stored at 45°C at different RHs were analyzed by photocolorimetry for pear and melon samples, and by image analysis for strawberry samples. Melon and pear were very slightly colored and presented a homogeneous color distribution. On the other hand, strawberries were mostly red, but they presented heterogeneous color distribution bearing a white and green center and different degrees of red from the center to the surface of the fruit. Therefore, in this case, color changes were analyzed globally and also dividing the disk in different sections.

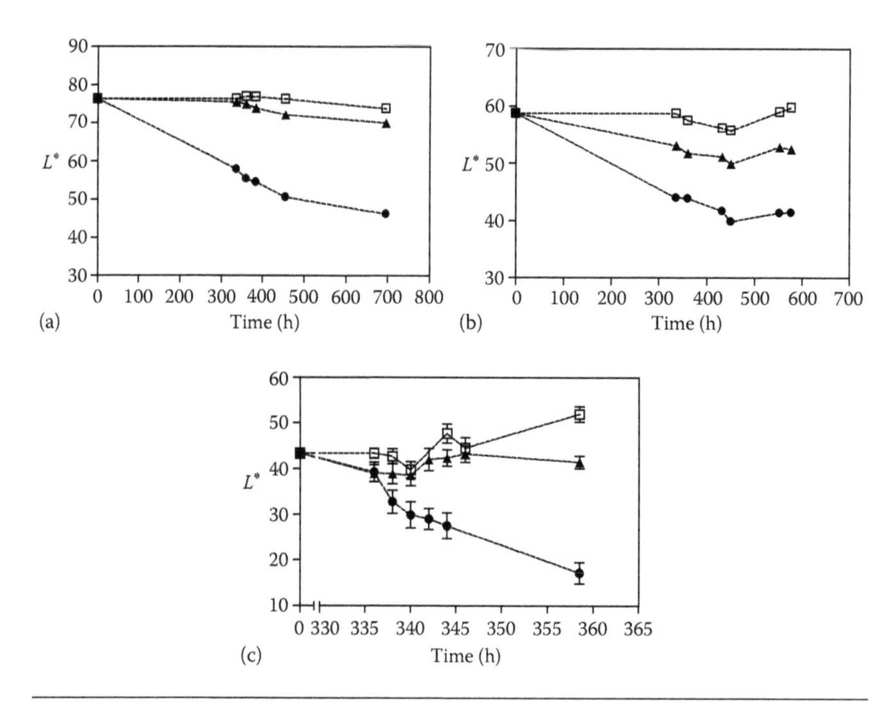

Figure 16.1 L^* changes as a function of storage time at 45°C for pear (a), melon (b), and strawberry (c) at 11% (□), 43% (▲), and 84% (●) RHs.

Figure 16.1 shows the changes of the L^* values observed for pear, melon, and strawberry during storage at 45°C. Both pear and melon developed homogeneous color changes along storage. The initial L^* value for melon was lower than that for pear. At RH of 11%, L^* values of the fruits did not show drastic changes; at 43% RH the highest L^* change was of 10 units, while at 84% RH the L^* values decreased faster due to sample browning development. Strawberry samples showed a similar behavior but displaced toward low L^* values.

Figure 16.2 shows the evolution of chromatic coordinates during storage time at different RHs; the arrows indicate the direction of time increase. Pear and melon showed a relative similar behavior. In these samples, at 11% RH small color changes were observed. The main change in melon samples was in the b^* coordinate, as a consequence of the browning development, as discussed in relation to the L^* coordinate. As increasing RH and storage time, the a^* and b^* values for pear and melon samples increased, while the opposite occurred in strawberry samples. The differences in relative browning development observed for melon and pear could be related to the chemical composition of each fruit (Agudelo Laverde et al. 2011). Previous studies performed in dehydrated apple reported the dependence of browning degree with RH, which is conditioned by the matrix–water interactions and the structural characteristics of the materials (Acevedo et al. 2006).

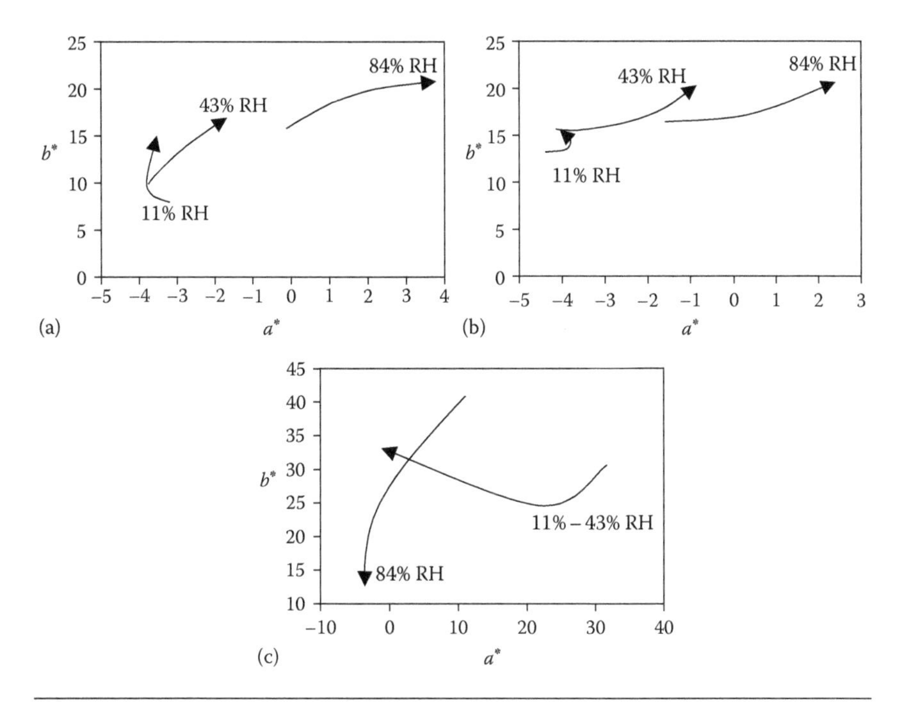

Figure 16.2 Evolution of chromatic coordinates for pear (a), melon (b), and strawberry (c) at 11%, 43%, and 84% RHs.

In the case of the strawberry samples, the initial L^* value was lower than those for pear and melon, as a consequence of the red coloration promoted by anthocyanins. At RHs lower than 43%, the a^* values were initially high and decreased markedly during storage time, with a slight change of b^*. At 84% RH, the a^* value was initially low due to the fast pigment degradation and the b^* values sharply decreased during storage. This decrease of b^* coordinate is associated to the strong darkening of the samples. It is to be noted that at the first stages of browning b^* increases due to sample "yellowing," but at the last stages b^* decreases due to the chromatic loss in the very dark samples.

In the case of strawberries, two reactions take place simultaneously: the destruction of anthocyanins and the formation of brown pigments. The stability of anthocyanins is affected by storage temperature, sugar, pH value, and ascorbic acid, water activity, and so on. Among these, water availability is essential for anthocyanin degradation (Tsai et al. 2004).

Table 16.1 shows the global color changes (ΔE_{00}) for pear, melon, and strawberry at 11%, 43%, and 84% RH, comparing the initial color coordinates and the color values after 24 h of storage at 45°C. Pear and melon presented small global color changes and in the same magnitude. The strawberry samples, however, presented much higher changes. This behavior could be related to the important changes observed in the a^* coordinate produced by the destruction of the red pigment in parallel with the formation of brown

TABLE 16.1 Global Color Changes (ΔE_{00}) for Pear, Melon, and Strawberry at 11%, 43%, and 84% RHs after 24 h

RH	Pear	Melon	Strawberry
11%	1.0 ± 0.3	2.2 ± 0.6	24 ± 3
43%	1.3 ± 0.2	1.53 ± 0.08	20 ± 5
84%	1.3 ± 0.1	2.5 ± 0.4	28 ± 2

products by the Maillard reaction. Pear and melon samples only presented formation of brown products.

As mentioned earlier, strawberry samples presented heterogeneous color distribution. This fruit showed different degrees of redness and some fractions were green or white. This characteristic suggests that color of heterogeneous samples should not be analyzed as it is usual for homogeneous materials. Therefore, image analysis was performed. Since the typical hue of strawberries is in the red region, quality changes were better represented by the a^* coordinate. The images of the strawberry slices were divided into four different zones, according to their a^* values and the pixels proportion in each zone was calculated (Figure 16.3). The pixels proportion of red areas in the pictures was greater at low RHs and short storage times; in general, the green and white areas were found in the center of the strawberry disks. By increasing RH and storage time at 45°C, the red areas started to disappear and the homogeneity of color distribution increased.

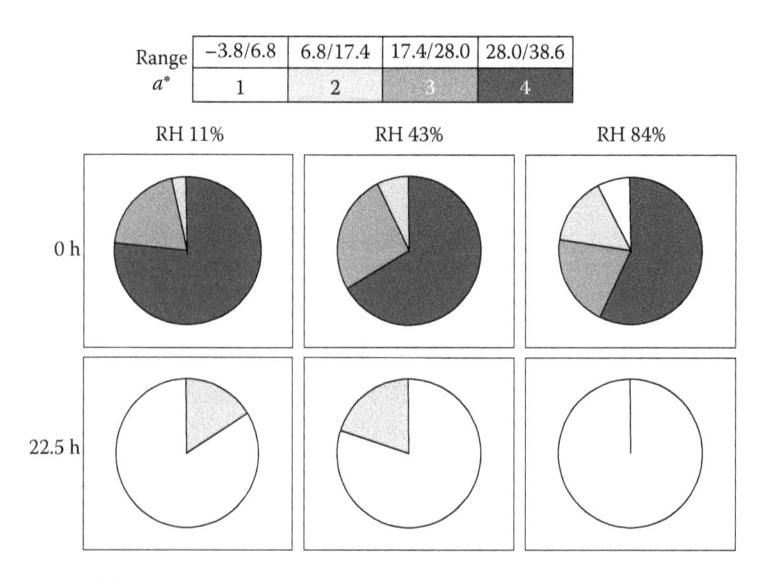

Figure 16.3 Percentage of pixels corresponding to different sections of a^* values for 0 and 22.5 h of storage at 45°C at 11%, 43%, and 84% RHs.

16.4 Conclusions

The chromatic displacement of the fruit samples depends on their water content, the presence and type of pigments, and the ability to develop browning reactions. Pigment deterioration and browning development were notably fast in strawberry samples. In general, in samples undergoing browning reactions a^* and b^* coordinates increase at the first stages, corresponding to a yellow coloration, until a red coloration is achieved. At the last stages, however, both a^* and b^* coordinates decrease due to the formation of dark brown melanoidins, which causes chromatic loss. The L^* coordinate continuously decreases during the browning of the samples. According to the analyzed stage, different behaviors can be detected.

Photocolorimetry is a simple and easy technique to evaluate global color changes, but it shows a limitation for the analysis of small areas or of heterogeneous samples. Computer vision analysis is a useful tool to assess optical properties of fruits with heterogeneous color distribution, based on its simplicity. It also allows the quantification of areas with different chromatic characteristics.

The segmented image analysis was appropriate to evaluate some relevant characteristics of the visual appearance changes occurring in dehydrated strawberries, related to the humidification level and storage time.

Acknowledgments

The authors acknowledge financial support from UBACyT Ex024, ANPCyT (PICT 0928), and Conicet (PIP 100846).

References

Acevedo, N., V. Briones, M. P. Buera, and J. M. Aguilera. 2008. Microstructure affects the rate of chemical, physical and color changes during storage of dried apple disc. *Journal of Food Engineering* 85 (2): 222–231.

Acevedo, N., C. Schebor, and M. P. Buera. 2006. Water–solid interactions, matrix structural properties and the rate of non-enzymatic browning. *Journal of Food Engineering* 77 (4): 1108–1115.

Agudelo Laverde, L. M., N. Acevedo, C. Schebor, and M. P. Buera. 2011. Integrated approach for interpreting browning rate dependence with relative humidity in dehydrated fruits. *LWT—Food Science and Technology* 44: 963–968.

Briones, V. and J. M. Aguilera. 2005. Image analysis of changes in surface color of chocolate. *Food Research International* 38 (1): 87–94.

Buera, M. P., J. Chirife, S. L. Resnik, and G. Wetzler. 1987. Nonenzymatic browning in liquid model systems of high water activity: Kinetics of color changes due to Maillard's reaction between different single sugar and glycine and comparison with caramelization browning. *Journal of Food Science* 52 (4): 1063–1064.

Greenspan, L. 1977. Humidity fixed points of binary saturated aqueous solutions. *Journal of Research* 8: 89–96.

Jayas, D., J. Paliwal, and N. Visen. 2000. Multi-layer neural network for image analysis of agricultural products. *Journal Agricultural Engineering Research* 77: 119–128.

León, K., D. Mery, F. Pedreschi, and J. León. 2006. Color measurement in $L^*a^*b^*$ units from RGB digital images. *Food Research International* 39: 1084–1091.

Mendoza, F., P. Dejmek, and J. M. Aguilera. 2006. Calibrated color measurements of agricultural foods using image analysis. *Postharvest Biology and Technology* 41 (3): 285–295.

Papadakis, S. E., S. Abdul-Malek, R. E. Kamdem, and K. L. Yam. 2000. A versatile and inexpensive technique for measuring color foods. *Food Technology* 54 (12): 48–51.

Sharma, G., W. Wu, and E. N. Dalal. 2005. The CIEDE2000 color difference formula: Implementation notes, supplementary test data, and mathematical observations. *Color Research and Application* 30 (1): 21–30.

Tsai, P.-J., Y.-Y. Hsieh, and T.-C. Huang. 2004. Effect of sugar on anthocyanin degradation and water mobility in a roselle anthocyanin model system using ^{17}O NMR. *Journal of Agricultural and Food Chemistry* 52 (10): 3097–3099.

Venir, E., M. Munari, A. Tonizzo, and E. Maltini. 2007. Structure related changes during moistening of freeze dried apple tissue. *Journal of Food Engineering* 81 (1): 27–32.

CHAPTER 17

Color Study at Storage of Lyophilized Carrot Systems

ALICIA DEL VALLE GALLO, MARÍA DEL PILAR BUERA, and CLAUDIO PETRIELLA

Contents

17.1 Introduction

Color is one of the appearance attributes that mainly influences food quality and acceptance. Due to the consumer preferences for natural foods in the last decades, vegetable colorants have been the subject of many research projects. The effects attributed to some vegetable pigments in the prevention of many chronic and degenerative diseases, which are in relation to their antioxidant properties, have also promoted several studies (Krinsky et al. 2003, Moeller et al. 2008). Carotenoids are a type of natural pigments which proved to develop antioxidant action. Among carotenoids, β-carotene and lycopene are some of the most studied. Carrot is an important source of the first, whereas tomato and watermelon are the sources of the second.

The isolated pigments of natural raw material and laboratory-synthesized analogs do not present the beneficial properties with equal intensity to those found in the original vegetable matrices (Liu 2003). This can be attributed to the different stability and absorption of phytochemicals or to the absence of other substances that accompany the pigments in their natural state and are necessary for their physiological action. Nevertheless, the fresh raw material is perishable, with seasonal availability and with a variable pigment concentration against the pure pigment preparations (Xiaquan et al. 2005). The development of preservation techniques may allow the availability of carotene-concentrated products in their natural matrices, which can be used either as innovative naturally colored healthy foods or as food ingredients with colorant properties (Vilstrup 2001). The encapsulation by freeze-drying is an excellent tool to preserve these compounds. Therefore, to study the color stability of a β-carotene-rich natural matrix, carrot pulp, employing different molecular weight maltodextrins as encapsulating agents, under different storage conditions is the proposed objective of this work.

17.2 Product Preparation and Stability Studies

17.2.1 Preparation of the β-Carotene-Rich Carrot Pulps

Figure 17.1 schematizes the preparation of the β-carotene-rich carrot systems. Carrots were washed, peeled, cut in pieces, steam-blanched, and ground. Blanching is a very well-known technological procedure that inactivates raw enzymes that can cause quality losses during processing or in the latter storage of food systems or ingredients. On the other hand, in the vegetable tissues, the pigments are located in chromoplasts, embedded in structures surrounded by the cell wall components. Thus, in order to release the carotene from the vegetable matrix and improve the homogeneity of the colorant-concentrated product, the obtained juice–pulp mixtures were treated with a commercial preparation of pectinase and hemicellulase enzymes. The conditions of time, temperature, and enzyme concentration were established in previous experiments.

17.2.2 Stabilization of the Carrot Pulps

After the enzymatic treatment, the carrot pulps were mixed either with calcium chloride (MCa), maltose (M), or maltodextrins of different molecular weights (MD-40, of average molecular weight 3600 and MD-150, of average

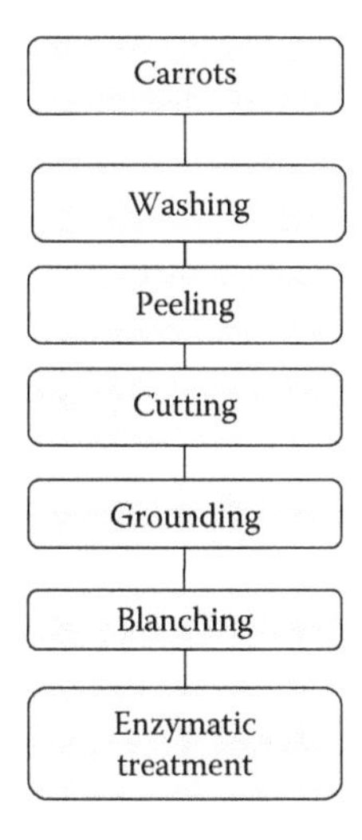

Figure 17.1 Scheme of the preparation of the β-carotene-rich carrot pulps.

molecular weight 1200) and then freeze-dried. Control samples were prepared of blanched carrots with and without enzymatic treatment (systems CE and C, respectively) with no other added substance.

17.2.3 Color Stability Studies

Freeze-dried samples were equilibrated at three different relative humidities (RHs = 11%, 43%, and 75%), employing saturated salt solutions of lithium chloride, potassium carbonate, and sodium chloride, respectively, and submitted to accelerated stability tests (55°C). At appropriate time intervals, the surface color of the samples was measured with a Minolta 508-D (Minolta Co. Ltd., Japan) integrating sphere spectrophotometer, and the CIELAB L^*, a^*, and b^* coordinates, calculated for illuminant D65 and 2° observer, were obtained. The chemical determination of β-carotene in the samples was performed spectrophotometrically in the hexane extracts of the aqueous suspension of the samples, by measuring the absorbance at 452 nm, and the percentage of β-carotene retention was calculated after each treatment (Desobry et al. 1997, Cinar 2004).

17.2.4 Microscopic Studies

The stabilized samples, equilibrated at different RHs, were observed by optical microscopy (OM) employing a Unico, MoticCam microscope and by scanning electron microscopy (SEM), with a Phillips electron microscope XL30 model TMP New Look. All samples for the SEM were mounted in aluminum holders and metalized with gold–palladium in a Termo VG Scientific SC 7620 metalizer.

17.3 Results

17.3.1 Effect of Transparency and Surface Phenomena on Luminosity

Luminosity, L^*, has been many times correlated to quality attributes of foods. Figure 17.2 shows L^* values of several studied samples as a function of storage time. As observed in Figure 17.2, the initial L^* values of samples (at time = 0) were affected by blanching, enzymatic treatment, and the matrix structure. Samples with added calcium salt and also those at low water activity had higher L^* values. The samples at higher RH values had a darker appearance due to a transparency effect caused by water on the structural biopolymers of vegetables (Agudelo Laverde et al. 2011), and this was reflected in the lower L^* values. Luminosity is highly dependent on surface phenomena, such as porosity, topography, and surface humidity (Prado et al. 2006). The opacity developed in the dry samples was caused by the different refraction indices of those interphases (MacDougall 2002). In the fresh tissues, water fills the pores and capillaries, while in the dry material the replacement of water by air

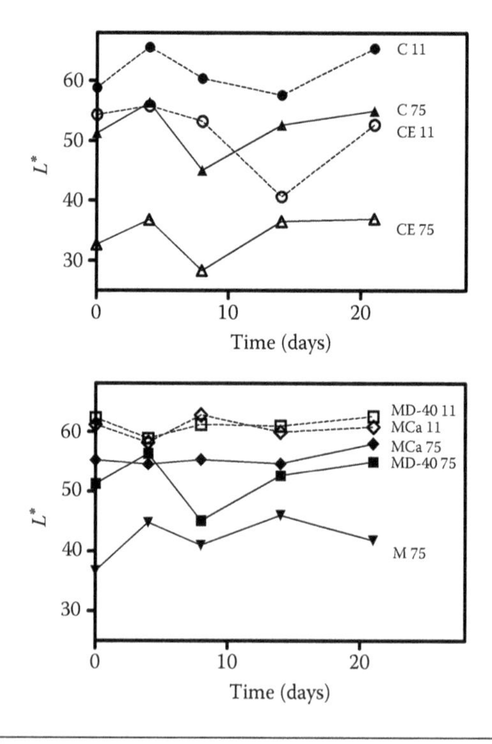

Figure 17.2 L^* behavior in dehydrated carrot samples during storage. C 11, C 75: control systems at 11% and 75% RHs, respectively. CE 11, CE 75: control systems with enzymatic treatment at 11% and 75% RHs, respectively. M 11, M 75: maltose systems at 11% and 75% RHs, respectively. MD-150 11 and MD-150 75; MD-40 11 and MD-40 75: maltodextrin MD-150 and MD-40 at 11% and 75% RHs, respectively. MCa 11, MCa 75: maltodextrin 150 with calcium chloride systems at 11% and 75% RHs, respectively.

generates several interphases. Since the refraction index of water is closer to that of the solid matter than the refraction index of air, the fresh samples are more transparent, and with darker appearance, since the pigments are more readily visible in a transparent medium. Samples at RH = 11% had very high L^* values and seemed to be lighter than the rest of the samples and clearly affected by the coverage with maltodextrins or maltose. Although initial values of L^* were very different according to the composition of the samples, L^* differences during storage were not significant. Tang and Chen (2000) encapsulated β-carotene in sucrose and gelatin matrices and also reported a small decrease of luminosity throughout the storage at 45°C, whereas the parameters a^* and b^* were not indicative of deterioration.

17.3.2 Microstructural Aspects of the Dehydrated Samples

OM images showed that the enzymatic treatment previous to freeze-drying enhanced tissue disruption and pigment extraction, which was also reflected in the visual appearance of the samples. Thus, the enzymatic treatment with pectinase and cellulose is a key point in the development of an adequate product appearance.

Figure 17.3a–c shows the micrographs of freeze-dried systems encapsulated with maltodextrins at the final stage of storage, after 30 days at

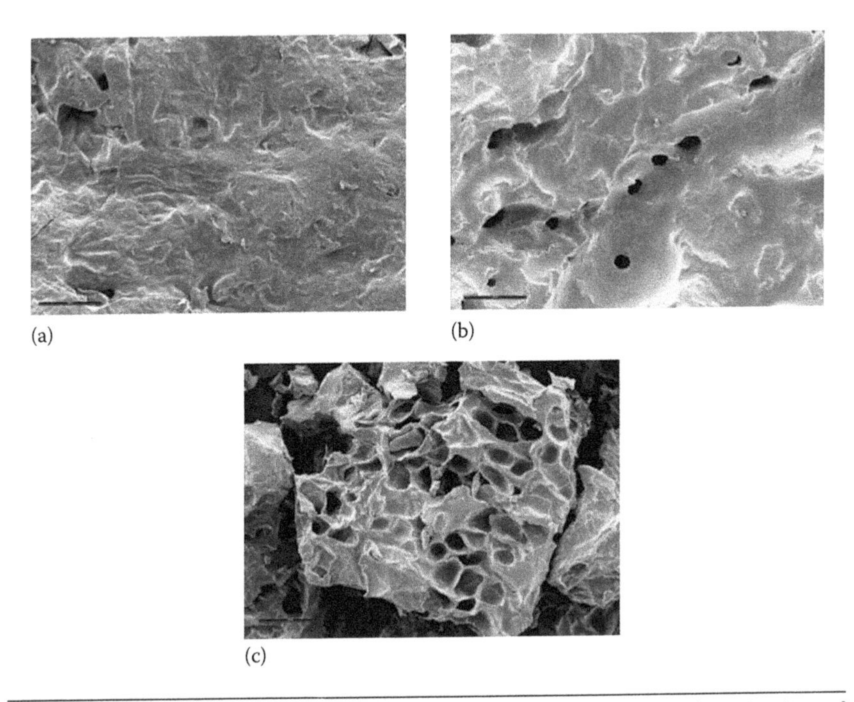

Figure 17.3 SEM micrographs showing the microstructural aspects of freeze-dried carrot systems of different relative humidities (RHs), encapsulated with maltodextrin MD-150 at the final stage of storage (21 days) at 50°C. (a) RH: 11%; (b) RH: 43%; and (c) RH: 75%.

the different RHs. At 11% RH, the coating properties of the maltodextrin were evident. The vegetable cells were totally covered by the matrix—the material surface was smooth and without pores, although some sample surfaces appeared slightly cracked—as can be seen in the left bottom region of Figure 17.3a.

At 43% RH, some pores were generated, which improved the contact with oxygen and water exchange (Figure 17.3b). At the condition of highest RH, 75% (Figure 17.3c), the porosity had increased notably, facilitating the gas exchange with the environment. These surface characteristics of the materials affected visual appearance, as discussed earlier, in relation to the L^* variable, but also influenced the kinetics of β-carotene loss. During storage, the samples at 75% RH developed yellow and then brown colors faster than the rest of the systems due to the exposure of the material when the maltodextrins did not offer an adequate coating to the tissues.

17.3.3 Correlation between β-Carotene Retention and Color Functions

Of the several color functions calculated in this work, the variable b^*/a^* correlated with the concentration of remaining β-carotene in the stored control samples and in maltodextrin-added samples equilibrated at different RHs, as determined by the chemical method. Figure 17.4 shows some examples of the observed correlations. The b^*/a^* ratio has been previously selected as a good indicator of vegetable color changes in the red–orange range hue (Little 1975), since it is a good variable to indicate the loss of redness and/or the increase of yellowness. As also reflected by the retention of β-carotene, the most remarkable change of the ratio b^*/a^* took place in the control samples with no enzymatic treatment and no added additives (systems C), while in the

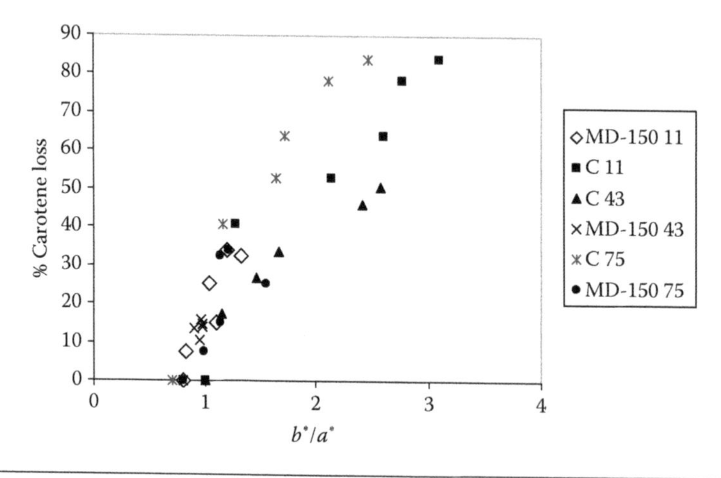

Figure 17.4 Correlation between the concentration of β-carotene and b^*/a^* in dehydrated carrot samples of different RHs and stored at 50°C.

enzymatically treated systems containing maltodextrins less drastic changes of b^*/a^* occurred. On the other hand, the color deterioration was greater and faster at high RHs: the samples became yellow and then brown at very short storage times.

17.4 Concluding Remarks

The lyophilized carrot systems described are more concentrated β-carotene sources than the raw vegetables. The enzymatic treatment previous to freeze-drying leads to a product of homogenous appearance and the presence of excipients such as maltodextrins improved β-carotene conservation in the dry systems. The β-carotene losses during processing and storage could be related to the b^*/a^* ratio. On the other hand, the luminosity values are highly dependent on the composition of the systems and on the surface characteristics of samples. Appearance was poorly related to the pigment concentration and it was highly influenced by the additives (maltose or maltodextrins or calcium salt) and by the water content. The correlation between pigment concentration and visual appearance was affected by transparency changes, mainly due to the generation or disappearance of interphases. In the most unfavorable storage conditions for β-carotene retention, which corresponded to the greatest RH studied, 75%, the presence of maltodextrin MD-150 provided stability, decreasing the color deterioration rate. In the freeze-dried carrot systems of intermediate and high water content (RHs = 43% and 75%), the encapsulation with MD-150 allowed the highest pigment retention during storage. The protecting coating action of maltodextrins was also evidenced by structural studies made with electronic microscopy. The encapsulation process herein described, which combines natural biopolymers and pigments of known antioxidant activity, could be a useful tool either in the development of innovative functional foods, of adequate appearance and health-promoting properties or as an ingredient with colorant properties.

Acknowledgments

The authors acknowledge financial support from Universidad de Buenos Aires, Universidad Nacional de Luján, ANPCyT (PICT 0928), and Conicet (PIP 100846).

References

Agudelo-Laverde, L. M., N. Acevedo, C. Schebor, and M. P. Buera. 2011. Integrated approach for interpreting browning rate dependence with relative humidity in dehydrated fruits. *LWT—Food Science and Technology* 44: 963–968.

Cinar, I. 2004. Carotenoid pigment loss of freeze-dried plant samples under different storage conditions. *LWT—Food Science and Technology* 37: 363–367.

Desobry, S. A., F. M. Netto, and T. P. Labuza. 1997. Comparison of spray-drying, drum drying and freeze-drying for β-carotene encapsulation and preservation. *Journal of Food Science* 62: 1158–1162.
Krinsky, N. I., J. T. Landrum, and R. A. Bone. 2003. Biological mechanisms of the protective role of lutein and zeaxanthin in the eye. *Annual Review of Nutrition* 23: 171–201.
Little, A. C. 1975. Off on a tangent. *Journal of Food Science* 40: 410–411.
Liu, R. H. 2003. Health benefits of fruit and vegetables are from additive and synergistic combinations of phytochemicals. *American Journal of Clinical Nutrition* 78(suppl): 517–520.
MacDougall, D. B. 2002. Discontinuity, bubbles, and translucence: Major error factors in food color measurement. In *AIC 2001, 9th Congress of the International Color Association, Proceedings SPIE 4421*, eds. R. Chung and A. Rodrigues. Bellingham, Washington: The International Society for Optical Engineering, pp. 685–688.
Moeller, S. M., R. Voland, and L. Tinker. 2008. Associations between age-related nuclear cataract and lutein and zeaxanthin in the diet and serum in the carotenoids in the age-related eye disease study, an ancillary study of the women's health initiative. *Archives of Ophthalmology* 126: 354–364.
Prado, S. M., M. P. Buera, and B. E. Elizalde. 2006. Structural collapse prevents β-carotene loss in a supercooled polymeric matrix. *Journal of Agricultural and Food Chemistry* 54: 79–85.
Tang, Y. C. and B. H. Chen. 2000. Pigment change of freeze-dried carotenoid powder during storage. *Food Chemistry* 69: 11–17.
Vilstrup, P. 2001. *Microencapsulation of Food Ingredients*. Surrey, U.K.: Leatherhead Food RA Publishing.
Xiaquan, S., J. Shi, Y. Kakuda, and J. Yueming. 2005. Stability of lycopene during food processing and storage. *Journal of Medical Food* 8: 413–422.

CHAPTER **18**

Temperature Abuses during Lettuce Postharvest

Impact on Color and Chlorophyll

MARÍA VICTORIA AGÜERO, ALICIA EVA BEVILACQUA, and SARA INÉS ROURA

Contents

18.1 Introduction

Lettuce is one of the most consumed leafy vegetables in the world (Martín-Diana et al. 2007). There are different varieties of lettuce each with particular morphological characteristics (Di Benedetto 2005). Among them, butter-head lettuce is one of the most spread varieties all over the world. The lettuce head is an assemblage of heterogenic morphological leaves that are packed together over the growing point of the plant. Its formation results from the accumulation of young leaves under the layers of leaves covering the growing point (Wien 1997). This growth pattern determines that leaves formed by different tissues, not only with different maturity degree, but also with different metabolic and physiological activities, and exposed to different environmental conditions coexist within a plant. So the lettuce plant is an interesting biological model where it is possible to evaluate these factors simultaneously in each unit.

In Argentine farms, lettuce is harvested using hand collection techniques, and as it is harvested, it is placed in package units, wooden crates, for its transportation to market. These crates are stacked on trucks and they are usually exposed to inadequate field conditions (usually high temperature) while they are waiting to be transported from the field to distribution centers (Mondino et al. 2007). Product exposed to sunlight could rapidly raise a temperature 4°C–6°C more than air temperature (Thompson et al. 2001). In the field, the combination between the heat of the sun and the respiration of the produce provokes the heat up of the produce, reducing its postharvest life. So the knowledge of the events occurring within the plant when temperature is uncontrolled during first hours after harvest is of fundamental importance to improve and optimize lettuce management. However, few researches have been developed to investigate the first few hours after harvest. Moreira et al. (2006) studied the effect of abusive temperatures after harvest over lettuce leaves and found that the first hours after harvest are crucial for vegetable shelf life because quality losses in this early stage could not been recovered. Jedermann and Lang (2007) proposed that the effect of short exposure of some few hours to inadequate conditions (too high or too low temperature and/or relative humidity) is sufficient to favor quality losses in the product.

The appearance of fresh vegetables strongly affects the purchase decision. Color is an appearance component and a transcendental property, since it impacts directly on consumer visual perception. The color of green leafy vegetables is determined by the chlorophyll concentration, which is the principal pigment of these photosynthetic tissues. Butterhead lettuce is characterized by light green color in outer leaves and yellow color in inner leaves.

The objective of the present research was to describe the response of lettuce color to the exposure at three isothermal conditions (0°C–2°C, 10°C–12°C, and 20°C–22°C), all at optimal relative humidity (RH) during the first 24 h after harvest. Leaf color behavior was evaluated through total chlorophyll content and $L^*a^*b^*$ color coordinates. Additionally, parameters were measured in three different lettuce sections: external (outer and older leaves), middle (mid leaves), and internal (inner and younger leaves) to evaluate the colorability of lettuce related to leaf age and position. Correlations between greenness indices were investigated in each section.

18.2 Materials and Methods

18.2.1 Plant Material and Sample Preparation

Heads of greenhouse butter lettuce (*Lactuca sativa* var. "Lores") were grown in Sierra de los Padres, Mar del Plata, Argentina, and were harvested at optimal maturity after reaching a marketable size (~24–30 leaves per head, corresponding to a weight of 500 ± 60 g). Once harvested, lettuce heads were immediately transported to the laboratory (~20 km) maintaining temperature and RH conditions at optimal levels (0°C–2°C and 97%–99% RH, respectively). Once arrived at laboratory, six whole plants were analyzed to obtain the initial value for each quality parameter. These experimental data represented the zero time of storage in each of the quality indicators analyzed.

Plants were not subjected to any preconditioning operation; they were just put in environmental chambers (SCT, Pharma, Argentina) at 0°C–2°C, 10°C–12°C, and 20°C–22°C, maintaining the RH levels in 97%–99%. Temperatures of 10°C–12°C and 20°C–22°C were chosen to simulate abusive refrigerated and room (warm) temperatures, respectively. High RH (97%–99%) was selected in all cases because the atmosphere within the crate is characterized by high RH due to respiration and transpiration of plants. Quality indicators evolution in samples at abusive temperatures was compared with the optimal temperature of lettuce management (0°C–2°C). Sampling was carried out at 0, 3, 6, and 24 h.

At each sampling time, three plants were taken from each storage chamber and were used to assess chlorophyll content and color. All parameters were measured in three different sections of the lettuce head called external (outer and older leaves), middle (mid leaves), and internal sections (inner and younger leaves). For each lettuce plant, sections were delimited visually, applying an organoleptic criterion, according to which the internal section was compact with yellow leaves and the middle and external sections corresponded to noncompact leaves of green and dark green color, respectively. Each section had a mean of approximately 6–9 leaves (Agüero et al. 2008). Three independent experimental runs were done.

18.2.2 Determination of Chlorophyll Content

The total chlorophyll (C) content of each section was determined following the methodology described by Moreira et al. (2003). All leaves in each section were homogenized with a commercial blender (Multiquick, MR 5550 CA Braun, Española SA, Barcelona, Spain), and two samples (1 g each) were taken from each homogenate. Each sample was then homogenized with 19 mL of a cold solution 18:1 propanone/ammonium hydroxide (0.1 N). This homogenate was filtered through sintered glass and water was removed from the filtrate with anhydrous sodium sulfate. Absorbance of the filtrate at 660.0 and 642.5 nm was measured with a UV 1601 PC UV–visible spectrophotometer (Shimadzu Corporation, Japan). C was calculated applying the formula $C = 7.12A660 + 16.8\ A642.5$ in which C is the total chlorophyll concentration (mg/L) and A660 and A642.5 are the absorbances at the corresponding wavelengths. C is reported as milligrams of chlorophyll/100 g fresh weight.

18.2.3 Determination of Color Coordinates

Color determination was carried out using a Minolta colorimeter CR 300 Series (Osaka, Japan) with an 8 mm diameter measuring area. The color of lettuce sections was measured by L^*, a^*, and b^* coordinates of the CIELAB space. The instrument was calibrated with a standard white plate ($L^* = 93.97$, $a^* = -21.85$, and $b^* = 1.21$). Color was measured in all leaves integrating each lettuce section. Ten different points were at least measured in each leaf. Mean values for L^*, a^*, and b^* coordinates were calculated for each leaf, and then for each section.

18.2.4 Statistical Analyses

Results reported in this chapter are LS-mean values (least square mean, mean estimators by the method of least squares) together with their standard deviations (Kuehl 2001).

Data were analyzed using SAS, software version 8.0 (SAS Inc. 1999). PROC general linear model (GLM) procedure was used for the analysis of variance (ANOVA). For all parameters, the factors employed as sources of variation were TIME (time after harvest), SECTION (section of the plant), TEMPERATURE, and interactions: TIME–SECTION, TIME–TEMPERATURE, and TIME–SECTION–TEMPERATURE. Differences between sections, temperatures, and time were determined by the Tukey–Kramer multiple comparison test ($p < 0.05$). PROC UNIVARIATE was used to validate ANOVA assumptions (Kuehl 2001).

18.3 Results and Discussion

18.3.1 Chlorophyll Content

Initial chlorophyll (C) content in harvested lettuce was 51.74 ± 1.93, 22.13 ± 2.11, and 10.60 ± 1.76 (mg chlorophyll/100 g fresh weight) for external, middle, and internal sections, respectively. Great differences ($p < 0.0001$) detected among lettuce sections could be related to the sunlight exposure, higher in outer leaves than in inner ones. Additionally, differences could be attributed to the physiological activity of each tissue. In this way, external leaves are photosynthetically active and they need chlorophyll molecules for this activity as they are the carbon sources of plant. Inner leaves are composed of young and growing tissues that import carbon from external leaves. Agüero et al. (2008), working with butterhead lettuce, reported lower C values in all lettuce sections than those informed in the present work. However, they found similar relationship between C contents in each section, being values in external section higher than those detected in internal one.

The exposure of almost all vegetables to inadequate temperatures (higher than the optimal one) brings about certain degradation on chlorophyll pigments (Zhang et al. 2008). ANOVA applied to C data showed a significant interaction between SECTION and TIME factors. This fact implies that each section have a particular postharvest behavior. In this way, while the external section exhibited chlorophyll pigments degradation throughout 24 h at any assayed temperature, the middle and internal sections showed no C changes during such period. Figure 18.1 depicts the evolution of C

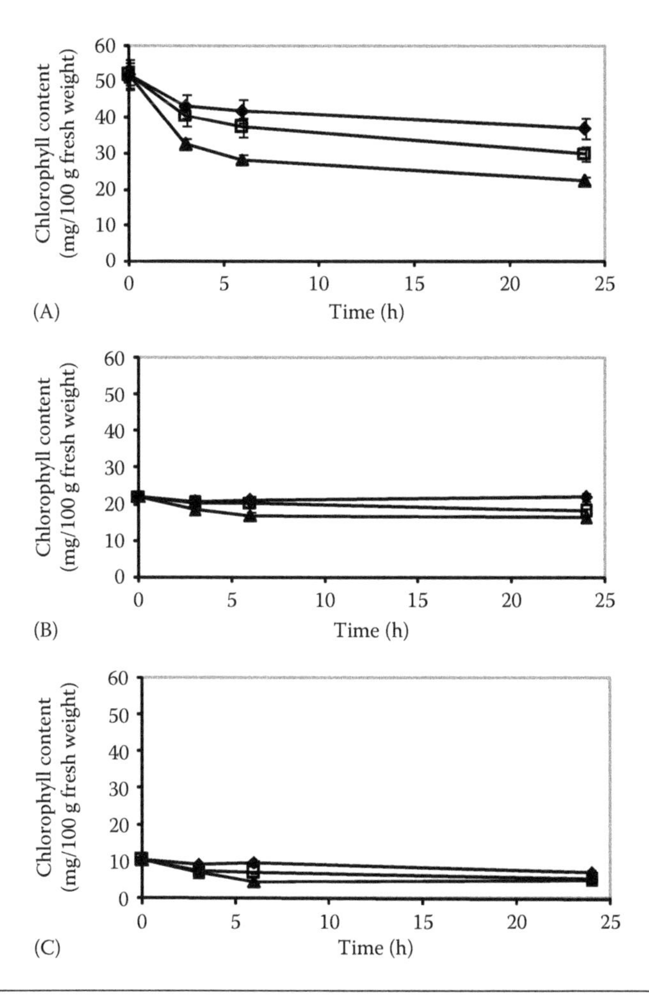

Figure 18.1 Chlorophyll content evolution in external (A), middle (B), and internal (C) lettuce sections during 24 h of exposure to isothermal conditions: (▲) 20°C; (□) 10°C; and (◆) 0°C.

in lettuce heads during 24 h after harvest. The chlorophyll degradation of external section was more pronounced at higher temperatures (36%, 20%, and 12% at 20°C–22°C, 10°C–12°C, and 0°C–2°C, respectively). Differences in chlorophyll degradation between sections could be related to the higher exposure of outer leaves to environmental factors such as light and oxygen, which could fasten pigment deterioration. It is assumed that during postharvest life, the leaf pigments undergo degradation that leads to leaf discoloration. Ferrante and Maggiore (2007) reported that *C* and carotenoid contents start to decline a few days after harvest and this phenomenon has been observed in many leafy vegetables. In the present research, *C* degradation was detected at 3 h after harvest. After this period, the rate of pigment loss was significantly lower. As a response to harvest stress, when water and nutrient supply was cut off, outer leaves (the most mature leaves of lettuce head, and leaves that are also near its physiological senescence) could trigger different physiological responses. Cran and Possinghani (1974) found different responses in *C* between mature and young spinach leaves when they were exposed to light or darkness during 7 days. During natural leaf senescence, nutrients are usually mobilized from the leaf to be used in other parts of the plant. At harvest, senescence process is induced artificially as a result of the removal of nutrient supplies (Page et al. 2001). As a rapid response, the degradation of *C* was evident in outer leaves, which started showing signs of chlorophyll losses.

Additionally, it is interesting to observe that chlorophyll degradation was only detected in older tissue suggesting that this process is mediated by leaf age associated with senescence more than by ambient conditions.

18.3.2 Color

Lightness coordinate (L^*) in lettuce at harvest showed significant differences among sections. The highest L^* value was observed in internal section (76.06 ± 2.00), decreasing toward the external one (65.71 ± 3.38 and 59.13 ± 1.95 for middle and external sections, respectively). The exposure of lettuce heads to different temperature conditions during 24 h did not introduce significant changes in L^* coordinate as a function of time or temperature (data not shown). Other authors (Ihl et al. 2003, Han et al. 2004, Martínez-Romero et al. 2008) also found for lettuce no significant changes in L^* coordinate throughout storage time.

Coordinate a^* in lettuce at harvest resulted negative in all sections indicating the predominance of green color in the product. Results showed significant differences between a^* values in each section at harvest, being more negative (greener) in external (−19.73 ± 0.83) and middle (−19.25 ± 1.28) sections (without significant differences between them) than in internal one (−16.30 ± 1.25). Table 18.1 presents a^* values obtained during sampling time. ANOVA applied to a^* data showed no significant interaction between factors considered in the analysis (neither triples nor doubles). However, each

Table 18.1 a^* Value in External (E), Middle (M), and Internal (I) Sections of Lettuce Heads during 24 h of Exposure to Three Different Isothermal Conditions (0°C–2°C, 10°C–12°C, and 20°C–22°C)

		a^* Values		
Time (h)	**Section**	**0°C–2°C**	**10°C–12°C**	**20°C–22°C**
3	E	−19.37 ± 0.32	−19.35 ± 0.56	−18.89 ± 0.24
	M	−19.49 ± 1.52	−18.85 ± 0.56	−18.90 ± 1.07
	I	−15.89 ± 0.88	−13.47 ± 3.86	−12.88 ± 1.52
6	E	−19.51 ± 1.09	−18.13 ± 2.37	−18.93 ± 0.03
	M	−18.43 ± 0.90	−16.84 ± 0.21	−18.03 ± 1.61
	I	−12.86 ± 1.04	−12.49 ± 0.58	−12.43 ± 2.41
24	E	−19.32 ± 0.34	−19.07 ± 0.35	−18.87 ± 0.06
	M	−19.14 ± 0.16	−17.94 ± 2.26	−17.44 ± 0.40
	I	−15.26 ± 0.76	−12.22 ± 1.86	−11.06 ± 2.04

factor individually resulted significant ($p < 0.0001$ for SECTION, $p = 0.0064$ for TEMPERATURE, and $p = 0.0006$ for TIME). For the SECTION factor, the significant differences detected between sections at harvest were maintained as time advanced at the three evaluated temperatures, that is, at each sampling time, outer and mid leaves resulted in more green (lower a^* values) than inner ones. For TEMPERATURE factor, samples exposed to the highest temperature registered the lowest absolute value of a^* coordinate in the three lettuce sections. Finally, for TIME factor, significant decreases in the absolute value of a^* (less green) were observed during 24 h in the three sections and the three assayed temperatures. A high correlation ($R^2 = 0.93$) was found between chlorophyll content and a^* in external section at the three assayed temperatures.

Coordinate b^* in lettuce at harvest resulted positive in all sections indicating the predominance of yellow in this scale. Significant differences between sections were found in b^* values at harvest, being higher (more yellow) in inner leaves (40.50 ± 0.78) than in mid (37.26 ± 0.85) and outer leaves (34.64 ± 0.88). ANOVA applied to b^* data showed neither double nor triple interactions among the factors considered in the analysis. Furthermore, TIME factor resulted nonsignificant indicating that no changes were registered in b^* values as time advanced at any analyzed temperature and section. SECTION and TEMPERATURE factors resulted significant ($p < 0.0001$, and $p = 0.0021$, respectively). For the SECTION factor, significant differences in b^* values detected among lettuce sections at harvest were maintained as time advanced at the three evaluated temperatures. For the TEMPERATURE factor, it was found that samples exposed to high temperatures registered lowest values of b^* coordinate.

Although chlorophyll concentration changes were detected only in the external section, color coordinates, especially the a^* coordinate, presented significant changes in the middle and internal sections too. This result could indicate that other pigments related to the red–green color scale may be experiencing changes in the early postharvest. Thus, the increase in a^* values, associated with a decrease in green color or an increase in red, could indicate an increase in the concentration of carotenoids in the tissue. Carotenoids possess a range of important and well-documented biological activities. They are potent antioxidants and free radical scavengers (Grassmann et al. 2002). When the tissue is subjected to stress, it is possible that cells synthesize carotenoids. Other authors have reported increases in carotenoid pigments during the postharvest period of some products (Lefsrud et al. 2007, Noichinda et al. 2007).

18.4 Conclusions

Leaf age had a significant effect on the initial greenness indices of butterhead lettuce. Older and senescent leaves (external section) showed higher chlorophyll concentration and lower L^* (lightness coordinate) and a^* values than younger leaves characterized by a clearest color (high L^* and a^* values and low chlorophyll content). Detriment in butterhead lettuce quality begins in the first hours after harvest and it was affected by both temperature and leaf age. Chlorophyll losses were detected at 3 h of storage at the three exposure temperatures and were only evident in outer leaves, presumably as a physiological response of mature leaves to artificial senescence induced by harvest. Mid and inner leaves (younger than outer leaves) did not show any change in its chlorophyll contents during 24 h of exposure at different temperatures. Maintaining the correct temperature from the first postharvest hours is an important factor in minimizing quality loss. The knowledge of the effect of leaf age on the evolution of quality indices is of fundamental importance for producers because they can take decisions based on this differential behavior related to the degree of development of tissue. In this way, they could use different lettuce leaves for alternative uses as fresh consumption, minimally processed products, among others.

Acknowledgments

This work was supported by Consejo Nacional de Investigaciones Científicas y Técnicas (Conicet), Agencia Nacional de Promoción Científica y Tecnológica (SECyT), and Universidad Nacional de Mar del Plata (UNMDP).

References

Agüero, M. V., M. V. Barg, A. Yommi, A. Camelo, and S. I. Roura. 2008. Postharvest changes in water status and chlorophyll content of lettuce (*Lactuca sativa* L.) and their relationship with overall visual quality. *Journal of Food Science* 73: 176–185.

Cran, D. G. and J. V. Possinghani. 1974. The effect of cell age on chloroplast structure and chlorophyll in cultured spinach leaf discs. *Protoplasma* 79: 197–213.

Di Benedetto, A. 2005. *Manejo de cultivos hortícolas: bases ecofisiológicas y tecnológicas*, 1st edn. Buenos Aires, Argentina: Orientadora Gráfica Editora.

Ferrante, A. and T. Maggiore. 2007. Chlorophyll a fluorescence measurements to evaluate storage time and temperature of *Valeriana* leaf vegetables. *Postharvest Biology and Technology* 45: 73–80.

Grassmann, J., S. Hippeli, and E. F. Elstre. 2002. Plant's defense mechanism and its benefits for animals and medicine: Role of phenolics and terpenoids in avoiding oxygen stress. *Plant Physiology and Biochemistry* 40: 471–478.

Han, J., C. L. Gomes-Feitosa, A. Castell-Perez, R. G. Moreira, and P. F. Silva. 2004. Quality of packaged romaine lettuce hearts exposed to low-dose electron beam irradiation. *LWT—Food Science and Technology* 37: 705–715.

Ihl, M., L. Aravena, E. Scheuermann, E. Uquiche, and V. Bifani. 2003. Effect of immersion solutions on shelf life of minimally processed lettuce. *LWT—Food Science and Technology* 36: 591–599.

Jedermann, R. and E. Lang. 2007. Semi passive RFID and beyond steps towards automated quality models tracing in the food chain. *International Journal of Radio Frequency Identification Technology and Applications* 1(3): 247–259.

Kuehl, R. O. 2001. *Diseño de experimentos*, 2nd edn. México: Thompson Learning International.

Lefsrud, M., D. Kopsell, A. Wenzel, and J. Sheehan. 2007. Changes in kale (*Brassica oleracea* L. var. *acephala*) carotenoid and chlorophyll pigment concentrations during leaf ontogeny. *Scientia Horticulturae* 112: 136–141.

Martín-Diana, A. B., D. Rico, C. Barry-Ryan, J. M. Frías, G. T. M. Henehan, and J. M. Barat. 2007. Efficacy of steamer jet-injection as alternative to chlorine in fresh lettuce. *Postharvest Biology and Technology* 45: 97–107.

Martínez-Romero, D., M. Serrano, G. Bailén, F. Guillén, P. J. Zapata, J. M. Valverde, S. Castillo, M. Fuentes, and D. Valero. 2008. The use of natural fungicide as an alternative to preharvest synthetic fungicide treatments to control lettuce deterioration during postharvest storage. *Postharvest Biology and Technology* 47: 54–60.

Mondino, M. C., J. Ferratto, I. Firpo, R. Rotondo, M. Ortiz Mackinson, R. Grasso, P. Calani, and A. Longo. 2007. Pérdidas poscosecha de lechuga en la región de Rosario, Argentina. *Horticultura Argentina* 26(60): 17–27.

Moreira, M. R., A. G. Ponce, C. E. del Valle, R. Ansorena, and S. I. Roura. 2006. Effects of abusive temperatures on the postharvest quality of lettuce leaves: Ascorbic acid loss and microbial growth. *Journal of Applied Horticulture* 8(2): 109–113.

Moreira, M. R., S. I. Roura, and C. E. del Valle. 2003. Quality of Swiss Chard produced by conventional and organic methods. *LWT—Food Science and Technology* 36: 135–141.

Noichinda, S., K. Bodhipadma, C. Mahamontri, T. Narongruk, and S. Ketsa. 2007. Light during storage prevents loss of ascorbic acid, and increases glucose and fructose levels in Chinese kale (*Brassica oleracea* var. *alboglabra*). *Postharvest Biology and Technology* 44: 312–315.

Page, T., G. Griffiths, and V. Buchanan-Wollaston. 2001. Molecular and biochemical characterization of postharvest senescence in broccoli. *Plant Physiology* 125: 718–727.

Thompson, J., M. Cantwell, M. L. Arpaia, A. Kader, and J. Smilanick. 2001. Effect of cooling delays on fruit and vegetable quality. *Perishables Handling Quarterly* 105: 1–4.

Wien H. C., ed. 1997. *The Physiology of Vegetable Crops*. New York: CAB International, pp. 479–509.

Zhang, M., Z. G. Zhan, S. J. Wang, and J. M. Tang. 2008. Extending the shelf-life of asparagus spears with a compressed mix of argon and xenon gases. *LWT—Food Science and Technology* 41: 686–691.

CHAPTER 19

Changes in Color and Anthocyanin Content of Different Dried Products Based on Sweet Cherries

LORENA FRANCESCHINIS, CAROLINA SCHEBOR, and DANIELA M. SALVATORI

Contents

19.1 Introduction

Cherry consumption has an increasing impact, due to the health benefits associated with regular intake of anthocyanins (Kirakosyan et al. 2009). Due to harvest losses and excess production, it is important to develop industrially processed cherry products.

Sweet cherries (*Prunus avium* L.) contain substantial amounts of anthocyanins and polyphenolic compounds (Tural and Koca 2008). Besides the interest because of their health-promoting properties, anthocyanins are responsible for the orange, red, and blue colors in many fruits and vegetables and have a critical role in the color quality of many fresh and processed products.

It is widely known that drying at high temperatures and long times may cause damage in the nutritive and sensorial characteristics, affecting flavor, color, and nutrients of the dried food (Lin et al. 1998). Among the best food dehydration techniques, freeze-drying is known to produce the highest quality dehydrated products (Khalloufi and Ratti 2003), being much more effective in preserving valuable food compounds than traditional methods. Color change during drying of a food product is indicative of the severity of the drying conditions, and is related to its pigment composition/concentration. Anthocyanin pigments are highly unstable and readily degrade during processing and storage of foodstuffs, which can have a dramatic impact on color quality and may also affect nutritional properties (Wrolstad et al. 2005b). A way of obtaining dried fruits of good quality is to use predrying treatments, such as osmotic dehydration, also termed as sugar infusion (Torregianni and Bertolo 2001).

Although there have been several investigations on the effect of processing on anthocyanin levels and color development in cherries (Serrano et al. 2009) and other fruits such as berries (Wojdyło et al. 2009), the available knowledge on changes after dehydration is still limited. Therefore, the objective of this work was to analyze the effect of different pretreatments and dehydration methods on color and anthocyanin content of two products obtained from sweet cherries (*Lapins* var.).

19.2 Materials and Methods

19.2.1 Sample Preparation, Pretreatments, and Drying Processes

Lapins cherry cultivar grown in Valentina Norte, Neuquén (Patagonia, Argentina) was used. Fruit characterization was performed according to AOAC (1990) methods: moisture 77.5% ± 1.8%, soluble solids 20 ± 2°Bx, acidity 0.82 ± 0.06 g% citric acid, pH 3.54 ± 0.07, ash 0.499% ± 0.009%. Water activity (a_w) measured at 25°C was 0.974 ± 0.005. Fruits were washed, peduncles were removed, and pitting was done with a manual cherry pitter. A group of cherries were cut into halves and another group were chopped into eight pieces in order to obtain different dried product geometries: disks and dices, respectively. Then they were subjected to the following pretreatments prior to drying:

Dry sugar infusion (SI): Fruits were placed in a mixture of solutes (sucrose and preservatives). The amount of sugar was calculated to attain a_w equilibration value of 0.87. Potassium sorbate (1000 μg/g) and sodium bisulfite (150 µg/g) were used. Two different systems were prepared: one for disks and other for dices.

Blanching (B): It was done by exposure of samples (disks only) to saturated steam during 90 s and then cooling by submerging them in water at 4°C during the same time.

Control (C): Cherry disks and dices without pretreatments were used as control.

Cherry disks and dices, with and without pretreatments, were dried by the following:

Freeze-drying (F): Condenser temperature of −84°C and chamber pressure of 0.04 mbar, during 48 h

Air-drying (A): In a forced convection oven, air temperature of 60°C, and 10% of HR, during 24 h

19.2.2 Color Analysis

Superficial color was measured by photocolorimetry, in the CIELAB color space, with C illuminant and 2° observer. Measurements were performed on skin and pulp in 40 disks for each condition. In the case of dices, 20 measurements were taken from a pull of samples randomly arranged in Petri dishes. The $L^*a^*b^*$ rectangular coordinates were used to calculate the following color functions in order to analyze the color changes caused by the different treatments:

1. Global color change ΔE^*_{ab}

$$\Delta E^*_{ab} = \sqrt{(\Delta L^*)^2 + (\Delta a^*)^2 + (\Delta b^*)^2}$$

   where $\Delta L^* = (\bar{L}^*_0 - L^*)$, $\Delta a^* = (\bar{a}^*_0 - a^*)$, and $\Delta b^* = (\bar{b}^*_0 - b^*)$

   $\bar{L}^*_0$, $\bar{a}^*_0$, and $\bar{b}^*_0$ are the average CIELAB coordinates for fresh cherry pulp, skin, or pull used for calculation of $\Delta E^*_{ab\ \text{pulp disks}}$, $\Delta E^*_{ab\ \text{skin disks}}$, and $\Delta E^*_{ab\ \text{dices}}$, respectively.
2. Hue angle (h_{ab}): $h_{ab} = \arctan(b^*/a^*)$.
3. Chroma or degree of color saturation: $C^*_{ab} = (a^{*2} + b^{*2})^{1/2}$.

19.2.3 Total Anthocyanin Content and Anthocyanin Degradation Index

Total anthocyanin content (T Acy) was determined using the pH differential method. Extracts were obtained with 95% ethanol/1.5 N HCl (85:15). Acy content (monomeric anthocyanin) was expressed as cyanidin-3-glucoside (MW: 445.2 and a molar extinction coefficient ε: 29,600 L/cm/mol) (Wrolstad et al. 2005a). Anthocyanin calculation was corrected by dry matter content.

The anthocyanin degradation index (ADI) is the ratio between total anthocyanin calculated by the single pH method and T Acy. This index is indicative of the proportion of degraded anthocyanin in the sample. Samples containing degraded pigment or other brownish colored compounds should give ADI higher than 1.0 (Fuleki and Francis 1968).

19.2.4 Water Content

Water content was determined by difference in weight before and after vacuum-drying over desiccant at 60°C.

19.2.5 Statistical Analysis

All statistical analyses were carried out using the Statgraphics Plus package (StatPoint Technologies, Inc., Warrenton, VA). Analysis of variance (ANOVA) was done to establish significant differences in parameters. Significance level was set at $p < 0.05$ and multiple comparisons were performed using the Tukey test.

19.3 Results and Discussion

19.3.1 Superficial Color of Cherry Products

Table 19.1 shows the L^*, a^*, and b^* average CIELAB coordinates of cherries disks. The applied pretreatments and drying methods caused different effects

Table 19.1 *$L^*a^*b^*$ Values and Global Color Change, ΔE^*_{ab}, for Cherry Disks after Drying*

		L^*		a^*		b^*		ΔE^*_{ab}	
Samples		**Mean†**	**SD**	**Mean†**	**SD**	**Mean†**	**SD**	**Mean†**	**SD**
Disk skin	Fresh	26.1^c	1.7	14^d	4	3.2^d	1.3		
	A-C	23^a	2	9.5^b	1.4	3.99^e	1.1	5.9^a	1.3
	A-B	25^b	2	10^b	3	5^e	2	5.8^a	1.8
	A-SI	36.9^e	1.8	5.6^a	1.8	2.19^c	1.0	13.9^d	1.7
	F-C	34.9^d	1.1	9^b	2	1.5^b	0.5	10.3^b	1.5
	F-B	36.4^d	1.5	12^c	3	2.34^c	1.0	11.2^c	1.8
	F-SI	35.2^e	1.3	4.44^a	1.2	0.8^a	0.4	13.5^d	1.0
Disk pulp	Fresh	32.8^c	1.7	5.3^a	1.7	0.8^a	0.6		
	A-C	15.6^a	1.6	7.6^c	1.8	$4.2^{c,d}$	1.6	17.9^e	1.2
	A-B	18.6^b	2.4	8.5^c	1.8	4.3^d	1.8	15^d	2
	A-SI	35.1^e	0.9	5.7^a	1.4	$3.6^{b,c}$	1.2	3.9^a	1.2
	F-C	$35.0^{d,e}$	1.8	13^d	2	4.6^d	1.6	9^c	3
	F-B	34.0^d	1.1	12.2^d	1.8	3.08^b	1.1	7^b	2
	F-SI	35.1^e	1.2	6.8^b	1.5	3.3^b	1.0	4^a	1.5

† For each group, means with the same letter superscript were not significantly different ($p < 0.05$).

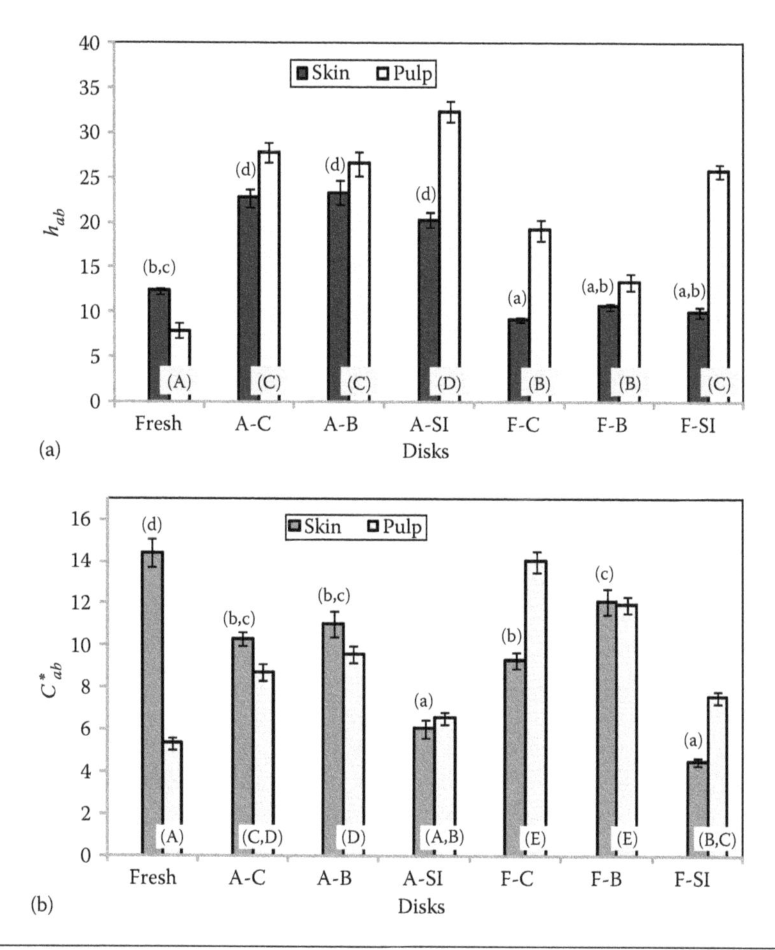

Figure 19.1 (a) Hue angle and (b) chroma for disks measured on skin and pulp. Bars with the same letter were not significantly different ($p < 0.05$).

on color when compared with the fresh fruit. Air-drying caused darkening (L^* decrease) in control cherries mainly due to nonenzymatic browning enhanced at high temperature, particularly on the pulp side of cherry disks. These samples exhibited a significant decrease of the a^* parameter in the skin and an increase of a^* and b^* values in the pulp.

Regarding hue angle (Figure 19.1), the values were in the red/yellow quadrant (0°–90°), and did not exceed 32°, which indicates that samples showed colors between red and orange. Air-drying produced a relatively large increase of hue angles when compared with the fresh disks. Although a heat pretreatment was applied to inactivate enzymes responsible for enzymatic browning, similar color (hue and chroma values) was obtained for cherries previously blanched and control ones. Only a slight but significant ($p < 0.05$) increase in L^* values was detected.

For all of the studied treatments, the chroma of disks decreased in the skin, but increased in the pulp, in which cherries experienced the highest color saturation due to concentration of pigments during drying. This behavior was mainly observed in freeze-dried samples with and without previous blanching.

In general, freeze-dried cherries showed the lowest increment in hue values, except for samples with previous SI pretreatment. FC and FB disks maintained the same hue angle of fresh cherries in the skin but increased in the pulp in a lesser extent than those observed for air-dried samples. In all the analyzed cases, freeze-dried cherries showed a clearer appearance ($>L^*$), which could be due to the porous structure generated in this process. Thus, freeze-drying inhibits color deterioration during drying, resulting in products with superior color compared with those dried by air-drying (Table 19.1 and Figure 19.1).

Previous studies have shown that infusion of sugars can stabilize color during/after drying of certain fruits such as apple and banana (Sosa et al. 2010). However, during infusion, some pigments can be transferred to the solution, with a significant loss of fruit color (Osorio et al. 2007). In dried cherries, the SI pretreatment caused discoloration, since samples showed higher L^* values, especially in the skin, and higher hue angle and no modification or slightly increase in chroma in the pulp of dried and freeze-dried cherries. This could be explained by three reasons: first, the diffusion of pigments into the solution during osmosis; second, some monomeric anthocyanins are in colorless sulfonic form due to acid addition of bisulfite; and finally, the presence of sugar crystals on fruit surface.

In the case of cherry dices, a more homogeneous behavior was observed because the color measurement was performed in bulk (mixture of pulp and skin). All samples were slightly lighter than the fresh ones and presented a decrease in the chromatic parameters (Table 19.2 and Figure 19.2). Hue angles for air-dried cherries were higher than those observed for fresh ones, and the values for freeze-dried samples were similar to the fresh ones. Changes were

Table 19.2 L^*, a^*, b^*, and Global Color Change, ΔE^*_{ab}, for Cherry Dices after Drying

		L^*		a^*		b^*		ΔE^*_{ab}	
Samples		**Mean†**	**SD**	**Mean†**	**SD**	**Mean†**	**SD**	**Mean†**	**SD**
Dices	Fresh	27.4^{a}	1.5	8^{d}	3	2.3^{c}	1.2		
	A-C	31.0^{c}	1.1	2.8^{a,b}	1.0	1.1^{a,b}	0.5	6.8^{b,c}	0.8
	A-SI	32.0^{d}	1.1	4.7^{c}	1.3	2.57^{c}	1.0	6.2^{a}	0.7
	F-C	29.5^{b}	0.6	1.98^{a}	0.7	0.5^{a}	0.2	6.9^{c}	0.7
	F-SI	31.42^{c,d}	1.1	3.9^{b,c}	1.4	1.4^{b}	0.5	6.2^{a,b}	0.6

† For each group, means with the same letter superscript were not significantly different ($p < 0.05$).

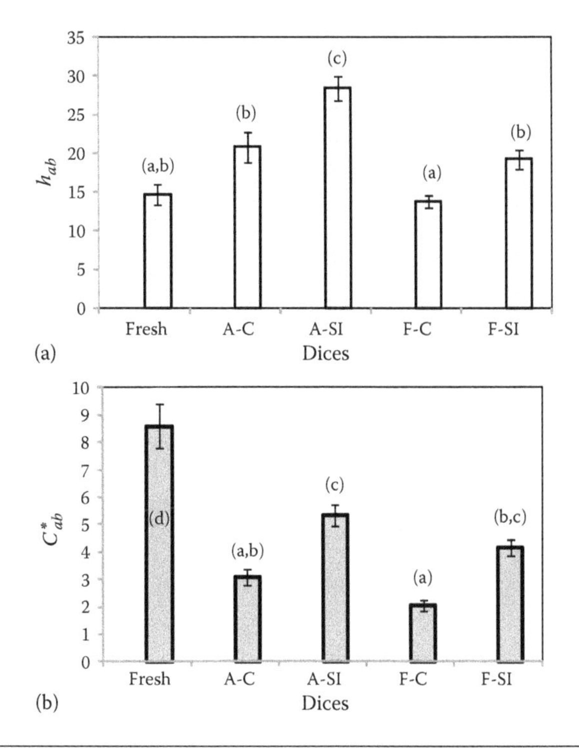

Figure 19.2 (a) Hue angle and (b) chroma for dices measured in pull. Bars with the same letter were not significantly different ($p < 0.05$).

more drastic when A-SI drying was applied, since the hue angles increased toward yellowish region of color wheel (Figure 19.2).

19.3.2 Anthocyanin Retention

The T Acy in all the samples was significantly lower than that in fresh cherries (Table 19.3). In general, Acy retention was higher in the freeze-dried fruits. However, SI pretreatment caused an important decrease in the anthocyanin content. Soaking during the osmotic pretreatment could allow the pigments leakage to the solution, which could cause anthocyanin loss before the drying step. The low values obtained for air-dried samples showed that more pigments could be destroyed by high temperature processing. Dices showed higher T Acy than disks for all the analyzed conditions (Table 19.3). A-SI samples presented the highest ADI values.

For both disks and dices, the increase in h_{ab} upon dehydration was in agreement with Acy pigment decrease. For a certain pretreatment, freeze-dried samples showed higher anthocyanin retention than air-dried samples (Figure 19.3).

As shown in Table 19.1, global color change function should not be used to predict color stability in cherry disks during drying, since the values proved

Table 19.3 Total Anthocyanin Content and ADI for Disks and Dices after Drying

	T Acy (mg/100 g Dry Weight)		ADI	
Disks	**Mean†**	**SD**	**Mean†**	**SD**
Fresh	236[a]	38	1.20[a]	0.01
A-C	38[d]	4	1.92[a]	0.06
A-B	50[d]	16	1.6[a]	0.3
A-SI	6.2[e]	1.7	6.7[b]	1.7
F-C	121[c]	8	1.46[a]	0.04
F-B	165[b]	22	1.29[a]	0.06
F-SI	31[d,e]	6	2.1[a]	0.3
Fresh	236[a]	38	1.20[a]	0.01
A-C	124[b]	6	1.37[a]	0.03
A-SI	22[c]	4	2.36[b]	0.02
F-C	212[a]	30	1.26[a]	0.03
F-SI	89[b]	8	1.36[a]	0.03

† Means with same letter superscript were not significantly different ($p < 0.05$).

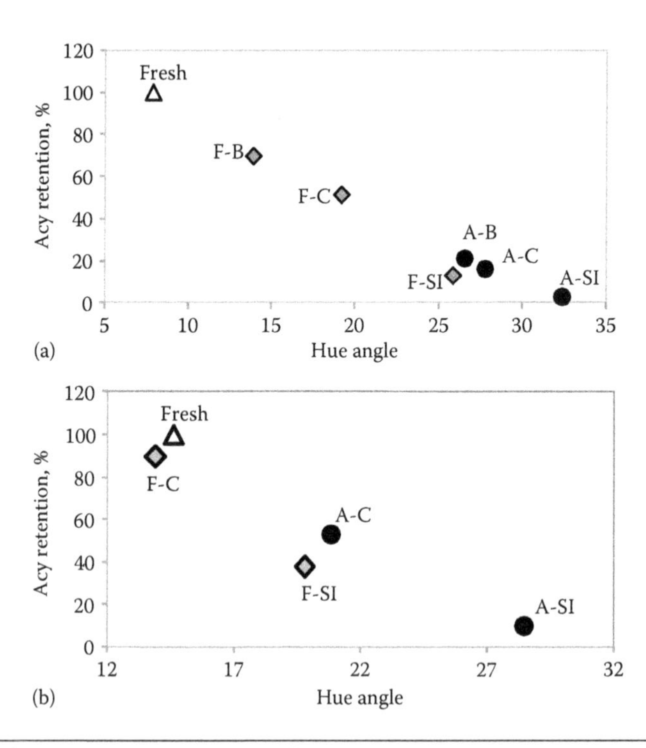

Figure 19.3 Anthocyanin retention vs. superficial red color (h_{ab}) on (a) cherry disk pulp and (b) dices after air- and freeze-drying.

to be less suitable to follow red color variations. It was observed that small ΔE^*_{ab} changes were exhibited by brownish samples and even by SI samples, the samples with lower red pigments concentration. Regarding cherry dices, no significant differences were observed in ΔE^*_{ab} values, and again this function appeared not to be recommended to make the appropriate technology selection in terms of surface color.

19.4 Conclusion

With respect to color and anthocyanin content, bettter quality products were obtained from cherry dices (small pieces) than disks (halves). The SI pretreatment caused an important decrease in anthocyanin pigments retention and a clearer appearance. Therefore, this procedure should be carefully performed in berries in order to retain the nutritional quality of the product.

Acknowledgments

The authors acknowledge the financial support from Universidad de Buenos Aires, Universidad Nacional del Comahue, Conicet, and Agencia Nacional de Promoción Científica y Tecnológica.

References

AOAC (Association of Official Analytical Chemists). 1990. *Official Methods of Analysis*, 15th edn. Arlington, VA: AOAC International.

Fuleki, T. and F. J. Francis. 1968. Quantitative methods for anthocyanins. Determination of total anthocyanins and degradation index for cranberry juice. *Journal of Food Science* 33: 78–83.

Khalloufi, S. and C. Ratti. 2003. Quality deterioration of freeze-dried foods as explained by their glass transition temperature and internal structure. *Journal of Food Science* 68: 892–903.

Kirakosyan, A., E. Seymour, D. Urcuyo Llanes, P. Kaufman, and S. Bolling. 2009. Chemical profile and antioxidant capacities of tart cherry products. *Food Chemistry* 115: 20–25.

Lin, T. M., T. D. Durance, and C. H. Scaman. 1998. Characterization of vacuum microwave, air and freeze dried carrot slices. *Food Research International* 31: 111–317.

Osorio, C., M. S. Franco, M. P. Castaño, M. L. González-Miret, F. J. Heredia, and A. L. Morales. 2007. Colour and flavour changes during osmotic dehydration of fruits. *Innovative Food Science and Emerging Technologies* 8(3): 353–359.

Serrano M., H. Díaz-Mula, P. Zapata, S. Castillo, F. Guillén, D. Martínez-Romero, J. Valverde, and D. Valero. 2009. Maturity stage at harvest determines the fruit quality and antioxidant potential after storage of sweet cherry cultivars. *Journal of Agricultural and Food Chemistry* 57: 3240–3246.

Sosa, N., D. Salvatori, and C. Schebor. 2010. Physico-chemical and mechanical properties of apple discs subjected to osmotic dehydration and different drying methods. *Food and Bioprocess Technology*. Published online: DOI 10.1007/s11947-010-0468-4.

Torregianni D. and G. Bertolo. 2001. Osmotic pre-treatments in fruit processing: Chemical, physical and structural effects. *Journal of Food Engineering* 49: 247–253.
Tural, S. and I. Koca. 2008. Physico-chemical and antioxidant properties of cornelian cherry fruits (*Cornus mas* L.) grown in Turkey. *Scientia Horticulturae* 116: 362–366.
Wojdyło A., A. Figiel, and J. Oszmianski. 2009. Effect of drying methods with the application of vacuum microwaves on the bioactive compounds, color, and antioxidant activity of strawberry fruits. *Journal of Agricultural and Food Chemistry* 57: 1337–1343.
Wrolstad, R., T. Acree, E. Decker, M. Penner, D. Reid, S. Schwartz, C. Shoemaker, D. Smith, and P. Sporns. 2005a. *Handbook of Food Analytical Chemistry*. Upper Saddle River, NJ: John Wiley & Sons.
Wrolstad, R., R. Durst, and J. Lee. 2005b. Tracking color and pigment changes in anthocyanin products. *Trends in Food Science & Technology* 16: 423–428.

CHAPTER 20

Color Measurement of Nova Mandarins Submitted to Heat and Degreening Treatments during Long Storage

SILVINA GUIDI, ANA MARÍA SANCHO, GUSTAVO POLENTA, GUILLERMO MEIER, DANIEL VÁZQUEZ, and CLAUDIA GONZÁLEZ

Contents

20.1 Introduction

The external color is one of the most important quality factors of citrus fruits that drive the purchasing decision of consumers. One problem, usually found in some early-season varieties of mandarins, is that the fruit reaches the required standard of internal maturity before the skin is fully colored. Thus, fruits have to be degreened with ethylene gas if they were to become commercially apt. To assess adequately this process, the objective measurement of color is the most important control activity to be carried out in any degreening research, in the packinghouse grading process, and in the inspection of citrus fruit. The application of ethylene gas, as an elicitor of the degreening process of fruits, has become a common commercial practice

(Brown and Miller 1999). This process stimulates the synthesis of yellow and orange carotenoids and accelerates, at the same time, the chlorophyll degradation, unmasking other colors in the flavedo (Wheaton and Stewart 1973, Goldschmidt 1974).

Similar to many other fruits, citrus production areas around the world are typically distant from their overseas markets, a fact that produces an important delay between harvesting time and the arrival of the fruit at the point of consumptions. This delay usually has negative effects on fruit quality because of all the modifications brought about by the uncontrolled ripening process. This problem also affects fruits devoted to domestic consumption, which have to be stored for long times in years with overproduction.

Considering all these aspects, it becomes evident that fruit producers need to rely on appropriate storage technologies capable of delaying the ripening process of the commodity, in order to maintain the quality of products and compete successfully in both local and export markets (Wills et al. 1989).

Based on the evidence that postharvest life of fruits strongly depends on the adequate control of the respiration rate, low-temperature storage is nowadays the most widely used technology to extend fruit shelf life, allowing the proper shipping of product to long distances, therefore contributing to regulate the supply of products in the market.

Citrus fruits, because of their tropical and subtropical origin, can suffer severe injuries when exposed to nonfreezing temperatures below 12°C (Lyons 1973), with the severity of the damage being dependent on the duration of the exposure and the temperature difference between this threshold and the actual storage temperature. This kind of disorder—of physiological origin—is collectively known as chilling injury (CI) and can cause significant deterioration of the overall quality (Lyons 1973). The critical temperature for CI varies among the different commodities; however, the damage was generally found to occur when fruits are stored typically at temperatures below 10°C–13°C.

In the case of citrus, CI is manifested as a brown-to-black peel pitting that blemishes when the damage becomes more severe. Although this alteration may result in commercial losses, the disorder has no effect on internal fruit quality. In recent years, it was found that the development of CI symptoms can be reduced by the application of heat treatments prior to fruit storage at chilling temperatures (Lurie 1998, Porat et al. 2000). Thereafter, several postharvest treatments have been designed to improve the low-temperature storage of chilling-sensitive fruit (Wang 1993). These treatments can be applied by dipping the fruit in hot water, by either vapor or hot air exposure, or by a short brushing in hot water. The application of these treatments constitutes a nonchemical mean to control postharvest diseases and reduce the sensitivity of fruits to cold storage, thus preventing the development of CI (Porat et al. 2000, Sanchez-Ballesta et al. 2000, Ben-Yehoshua 2005). Porat el al. (2000), have demonstrated that some postharvest heat treatments such as curing at

36°C for up to 3 days, hot-water dipping for 2 min at 52°C, and a hot-water brushing for 20 s at 62°C were successful in decreasing decay and CI in grapefruits stored for up to 8 weeks at 2°C.

Interestingly, some mandarin cultivars such as Clementine and Clemenules are tolerant to low-temperature storage (Martínez-Jávega and Cuquerella 1981, Puppo et al. 1988, Martínez-Jávega et al. 1991), while other cultivars such as Nova and Fortune are typically susceptible to this condition (Cuquerella et al. 1990, Martinez-Jávega et al. 1991). In spite of the economical consequences of this problem, studies addressing the application of heat treatments to citrus fruits are still scarce.

Therefore, the aim of the present study was to investigate the effect of postharvest treatments applied to reduce CI (heat shock and degreening) on the color attributes of "Nova" mandarins stored for long period at low temperature.

20.2 Materials and Methods

20.2.1 Plant Material and Treatments Applied

"Nova" mandarins (*Citrus clementina* Hort. ex. Tanaka x *Citrus paradisi* Macf. x *Citrus reticulata* Blanco) grown in Entre Rios (Argentina) were harvested and divided into four lots: one was incubated at 20°C–21°C in a chamber for 48 h in the presence of ethylene atmosphere (5 ppm, ethylene degreening, D); the second was incubated in a chamber at 36°C for 48 h (heat shock, HS); the third was first heat treated at 36°C for 48 h and then incubated at 20°C–21°C for 48 h with ethylene 5 ppm (HS + D); and, finally, nontreated fruit was used as control (C). After the applications of the treatments, the fruit was evaluated every 15 days during 60 days' storage at two different temperatures: 2°C (chilling temperature) and 9°C (nonchilling temperature), and under two different conditions: immediately after cold withdrawal and after storage for 7 days at 20°C in chamber to simulate commercial conditions.

20.2.2 Color Measurements

Color measurements (L^*, a^*, and b^* parameters) were performed using a Minolta Chroma Meter with an illuminant D65 and 2° observer geometry. Average values (three measurements per sample) were calculated for each fruit and converted into color index values $IC = (1000 \times a^*)/(L^* \times b^*)$ (Jiménez-Cuesta et al. 1981). For each temperature of conservation, the evaluations were performed at harvest, immediately after the application of treatments ($M1_0$), and periodically every 15 days under the two conditions mentioned earlier: immediately after cold storage withdrawal ($M15_0$, $M30_0$, $M45_0$, $M60_0$) and after submitting the fruit—withdrawn from cold storage—to 7 days at 20°C, in order to simulate commercial conditions ($M15_7$, $M30_7$, $M45_7$, $M60_7$).

20.2.3 *Statistical Analysis*

The experiment was designed as a 4 × 5 factorial. Main effects were treatments (C, D, HS, HS + D) and times (1, 15, 30, 45, and 60), and the statistical analysis was accomplished independently for each conservation temperature evaluated. The result of the initial measurement (at harvest) was used as a covariable to analyze both storage conditions. ANOVA and covariance analysis were analyzed by using SPSS® software (version 12.0 Illinois, United States). LS means were utilized for means comparison.

20.3 Results and Discussion

Table 20.1 shows IC values obtained immediately after the application of the treatments and periodically every 15 days during the conservation at 2°C. Control fruits (C) showed an increase in IC values after 15 days at 2°C ($M15_0$; 9.4) and remained steadily constant during the rest of the storage period. None of the treated fruits showed any color modification throughout 60 days of cold storage.

By performing a comparison among treatments, it can be seen that immediately after the application ($M1_0$), fruits submitted to treatment D showed the highest IC value (16.6, $p < 0.05$). The simultaneous application of heat and degreening (HS + D) caused a significant ($p < 0.05$) reduction of IC values from 16.6 to 12.7. The heat treatment alone (HS) rendered an IC score smaller than the one mentioned earlier, which was only slightly higher than the value determined for C (9.0 and 7.0, respectively), though still significantly different. These results lead us to speculate that the application of heat could affect the metabolism of the pigments responsible for the color change of fruit, therefore attenuating the effect of the degreening treatment.

Fruits analyzed during the storage period showed that the IC value attained immediately after treatment application remained unchanged thereafter. The exception was HS treatment, which showed no significant differences with C. This result highlights the advantage of applying ethylene,

Table 20.1 IC Evolution in Mandarin "Nova" during the Conservation at 2°C

	Periods of Sampling				
Treatment	$M1_0$	$M15_0$	$M30_0$	$M45_0$	$M60_0$
C	7.0 ± 0.3 d	9.4 ± 0.3 c	9.7 ± 0.3 c	10.2 ± 0.3 c	10.5 ± 0.3 c
D	16.6 ± 0.3 a	17.2 ± 0.3 a	16.2 ± 0.3 a	16.0 ± 0.3 a	17.3 ± 0.3 a
HS	9.0 ± 0.3 c	10.2 ± 0.3 c	9.7 ± 0.3 c	9.7 ± 0.3 c	9.7 ± 0.3 c
HS + D	12.7 ± 0.3 b	13.8 ± 0.3 b	13.8 ± 0.3 b	13.7 ± 0.3 b	13.7 ± 0.3 b

Notes: Measurements were accomplished in control (C) or treated fruit (D, HS, and HS + D) after treatment applications ($M1_0$) and periodically each 15 days ($M15_0$, $M30_0$, $M45_0$, $M60_0$) during the conservation. Means with different letters are significantly different ($p < 0.05$ LS means).

which is widely used in packinghouses to accelerate fruit color development by inducing both the synthesis of carotenoids and the catabolism of chlorophylls. With regard to the combined treatment (HS + D), which rendered IC values lower than D, it can be hypothesized that the heat applied prior to ethylene could have caused a delay in the color evolution. Supporting this thought it has been reported that the color change evidenced during the degreening process involves both the destruction of chlorophylls (revealing carotenoids already present in the cell) and the synthesis of carotenoid pigments, which is externally reflected as an orange color development of the fruit (Wheaton and Stewart 1973). Interestingly, the enzymatic reactions involved in these processes have different optimal temperatures. For chlorophyll degradation, optimal and inhibitory temperatures are respectively 28°C and 40°C, while for carotenoid synthesis, these temperatures are 18°C and 30°C, respectively (Cuquerella 1997). Therefore, the temperature applied in the heat treatment (36°C, 48 h) would induce the inhibition of carotenoid synthesis, impairing at the same time the chlorophyll degradation and rendering in consequence a lower IC value.

The accelerated evolution of color induced by the exposure of fruits to ethylene had been previously described in citrus fruits by Martínez-Javega (2002). Regarding the color inhibition caused by heat treatments, a similar effect was also observed by Plaza et al. (2004) in mandarins. These authors found that fruits submitted to ethylene alone (5–10 μL at 20°C) had a higher color development compared to fruits treated simultaneously with ethylene gas and high temperature (5–10 μL at 40°C for 24 h + 5–10 μL at 20°C).

Table 20.2 shows IC values obtained in fruits withdrawn from cold storage and submitted for 7 days at 20°C to simulate commercial conditions. Analysis of the color evolution during the storage period showed that control fruits (C) did not show significant differences among the different sampling periods, reaching the highest IC value (nonsignificant) after 15 days of storage ($M15_7$). In general, IC values of fruits submitted to the other treatments and

Table 20.2 IC Evolution in Mandarin "Nova" during the Conservation at 2°C and Subsequent 7 Days Storage at Room Temperature to Simulate Commercial Conditions

	Periods of Sampling				
Treatment	**$M1_0$**	**$M15_7$**	**$M30_7$**	**$M45_7$**	**$M60_7$**
C	6.9 ± 0.3 e	7.9 ± 0.3 cde	9.1 ± 0.3 cd	10.1 ± 0.3 c	10.1 ± 0.3 c
D	16.4 ± 0.3 a	16.1 ± 0.3 a	15.6 ± 0.3 a	15.6 ± 0.3 a	15.7 ± 0.3 a
HS	9.0 ± 0.3 cd	9.6 ± 0.3 c	9.5 ± 0.3 c	9.9 ± 0.3 c	9.9 ± 0.3 c
HS + D	12.7 ± 0.3 b	13.1 ± 0.3 b	13.2 ± 0.3 b	13.3 ± 0.3 b	14.0 ± 0.3 b

Notes: Measurements were accomplished in control (C) or treated fruit (D, HS, and HS + D) after treatment applications ($M1_0$) and periodically each 15 days + 7 days to room temperature ($M15_7$, $M30_7$, $M45_7$, $M60_7$). Means with different letters are significantly different ($p < 0.05$ LS means).

incubated for 7 days were similar within each treatment for the different storage times. Hence, D treatment had the highest IC values while the rest of the treatments showed lower IC values in the following order: HS + D > HS > C. This result suggests that the exposure of fruits for 7 days at room temperature after the cold withdrawal caused no further change when compared to fruits analyzed immediately after treatments.

Table 20.3 shows IC values obtained immediately after treatment application and periodically every 15 days during the conservation at 9°C, which represents the optimal storage temperature for citrus fruit. At this temperature, the color index increased gradually in C samples ($p < 0.05$). The degreening treatment alone induced a faster color index development compared to C, with no further significant increment being detected during the storage period. Again, the exposure of fruit to heat reduced the IC value compared to D, with no color index modification being observed after treatment application. HS treatment showed a similar pattern as the one found in Tables 20.1 and 20.2, with the color index being lower than those found in fruit submitted to the other treatments. After 30 days, the color index remained unchanged, with a slight increase being detected around day 45. This value represents the maximum attained and was evidenced toward the end of the storage period.

Table 20.4 shows the IC values analyzed in fruit stored at 9°C and incubated, after cold withdrawal, for 7 days at 20°C to simulate commercial conditions. In general, IC values of treated fruit measured along the storage were coincident, within each treatment, with those determined in fruits analyzed immediately after withdrawal from 9°C.

No statistical comparison of IC values was conducted between the different storage temperatures, although fruit stored at 9°C had higher IC values compared to fruits stored at a lower temperature. Coincident with this result, fruits analyzed at 9°C also presented a more regular coloration on the whole surface.

Table 20.3 IC Evolution in Mandarin "Nova" during the Conservation at 9°C

	Periods of Sampling				
Treatment	**$M1_0$**	**$M15_0$**	**$M30_0$**	**$M45_0$**	**$M60_0$**
C	6.3 ± 0.4 i	11.6 ± 0.4 fg	12.9 ± 0.4 efg	16.6 ± 0.4 bc	17.5 ± 0.4 abc
D	16.2 ± 0.4 bcd	17.0 ± 0.4 bc	17.9 ± 0.4 ab	17.8 ± 0.4 ab	19.2 ± 0.4 a
HS	9.1 ± 0.4 h	9.5 ± 0.4 h	10.9 ± 0.4 gh	13.1 ± 0.4 ef	15.5 ± 0.3 cd
HS + D	12.3 ± 0.4 fg	12.8 ± 0.4 efg	13.5 ± 0.4 ef	13.5 ± 0.4 def	14.4 ± 0.4 de

Notes: Measurements were accomplished in control (C) or treated fruit (D, HS, and HS + D) after treatment applications ($M1_0$) and periodically each 15 days ($M15_0$, $M30_0$, $M45_0$, $M60_0$) during the conservation. Means with different letters are significantly different ($p < 0.05$ LS means).

Table 20.4 IC Evolution in Mandarin "Nova" during the Conservation at 9°C and Subsequent 7 Days to Room Temperature to Simulate Commercial Conditions

Treatment	Periods of Sampling				
	$M1_0$	$M15_7$	$M30_7$	$M45_7$	$M60_7$
C	6.3 ± 0.4 i	11.1 ± 0.4 gh	13.9 ± 0.4 de	17.2 ± 0.4 abc	18.1 ± 0.4 ab
D	15.6 ± 0.4 cd	15.8 ± 0.4 cd	18.9 ± 0.4 ab	19.1 ± 0.4 ab	19.4 ± 0.4 a
HS	8.9 ± 0.4 h	9.2 ± 0.4 h	11.5 ± 0.4 fg	14.4 ± 0.4 de	16.9 ± 0.3 bc
HS + D	12.2 ± 0.4 efg	12.4 ± 0.4 efg	13.7 ± 0.4 def	14.5 ± 0.4 de	15.7 ± 0.4 cd

Notes: Measurements were accomplished in control (C) or treated fruit (D, HS, and HS + D) after treatment applications ($M1_0$) and periodically each 15 days + 7 days to room temperature ($M15_7$, $M30_7$, $M45_7$, $M60_7$). Means with different letters are significantly different ($p < 0.05$ LS means).

The results found in present work indicate that the application of ethylene (D) rendered more colored fruits (based on IC values), and this color remained unchanged throughout the cold storage. On the other hand, additional storage for 7 days at 20°C to simulate commercial conditions had no important effect on color change.

The adverse effect of high temperature (HS treatment) on the color development during the degreening of citrus fruit has been previously described (Jahn et al. 1973, Wheaton and Stewart 1973). In this regard, it is well known that during the first hours of ethylene exposure, the yellowness of fruit is produced as a result of chlorophyll destruction, which unmasks carotenoids already present. After this phenomenon, carotenoids synthesized de novo begin to accumulate, which are externally visualized as an orange color. Temperatures around 30°C (typically used for curing treatments) accelerate the chlorophyll destruction and, at the same time, impair carotenoid synthesis.

The conclusions drawn from the present study by considering all findings are an important contribution toward the production, conditioning and commercialization of high-quality citrus fruits. It becomes evident that when curing treatments are to be applied in combination with degreening, the rest of the conditions should be carefully determined for the successful application of this combined treatment at a commercial scale. Similarly to the production of other fruits, the main challenge of the citrus industry is to develop technologies able to adequately maintain the market quality of product all throughout the commercialization chain.

20.4 Conclusion

The present investigation highlights the importance of the use of ethylene gas in the treatment of mandarin to artificially induce the degreening process. Among other advantages, this treatment offers the benefit of a shorter time required for the fruit to reach a commercially apt color. However, if this technology is applied in combination with heat treatments, conditions should be carefully determined, considering the reduction in the effect brought about by the exposure to high temperatures. The potential use and the applicability of the combined treatment will strongly depend on the final destination of the fruit and the intended condition of storage.

References

Ben-Yehoshua, S. 2005. *Environmental Friendly Technologies for Agricultural Produce Quality*. Boca Raton, FL: CRC Press, Taylor & Francis.

Brown, G. E. and W. R. Miller. 1999. Maintaining fruit health after harvest. In *Citrus Health Management*, eds. L.W. Timmer and L.W. Duncan. St. Paul, MN: APS Press.

Cuquerella, J. 1997. Técnicas y prácticas de desverdización de cítricos producidos en condiciones mediterráneas. *Phytoma España* 90: 106–110.

Cuquerella, J., C. Saucedo, J. M. Martínez-Jávega, and M. Mateos. 1990. Influencia de la temperatura y envoltura plástica en la conservación de mandarina Fortune. In *Congreso Nacional Soc. Esp. Hort. 2*. Puerto de La Cruz, Tenerife, España, pp. 410–416.

Goldschmidt, E. E. 1974. Hormonal and molecular regulations of the chloroplast senescence in citrus peel. In *Plant Growth Substances*, ed. Y. Sumiki. Tokyo, Japan: Hirokawa Press.

Jahn, O. L., W. G. Chace, and R. H. Cubbedge.1973. Degreening response of "Hamlin" oranges in relation to temperature, ethylene concentration, and fruit maturity. *J. Am. Soc. Hort. Sci.* 98: 177–181.

Jiménez-Cuesta, M., J. Cuquerella, and J. M. Martínez-Jávega. 1981. Determination of a color index for citrus degreening. *Proc. Natl. Acad. Sci. USA* 78: 3526–3530.

Lurie, S. 1998. Postharvest heat treatments of horticultural crops. *Hort. Rev.* 22: 91–121.

Lyons, J. M. 1973. Chilling injury in plants. *Ann. Rev. Plant Physiol.* 24: 445–446.

Martínez-Javega, J. M. 2002. Estado actual de las aplicaciones del frío en la poscosecha de cítricos. In *Actas del I Congreso Español de Ciencia y Tecnología del Frío*, eds. A. López, A. Esnoz, and F. Artés. Cartagena, Spain: Universidad Politécnica de Cartagena, pp. 433–442.

Martinez-Jávega, J. M. and J. Cuquerella. 1981. Factors affecting cold storage of Spanish oranges and mandarins. *Proc. Int. Soc. Citric.* 1: 511–514.

Martínez-Jávega, J. M., J. Cuquerella, and P. Navarro. 1991. Influencia de la temperatura de conservación y condiciones de almacenamiento en la calidad final de mandarina Nova. In *III World Congress Food Technology, Abstracts*. Barcelona, Spain, p. 62.

Plaza, P., A. Sanbruno, J. Usall, N. Lamarca, R. Torres, J. Pons, and I. Viñas. 2004. Integration of curing treatments with degreening to control the main postharvest diseases of clementine mandarins. *Postharvest Biol. Technol.* 34: 29–37.

Porat, R., D. Pavoncello, J. Peretz, S. Ben-Yehoshua, and S. Lurie. 2000. Effects of various heat treatments on the postharvest qualities of "Star Ruby" grapefruit. *Postharvest Biol. Technol.* 18: 159–165.

Puppo, A. H., E. Zefferino, L. Bisio, J. C. Codina, and E. Supino. 1988. Study of ten different cultivars of tangerine in Uruguay. *Proc. Int. Soc. Citric.* 6: 1499–1504.

Sanchez-Ballesta, M. T., L. Zacarias, A. Granell, and M. T. Lafuente. 2000. Accumulation of Pal transcript and Pal activity as affected by heat-conditioning and low-temperature storage and its relation to chilling sensitivity in mandarin fruit. *J. Agric. Food. Chem.* 48: 2726–2731.

Wang, C. Y. 1993. Approaches to reduce chilling injury of fruits and vegetables. *Hort. Rev.* 15: 63–132.

Wheaton, T. A. and I. Stewart. 1973. Optimum temperature and ethylene concentrations for postharvest development of carotenoid pigments in citrus. *J. Am. Soc. Hort. Sci.* 98: 337–340.

Wills, R. B. H., W. B. McGlasson, D. Graham, T. H. Lee, and E. G. Hall. 1989. *Postharvest: An Introduction to the Physiology and Handling of Fruits and Vegetables*. Oxford, U.K.: BSP Professional Books.

CHAPTER 21

Color of Vacuum-Packed Squid (*Illex argentinus*) Mantle Rings Treated with Gamma Radiation

ALEJANDRA TOMAC and MARÍA ISABEL YEANNES

Contents

21.1 Introduction

Food irradiation is perhaps the most intensively studied technology for toxicological safety in the history of food preservation. Studies on nutritional adequacy, toxicological innocuity, and microbiological safety of irradiated foods were carried out for more than 50 years, and all scientific evidence supports its safety (Diehl 2002, Sommers et al. 2006, WHO 1994, 1999).

Food irradiation has proven to be not only a useful technology but also an efficient one to reduce microbial counts and to increase food shelf life (Kilcast 1995, Kodo 1990).

Shelf life of sea bream (Chouliara et al. 2004), *Merluccius hubbsi* (Lescano et al. 1990), and whole anchovies (Lakshmann et al. 1999) was increased by gamma irradiation.

Illex argentinus is the most abundant cephalopod in the Southwest Atlantic Ocean and the second fishery of economic interest in Argentina (Brunetti et al. 1999). Its captures were more than one million tons in 2006, representing the second among marine captures with 27.3% of the total (MINAGRI 2007).

Squid is processed for its consumption in many different ways. Fried floured squid rings (*rabas*) are consumed in North America, northern Europe, and Latin America. Squid tubes and rings can be treated with commercial solutions containing polyphosphates, which are largely used in the fishery industry to improve water-holding capacity of proteins. This fact benefits the final quality of the product by retaining natural moisture, flavor, and nutrients, improving texture and reducing the cooking loss. In addition, phosphates delay lipid oxidation and stabilize color by quelling enzyme (metal) cofactors (Gonçalves and Duarte Ribeiro 2008, 2009, Knipe 2004, Lampila 1993).

After capture, sensory quality of squid decreases due to chemical and microbiological deterioration reactions that promote changes in squid skin color and favor muscle stain formation. These color changes associated with quality loss have been studied by Lapa-Guimarâes et al. (2001) in *Loligo plei*, by Sungsri-in et al. (2011) in *Loligo formosana*, and by Thanonkaew et al. (2006) in *Loligo peali*. These changes have been related to modifications in the color parameters of the CIELAB color space (L^*, a^*, b^*) by the increase of a^* due to the generation of pink spots (Sungsri-in et al. 2011) and the increase of b^*, mostly related to lipid oxidation and deterioration reactions (Thanonkaew et al. 2006).

Irradiation of squid has been studied by Byun et al. (2000) in fermented and salted *Todarodes pacificus*, but no analyses on color changes were included. Color is considered one of the most important attributes of food appearance and is usually used to determine different aspects of food quality such as ripeness and spoilage grade (Francis 1995, Potter and Hotchkiss 1995). Color directly influences consumer purchase choices, so it is important to evaluate how the color of squid changes due to irradiation and during storage, in order to understand how the product is affected by this technology. This work is part of research intended to extend shelf life of minimally processed squid by gamma radiation, taking into account radiation effects on color. The objective of this work was to analyze the effect of different gamma radiation doses on the color of squid (*Illex argentinus*) rings, vacuum-packed, during storage at 4°C–5°C.

21.2 Materials and Methods

21.2.1 Raw Material Source, Treatment, and Storage

Peeled squid mantle rings of *Illex argentinus* specimens were acquired in the port of Mar del Plata (Argentina). They had been pretreated according to manufacturer instructions with a polyphosphate commercial solution.

Samples of approximately 110 ± 2 g were vacuum-packed in Cryovac® bags of LDPE and nylon (125 μm) using a vacuum packaging machine Minimax 430 M (Servivac, Argentina). Samples were transported at 4°C ± 3°C to the semi-industrial Ezeiza Atomic Centre facilities of Argentina (National Atomic Energy Commission (CNEA); activity: 600,000 Ci). They were gamma irradiated with a ^{60}Cobalt source at doses of 1.8, 3.3, and 5.8 kGy (minimum doses absorbed). Doses were determined with Amber Perspex dosimeters. Irradiated and nonirradiated samples (control, 0 kGy) were stored at 4°C–5°C for 22 days and analyzed days 1, 5, 8, 12, 15, 19, and 22.

21.2.2 Color Analysis

A portable colorimeter (NR-3000, Nippon Denshoku Kogyo Co. Ltd., Tokyo, Japan) with D65 standard illuminant and 2° standard observer was used to determine CIELAB color space system parameters (L^* = lightness, a^* = positive values for red color intensity and negative values for green color intensity, and b^* = positive values for yellow color intensity and negative values for blue color intensity) of squid rings. Two measurements were made on five rings of each sample. Color difference (DE2000) with reference to the first day of storage was calculated with CIE 2000 color difference formulas (CIE 2001).

21.2.3 Statistical Analysis

Results were analyzed by two main factors on a completely aleatorized design. Factors analyzed were radiation dose (0, 1.8, 3.3, and 5.8 kGy) and days of storage (1, 3, 5, 9, 12, 15, 19, and 22 days). A two-way ANOVA test was used to evaluate main effects and interaction. In further analysis, the Tukey test was used to compare means ($p < 0.05$). The experiment was repeated twice. The statistical analysis was carried out using the R-Project software (R Development Core Team 2008).

21.3 Results and Discussion

Color changes were analyzed for any association with spoilage reactions. Figures 21.1 through 21.3 show the evolution of L^*, b^*, and a^* color parameters, respectively, during storage of squid mantle rings at 4°C–5°C. Color difference was calculated and evaluated for each day of analysis with respect to the first day, using CIEDE2000 equation. Results, standard errors, and Tukey test results ($p < 0.05$) are shown in Figure 21.4.

No significant differences due to radiation dose were detected in L^*, a^*, b^*, and DE2000 on day 1. Radiation did not induce color changes in squid rings after its application.

Values of L^* significantly increased in control sample during storage; meanwhile in radiated samples, it did not significantly change during the whole storage period (Figure 21.1). After 5 days of storage, L^* of control was significantly higher than L^* of radiated samples ($p < 0.05$).

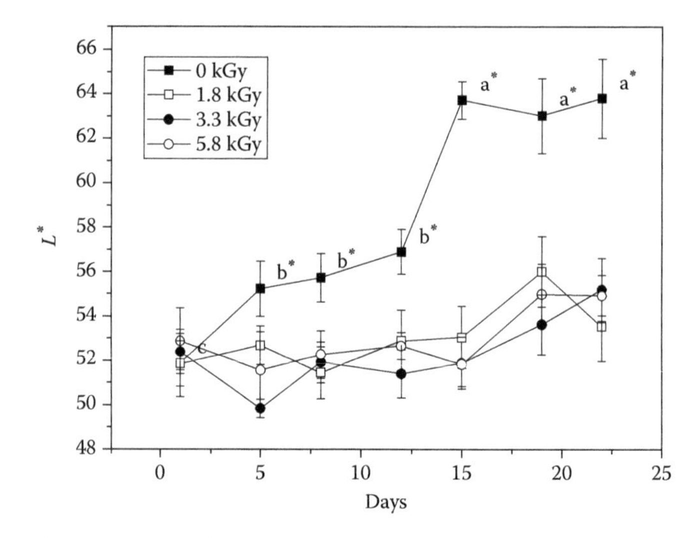

Figure 21.1 Evolution of L^* in squid rings during storage at 4°C–5°C. Standard error represented by bars. Different letters (a, b, c) indicate significant differences due to storage time. Values with * are significantly different between doses.

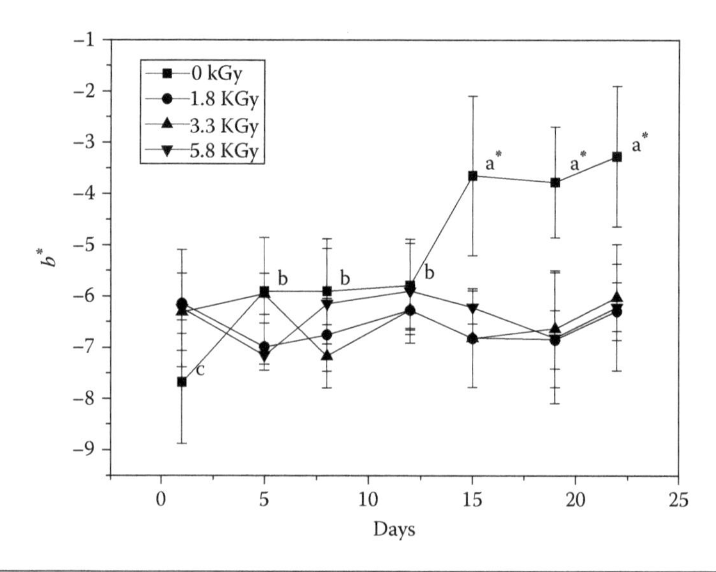

Figure 21.2 Evolution of b^* in squid rings during storage at 4°C–5°C. Standard error represented by bars. Different letters (a, b, c) indicate significant differences due to storage time. Values with * are significantly different between doses.

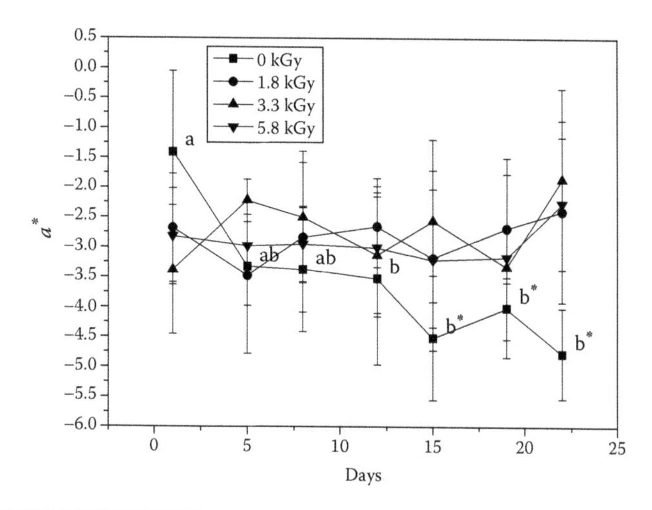

Figure 21.3 Evolution of *a** in squid rings during storage at 4°C–5°C. Standard error represented by bars. Different letters (a, b, c) indicate significant differences due to storage time. Values with * are significantly different between doses.

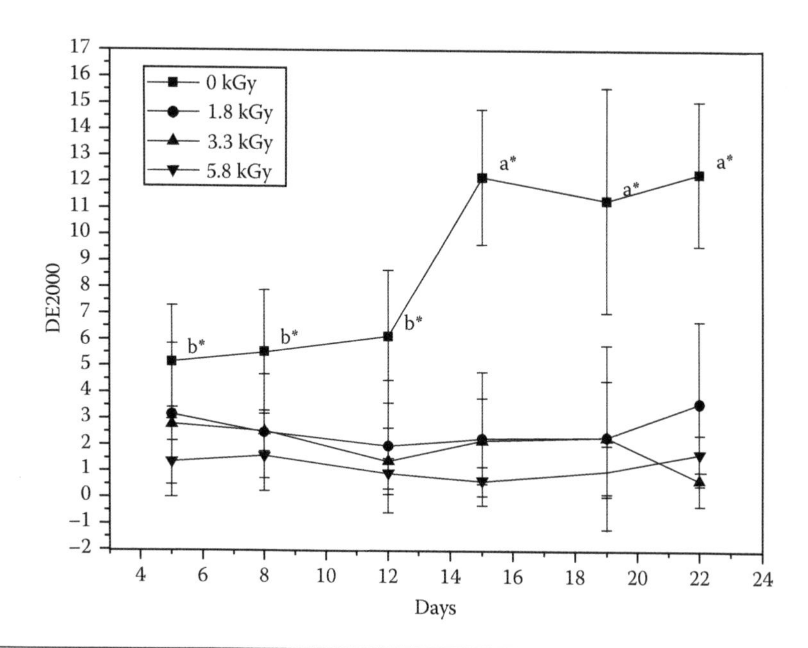

Figure 21.4 Squid ring color difference (DE2000) with respect to day 1 during storage at 4°C–5°C. Standard error represented by bars. Different letters (a, b, c) indicate significant differences due to storage time. Values with * are significantly different between doses.

Values of b^* increased toward yellow in control, being significantly higher than radiated samples on day 12 (Figure 21.2). Values of b^* of radiated samples remained unchanged, ranging between −7 and −6 during 22 days ($p < 0.05$). There were no significant differences between samples radiated with 1.8, 3.3, and 5.8 kGy, indicating that b^* was independent form the dose applied during 22 days. Values of a^* did not significantly change in radiated samples during 22 days of refrigerated storage; meanwhile a slight decrease was noticed in control ($p < 0.05$) (Figure 21.3).

DE2000 of control significantly increased during storage, reaching a value of 12.33 on day 22. After day 8, DE2000 of control was significantly higher than irradiated samples. DE2000 was not significantly different for all radiated samples, ranging between 1.96 and 3.58 for 1.8, 3.3, and 5.8 kGy during the whole storage period ($p < 0.05$) (Figure 21.4).

Gamma irradiation permitted to keep initial color characteristics of squid rings during 22 days of storage at 4°C–5°C, while color parameters of nonirradiated sample significantly changed during storage, showing increases in b^* and L^* and a slight decrease in a^* values.

Results of control sample in this work were in agreement with what was found by Lapa- Guimarâes et al. (2001), who studied color changes in *Loligo plei* and observed increases in b^* and L^* and small changes in a^* that varied according to the storage conditions (ice contact or noncontact). Values of a^* showed small alterations. A decreasing tendency was informed if the squid was stored with ice contact and a slight increase if it was not in contact with ice (Lapa-Guimarâes et al. 2001).

During spoilage, the color of the squid mantle is altered, modifying the appearance and quality of the product. Thanonkaew et al. (2006) worked with *Loligo peali* and noticed an increase of yellowness represented by an increase of b^*. They associated it to deterioration and lipid oxidation product reactions and suggest that formation of off-color in squid mantle could be due to nonenzymatic browning reactions. Sugiyama et al. (1989) found that after capture, the transparency is gradually lost, and squid meat becomes tinged white. This white tinge could be associated to an L^* increase, as observed in control in this work. Sungrisi-in et al. (2011) worked with *Loligo formosana* and observed pink stains formed during spoilage, represented by an increase in a^*. They also noticed an increase of b^* related to yellow pigment formation.

In this work, no pink-spot formations were detected, neither in control nor in radiated samples. Values of a^* tended to decrease in control during storage, a fact that could be explained by two main reasons: (a) the previous skinning of squid, which has been found to be effective in preventing pink discoloration (pink discoloration was associated with skin chromatophore pigments released when squid was not skinned [Sungrisi-in et al. 2011]), and (b) the improvement and stabilization of color due to the previous treatment with polyphosphates (Gonçalves and Ribeiro 2008, 2009, Knipe 2004, Lampila 1993).

According to Tükenmez et al. (1997), radiation-induced free radicals are known to favor lipid oxidation, and so it could have been expected to find higher b^* values in irradiated samples. In this work, radiated sample b^* values were practically constant while b^* increased in control. This can be explained by the lack of oxygen due to vacuum packaging; as reported by Brewer (2009), the absence of oxygen affects aldehyde generation, reducing lipid oxidation. Furthermore, polyphosphate treatment was effective in stabilizing squid color by quelling metals that act as enzyme cofactors, inhibiting lipid oxidation and thus changes of b^*. The yellowness increase in control sample, also vacuum-packed, could be explained by the loss of freshness related to bacterial spoilage reactions. The effectiveness of radiation in keeping color parameters unchanged during storage can be explained by the fact that gamma radiation reduced bacterial spoilage by reducing initial mesophilic and psychrotrophic counts in a dose-dependent way (data not shown).

Gamma irradiation was effective in preventing color changes associated with spoilage and did not induce modifications of initial color parameters of *Illex argentinus* rings.

21.4 Conclusions

Gamma irradiation avoided color changes of vacuum-packed squid rings during 22 days of refrigerated storage. It was useful for reducing yellow pigment formation, pink-spot formation, and white tinges, diminishing color-associated deterioration reactions.

Gamma irradiation improved the quality of a minimally processed squid product by stabilizing color without inducing color changes, indicating that shelf life of this product could be improved by this technology.

Acknowledgments

This work was supported by UNMDP (Projects 15/G206/07 and 15/G264/09). The authors are grateful to this institution.

References

Brewer, M. S. 2009. Irradiation effects on meat flavour: A review. *Meat Science* 81: 1–14.

Brunetti, N. E., M. L. Ivanovic, and M. Sakai. 1999. *Calamares de importancia comercial en la Argentina. Biología, distribución, pesquerías, muestreo biológico*. Mar del Plata, Argentina: Instituto Nacional de Investigación y Desarrollo Pesquero (INIDEP).

Byun, M. W., K. H. Lee, D. H. Kim, J. H. Kim, H. S. Yook, and H. J. Ahn. 2000. Effects of gamma radiation on sensory qualities, microbiological and chemical properties of salted and fermented squid. *Journal of Food Protection* 63 (7): 934–939.

Chouliara, I., I. N. Savvaidis, N. Panagiotakis, and M. G. Kontominas. 2004. Preservation of salted, vacuum-packed, refrigerated sea bream (*Sparus aurata*) fillets by irradiation: Microbiological, chemical and sensory attributes. *Food Microbiology* 21: 351–359.

CIE (Commission Internationale de l'Éclairage). 2001. *Improvement to Industrial Colour-Difference Evaluation*. Vienna, Austria: CIE Central Bureau, Publication CIE 142-2001.

Diehl, J. F. 2002. Food irradiation—Past, present and future. *Radiation Physics and Chemistry* 63: 211–215.

Francis, F. J. 1995. Quality as influenced by color. *Food Quality and Preference* 6: 149–155.

Gonçalves, A. A. and J. L. Duarte Ribeiro. 2008. Optimization of the freezing process of red shrimp (*Pleoticus muelleri*) previously treated with phosphates. *International Journal of refrigeration* 31: 1134–1144.

Gonçalves, A. A. and J. L. Duarte Ribeiro. 2009. Effects of phosphate treatment on quality of red shrimp (*Pleoticus muelleri*) processed with cryomechanical freezing. *LWT—Food Science and Technology* 42: 1435–1438.

Kilcast, D. 1995. Food irradiation: Current problems and future potential. *International Biodeterioration and Biodegradation* 36: 279–296.

Knipe, L. 2004. Use of phosphates in meat products. In *Meat Industry Research Conference*. Nashville, TN.

Kodo, J.-L.1990. *L'ionisation des produits de la pêche*. Collection Valorisation des produits de la mer. Issy-les-Moulineaux, France: Ifremer. Available at http://archimer.ifremer. fr/doc/1990/rapport-649.pdf

Lakshmanan, R., V. Venugopal, K. Venketashvaran, and D. R. Bongirwar. 1999. Bulk preservation of small pelagic fish by gamma irradiation: Studies on a model storage system using Anchovies. *Food Research International* 32: 707–713.

Lampila, L. E. 1993. Polyphosphates: Rationale for use and functionality in seafood and seafood products. In *Proceedings of the 18th Annual Tropical and Subtropical Fisheries Technological Conference of the Americas*. Williamsburg, VA, pp. 13–20.

Lapa-Guimarâes, J., M. A. Acevedo da Silva, P. Eduardo de Felicio, and E. Contreras Guzmán. 2001. Sensory, colour and psychrotrophic analyses of squids (*Loligo plei*) during storage in ice. *Lebensmittel Wissenschaft und Technologie* 35: 21–29.

Lescano, G., E. Kairiyama, P. Narvaiz, and N. Kaupert. 1990. Studies on quality of radurized (refrigerated) and non-radurized (frozen) Hake (*Merluccius merluccius hubbsi*). *Lebensmittel Wissenschaft und Technologie* 23: 317–321.

MINAGRI. 2007. *Pesquerías de calamar y langostino. Situación actual*. Ministerio de Agricultura, Ganadería y Pesca, Presidencia de la Nación Argentina. http://www.minagri.gob.ar/SAGPyA/pesca/pesca_maritima/04=informes/05-economia_pesquera/index.php. Accessed on January 2010.

Potter, N. N. and J. H. Hotchkiss. 1995. *Food Science*, 5th edn. London, U.K.: Chapman & Hall, International Thomson Publishing.

R Development Core Team. 2008. R: A language and environment for statistical computing. Vienna, Austria: R Foundation for Statistical Computing. http://www.R-project.org. Accessed on March 2009.

Sommers, Ch. H., H. Declincée, J. S. Smith, and E. Marchioni. 2006. Toxicological safety of irradiated foods. In *Food Irradiation Research and Technology*, eds. Ch. H. Sommers and Xuetong Fan. Ames, IA: IFT Press, Blackwell Publishing, Chapter 4.

Sugiyama, M., S. Kòusu, M. Hanabe, and Y. Okuda. 1989. *Utilization of Squid*. Rotterdam, the Netherlands: A. A. Balkema.

Sungsri-in, R., S. Benjakul, and K. Kijroongrojana. 2011. Pink discoloration and quality changes of squid (*Loligo formosana*) during iced storage. *LWT—Food Science and Technology* 44 (1): 206–213.

Thanonkaew, A., Z. Benjakul, W. Visessanguan, and E. A. Decker. 2006. Development of yellow pigmentation in squid (*Loligo peali*) as a result of lipid oxidation. *Journal of Agricultural and Food Chemistry* 54: 956–962.

Tükenmez, I., M. S. Ersen, A. T. Bakioglu, A. Biçer, and V. Pamuk. 1997. Dose dependent oxidation kinetics of lipids in fish during irradiation processing. *Radiation Physics and Chemistry* 50 (4): 407–414.

WHO (World Health Organization). 1994. *Safety and Nutritional Adequacy of Irradiated Food*. Geneva, Switzerland: WHO, pp. 81–107.

WHO (World Health Organization). 1999. *High-Dose Irradiation: Wholesomeness of Food Irradiated with Doses above 10 kGy*. Geneva, Switzerland: WHO, Technical Report Series 890.

CHAPTER 22

Evaluation of Blueberry Color during Storage by Image Analysis

SILVIA MATIACEVICH, PATRICIA SILVA, FERNANDO OSORIO, and JAVIER ENRIONE

Contents

22.1 Introduction

Consumption of blueberry across the world has risen mainly due to its well-known health benefits, such as low calorie and the presence of anticancer and antioxidant properties that prevent various diseases, becoming an important component of a healthy diet (Sinelli et al. 2008). Blueberries are blue little fruits from the genus *Vaccinium* with a short shelf life. It has been stated that under refrigeration temperatures (~0°C), blueberries' shelf life is about 14–20 days (Yommi and Godoy 2002, Nunes et al. 2004). Penetration of global markets and retention of the existing markets depend on the ability to deliver consistently high-quality products. The color of this fruit ranges from light blue to deep black depending on the cultivar and presence of an epicuticular wax on the skin, which gives it an attractive appearance (Nunes et al. 2004). The main quality indicators of blueberry are fruit appearance (color, size, and shape), firmness or texture, flavor (soluble solids, titratable acidity, and pH), and nutritive value (vitamins A and C and antioxidants) (Duarte et al. 2009). However, changes in color during storage can have a profound effect on consumer acceptability, the darkening of the color being the limiting factors for berries stored at 10°C or 15°C (Nunes et al. 2004). Computer vision (CV) is a nondestructive technology used for acquiring and analyzing digital images to obtain information of a product or to control processes in manufacturing (Brosnan and Sun 2004). It has been regarded as a valuable tool which helps to improve the automatic assessment of food quality (Pedreschi et al. 2006, Mery et al. 2010). CV has been used in the food industry for quality and color evaluation, detection of defects, grading and sorting of fruits and vegetables, bakery products, and potato chips, among other applications (Gunasekaram and Ding 1994, Leemans et al. 1998, Pedreschi et al. 2006, Mery et al. 2010). Since color measured by CV can easily be compared to that obtained from instruments, the instrumental color spaces offer a possible way of evaluating the performance of CV systems in measuring color of objects. Commercial colorimeters measure in CIELAB space standard only over a very few square centimeters, and thus their measurements are not very representative of heterogeneous materials such as most food products (Segnini et al. 1999, Papadakis et al. 2000).

Therefore, the objective of this work, as part of a comprehensive study of blueberry conservation, was to study color changes of different blueberry cultivars hand-harvested in Chile and stored under various conditions through image analysis obtained using CV with the aim to obtain an objective and quantitative measurement of blueberry color.

22.2 Materials and Methods

22.2.1 Blueberry Cultivars and Storage Conditions

Blueberry cultivars (*Duke, Brigitte, Elliot, and Jewel*) of Northern highbush variety, cultivated at Curacaví, near Santiago (Chile), were donated by the Berry Committee of the National Association of Exporters (ASOEX). Blueberries were manually picked at full maturity (100% blue) and transported to the laboratory on the same day. High-quality blueberries were presorted by hand, discarding the excessively small, soft, visually damaged fruits and those without presence of pedicel and floral remains for all experiments ($n = 50$). The fruits were stored for 0, 7, 14, and 21 days at temperatures of 4°C and 15°C and relative humidities (RHs) of 75% and 90%. The cv. *Jewel*, with and without epicuticular wax, was also stored for the same period of time and temperatures as mentioned earlier but only at 75% RH.

22.2.2 Image Analysis Using Computer Vision

Digital images from each blueberry (both front and back) were captured at each storage time through a CV system setup, which consisted of a black box with four natural daylight (D65) tubes of 18 W (Philips) and a camera (Canon 4 MP Powershot G3) placed in vertical position at 22.5 cm from samples (the camera lens angle and light was 45°, according to Pedreschi et al. [2006]). All images were acquired at the same conditions; the camera was remotely controlled by ZoomBrowser software (v6.0 Canon). Surface color data were measured in the CIELAB space standard, and the images were analyzed using the Balu Toolbox in MATLAB® software (v7) (Mery and Soto 2008), which can extract intensity characteristics of regions. Parameters of the camera and Balu software were calibrated using 30 color charts with a Minolta colorimeter. This software is composed by routines in MATLAB, specially developed to segment the fruit images and to convert RGB images into L^*, a^*, and b^* units (León et al. 2006, Mery and Soto 2008). Therefore, L^*, a^*, and b^* values obtained from image analysis were equal to those values obtained from the colorimeter.

Fungal growth (Equation 22.1) was also obtained through image analysis, visually considering the presence of fungal filaments:

$$\%\ \text{fungal growth} = \frac{\text{Number of images with presence of fungal filaments}}{\text{Total number total of digital images}} \times 100 \quad (22.1)$$

This technique was validated by a linear correlation of fungal growth measured by absorbance at 720 nm (data not shown). Pearson correlation coefficient was 0.99, indicating a good positive correlation between the values reported by absorbance and image analysis using Equation 22.1.

22.2.3 Sensory Evaluation

Sensory evaluation of color and acceptability were performed using a nine-level hedonic scale (Peryam and Pilgrim 1957). A consumer panel consisting of 15 individuals was recruited from the university community. Four coded blueberries of each storage time condition were presented in random order. Age distribution of panelists was in the range of 24–57 years with a mean age of 35.4 years. Gender distribution was 60% female and 40% male. Other data indicated that the panelists were active consumers of fresh blueberries (100%).

22.2.4 Statistical Analysis

Statistical analysis was performed by one-way ANOVA, and Tukey post-test was applied if significant differences appeared. Significant differences are reported at $p \leq 0.05$. Pearson correlation coefficient between two independent parameters was calculated, where $p = 1$ indicates a positive correlation between them.

22.3 Results and Discussion

The use of image analysis allowed differentiating blueberry color from different cultivars and with storage time. Fresh blueberry colors were significantly different ($p < 0.05$) among cultivars (Figure 22.1), where cv. *Duke* showed lower lightness (L^*). The proportion of fruit surface covered by epicuticular wax was 28 ± 8%, and no significant differences ($p > 0.05$) were observed among cultivars. Therefore, L^* value observed for cv. *Duke* could not be associated to a lower presence of epicuticular wax on the fruit surface. Moreover, high dispersions in color parameters (a^* vs. b^*) were obtained within the same cultivar (Figure 22.2), indicating that the characteristic blue color is not a unique value and it slightly differs for each fruit. The acceptable color range for consumers, in the case of intact blueberries, was obtained in the range from −10 to 0 for a^* and from −3 to 5 for b^* (Figure 22.2). However, a change of this range was observed for blueberries without epicuticular wax, obtaining a range from −6 to 0 for a^* and from 0 to 5 for b^*, indicating that the characteristic blueberry color changed by the removal of the epicuticular wax from blue and dull to dark blue and brighter color (Figure 22.3a and b).

During storage, fruit color changed from blue to red (Figures 22.2 and 22.3c). The a^* parameter increased from −6/0 to 0/20 and b^* from 0/5 to −1/10 after 14 days at both storage temperatures. This change was more profound in blueberries without the protective epicuticular wax ($p < 0.01$) (as shown in Figure 22.2) where 100% of fruits changed color measured as a^* and b^* parameters when compared to the 34% of intact fruits. Storage temperature did not significantly affect ($p > 0.05$) the color parameters, but color dispersion increased as storage time increased, independently of the analyzed cultivars.

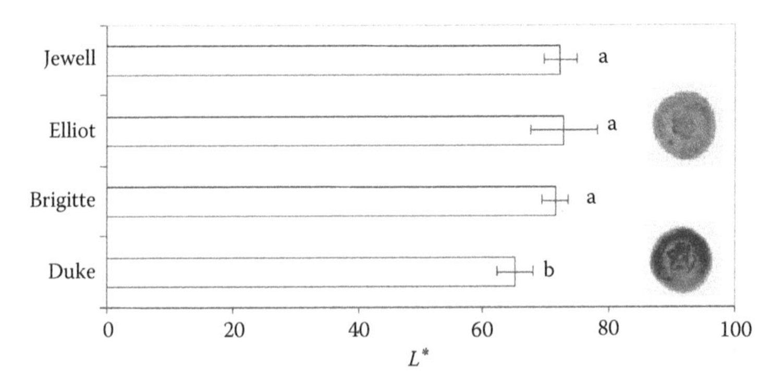

Figure 22.1 Total lightness (L^*) for fresh blueberry cultivars. Different characters indicate significant differences ($p < 0.05$ obtained by Tukey test).

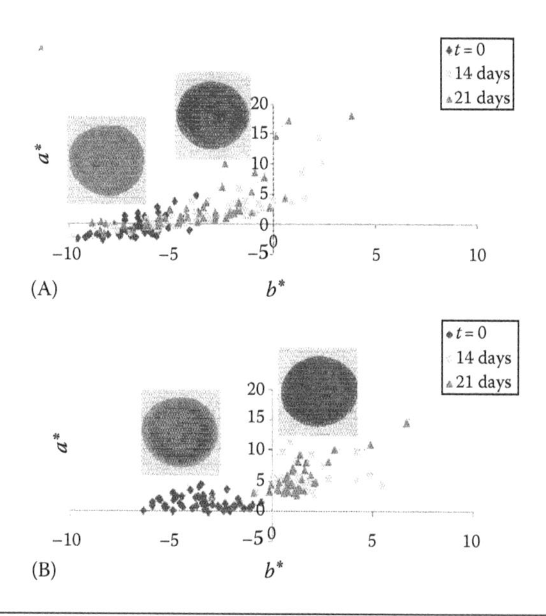

Figure 22.2 Color parameter changes by image analysis (a^* vs. b^*) during 21 days of storage at 15°C and 75% HR. (A) Intact blueberries (with epicuticular wax). (B) Blueberries without epicuticular wax. Similar behavior was obtained for all cultivars.

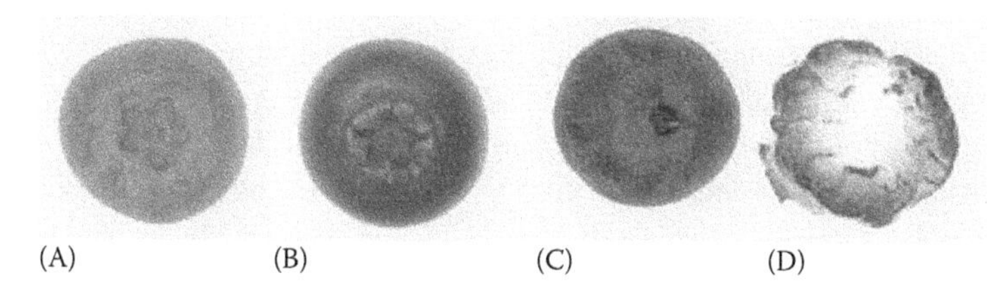

Figure 22.3 Image analysis of blueberry cultivars during storage time. (A) Intact blueberry (with epicuticular wax) at initial time. (B) Blueberry without epicuticular wax at initial storage time. (C) Intact blueberry after storage for 21 days at 15°C and 90% HR. (D) Blueberry without epicuticular wax after storage for 21 days at 15°C and 90% HR.

Total lightness (L^*) increased during storage time (Figure 22.4). This increase was related to fungal growth in all cultivars (Figures 22.3d and 22.4), which was attributed to *Botrytis cinerea* identified taxonomically (data not shown). Pearson correlation coefficient between lightness and fungal growth was lower than 0.9 for all cultivars, indicating that total lightness increased principally due the characteristic white-gray color of *Botrytis* filaments. This fungal growth increased as storage time increased being also affected by temperature, showing a lower growth kinetic at 4°C (2%) than at 15°C (up to 14%) at both RHs after 21 days. Interestingly, blueberries without the epicuticular wax showed a significant increase in fungal growth (up to 53%) at 90% RH and 15°C when compared to samples stored at 4°C (up to 4%) at both RHs after 21 days of storage.

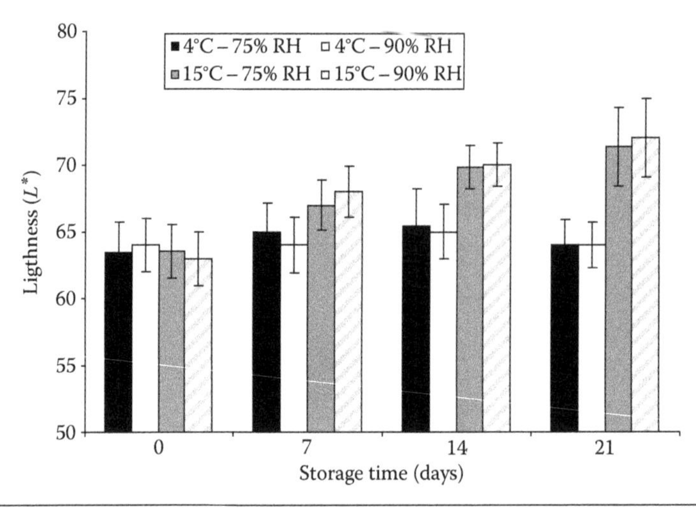

Figure 22.4 Total lightness (L^*) of cv. *Brigitte* at different temperatures (°C) and RHs of storage conditions. Similar behavior was observed for other cultivars.

Therefore, color changes (Figures 22.2 and 22.4) and fungal growth were observed during storage for all cultivars, significantly influenced by temperature, although no significant differences ($p > 0.05$) were obtained between both RHs. These important quality changes occurred more remarkably on blueberries without epicuticular wax than in intact blueberries.

Sensorial evaluation showed that Chilean consumers preferred ($p < 0.05$) blue- than red-colored blueberries and accepted ($p > 0.05$) the color of this fruit without epicuticular wax at initial storage time (Figure 22.5). However,

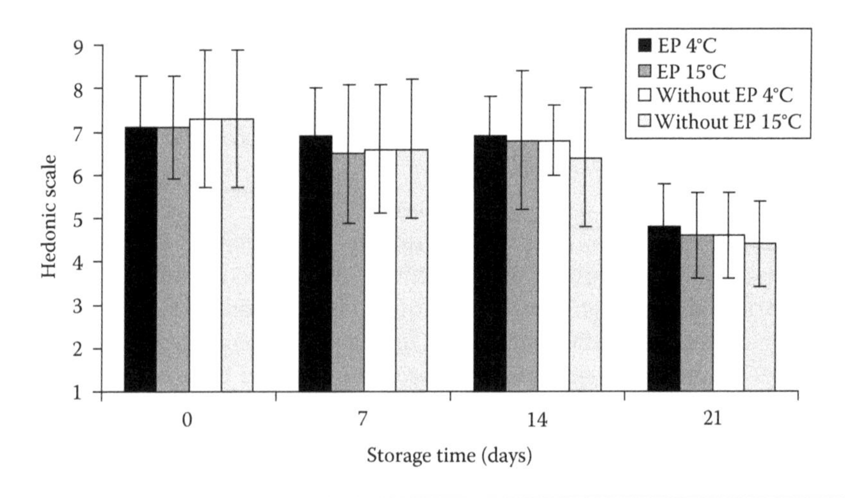

Figure 22.5 Ranking generated by a sensorial evaluation of blueberries in presence (EP) and absence (without EP) of epicuticular wax using an hedonic scale for cv. *Jewel* fruits stored at 4°C and 15°C. Similar behavior was obtained for other cultivars. * indicates significant difference ($p < 0.05$).

it is well known that U.S. market does not accept blueberries without epicuticular wax. Acceptability of the fruit on storage was mainly affected by both variations in color and fungal growth at 14 days of storage, being rejected after 21 days for both cases, intact or without epicuticular fruits. From these experimental results, there were no significant differences in color ($p > 0.05$) among cultivars during storage.

22.4 Conclusions

Differences among studied fresh cultivars were observed, particularly in blue color (L^* was lower in cv. *Duke* than other cultivars); however, during storage, its variations were similar among cultivars.

Important quality factors such as blueberry color (L^*, a^*, and b^* parameters) during storage were significantly influenced by temperature but not by storage RH. Color changes from blue to red were observed during storage. Fungal growth increased as storage time and temperature increased. Unprotected blueberries (without epicuticular wax) underwent high deteriorative processes (fungal growth, color change, and acceptability) with storage time. Consequently, it is necessary to maintain low storage temperature and the epicuticular wax on blueberries, which can be easily removed by handling (e.g., during harvesting). Image analysis obtained through CV system is a nondestructive method to measure objectively food color patterns. The results of this work showed that CV is a useful tool to assess optical properties of blueberries such as color and the appearance due to the fungal growth during storage based on its simplicity, allowing the analysis of heterogeneous materials. It is important to note that this evaluation and results could be used to improve the automatic evaluation of blueberries' quality for exporters.

Acknowledgments

The authors acknowledge the financial support from INNOVA-CORFO Project CT11 PUT-20, FONDECyT 11100209, and Bicentennial Program PBCT-PSD-62 from CONICyT.

References

Brosnan, T. and D. W. Sun. 2004. Improving quality inspection of food products by computer vision—A review. *Journal of Food Engineering* 61: 161–166.

Duarte, C., M. Guerra, P. Daniel, A. López Camelo, and A. Yommi. 2009. Quality changes of highbush blueberries fruit stored in CA with different CO_2 levels. *Journal of Food Science* 74: 154–159.

Gunasekaram, S. and K. Ding. 1994. Using computer vision for food quality evaluation. *Food Technology* 48: 151–154.

Leemans, V., H. Maein, and M. F. Destain. 1998. Defects segmentation on 'Golden Delicious' apples by using color machine vision. *Computers and Electronic in Agriculture* 20: 117–130.

León, K., D. Mery, F. Pedreschi, and J. León. 2006. Color measurements in *L**a*b* units from RGB digital images. *Food Research International* 39: 1084–1091.

Mery, D., J. J. Chanona-Perez, A. Soto, J. M. Aguilera, A. Cipriano, N. Veléz-Rivera, I. Arzate-Vázquez, and G. F. Gutiérrez-López. 2010. Quality classification of corn tortillas using computer vision. *Journal Food Engineering* 101: 357–364.

Mery, D. and A. Soto. 2008. Features: The more the better. In *New Aspects of Signal Processing, Computational, Geometry and Artificial Vision; Proceedings of the 8th International Conference on Signal Processing, Computational Geometry and Artificial Vision*, eds. N. Mastorakis, M. Demiralp, V. Mladenov, and Z. Bojkovic, Rhodes, Greece, pp. 46–50.

Nunes, M. C., J. P. Emond, and J. K. Brecht. 2004. Quality curves for Highbush blueberries as a function of the storage temperature. In *Proceedings of the 9th North American Blueberry Research and Extension Workers Conference; and In Small Fruits Review*. Binghamton, NY: Food Product Press, Haworth Press, pp. 423–438.

Papadakis, S. E., S. Abdul-Malek, R. E. Kamdem, and K. L. Yam. 2000. A versatile and inexpensive technique for measuring color of foods. *Food Technology* 54 (12): 48–51.

Pedreschi, F., J. León, D. Mery, and P. Moyano. 2006. Development of a computer vision system to measure the color of potato chips. *Food Research International* 39 (10): 1092–1098.

Peryam, D. and F. Pilgrim. 1957. Hedonic scale method of measuring food preferences. *Food Technology* 1 (9): 14.

Segnini, S., P. Dejmek, and R. Öste. 1999. A low cost video technique for color measurement of potato chips. *Lebensmittel Wissenschaft und Technologie* 32: 216–222.

Sinelli, N., A. Spinardi, V. Di Egidio, Ll. Mignani, and E. Casiraghi. 2008. Evaluation of quality and nutraceutical content of blueberries (*Vaccinium corymbosum* L.) by near and mid-infrared spectroscopy. *Postharvest Biology Technology* 50: 31–36.

Yommi, A. and C. Godoy. 2002. Arándanos: fisiología y tecnologías de postcosecha. http://www.inta.gov.ar/balcarce/info/documentos/agric/posco/fruyhort/arandano.htm. Accessed April 19, 2011.

CHAPTER 23

Effect of Washing-Disinfection Conditions on Total Anthocyanin Retention and Color of Fresh-Cut Strawberries

FRANCO VAN DE VELDE, MARÍA PIROVANI, DANIEL GÜEMES, and ANDREA PIAGENTINI

Contents

23.1 Introduction

Fresh-cut produce is defined as "any fruit or vegetable or combination that has been physically altered from its original form, but remains in a fresh state" by the International Fresh-Cut Produce Association (IFPA) (Ölmez and Kretzschmar 2009). Recently, there has been an increasing demand for fresh-cut fruits and vegetables, mainly due to their convenience as ready-to-eat products as well as for the health benefits associated with their consumption. Strawberries are a good source of vitamin C and anthocyanins.

Anthocyanins are the most important pigments of the vascular plants. These pigments are responsible of the shiny orange, pink, red, violet, and blue colors in the flowers and fruits of some plants. They are responsible for the appealing, bright red color of strawberries. Another significant property is their antioxidant activity, which plays a vital role in the prevention of neuronal and cardiovascular illnesses, cancer, and diabetes (Castañeda-Ovando et al. 2009). Disinfection of fresh-cut products by washing is an essential operation to eliminate foreign matter and cellular fluids produced by cutting (Pirovani et al. 2004). Some technological problems that may take place during washing operations of fresh-cut fruits and vegetables are the loss of pigments, vitamins by oxidation and/or leaching, and other compounds by dissolution in the wash water (Van de Velde et al. 2010). If operative conditions are not adequate, the washing-disinfection may spoil appearance, and reduce nutrients and bioactive compounds from these products. The color of a food product is an important freshlike attribute for the consumer to evaluate the quality of the product. Several disinfection agents possess strong oxidizing properties causing deleterious effects on the color of vegetables by inducing browning or bleaching of the vegetable tissue (Vanderkinderen et al. 2009). Compounds derived from chlorine are disinfectants widely used in the fresh-cut vegetables industry (Pirovani et al. 2010). Moreover, reaction of chlorine with organic matter may result in the formation of carcinogenic halogenated disinfection by-products, like trihalomethanes and haloacetic acids (Ölmez and Kretzschmar 2009). Recently, the industry and the academy are trying to replace the chlorine in the washing-disinfection operations of fresh-cut vegetables and incorporate alternative disinfectants to maintain quality, food security, and shelf life as if they had been washed with chlorine. Commercial peracetic acid is a quaternary equilibrium mixture of acetic acid, hydrogen peroxide, peracetic acid (PAA) and water. The PAA is a sanitizing agent which does not react with proteins to cause toxic or carcinogenic compounds, and its only decomposition products that have been reported are oxygen and acetic acid (Silveira et al. 2008, Vandekinderen et al. 2009). The U.S. Code of Federal Regulations (CFR) allows the use of PAA in the washing water of fruits and vegetables at 80 ppm concentration.

In this study, it was proposed to quantify and to model changes in the total anthocyanin content and color change as a consequence of washing-disinfection of fresh-cut strawberries using solutions of PAA at different contact times and temperatures.

23.2 Materials and Methods

23.2.1 Plant Material

Strawberry fruits (*Fragaria ananassa* Duch var. *Camarosa*) were received directly from a local producer from Arroyo Leyes, Santa Fe, Argentina. The fruits were stored overnight at 3°C before processing.

23.2.2 Minimal Processing

The strawberries were sorted, eliminating those fruits with signs of damage, as well as parts of the plant. Calyxes and peduncles were cut off using a sharp knife, then washed with tap water, and drained on absorbent paper. Subsequently, processing was continued into a "clean zone" of the pilot plant of the Instituto de Tecnología de Alimentos, FIQ, UNL, at 18°C. The strawberries were cut longitudinally into quarts using a stainless-steel sharp knife. Then, they were washed according to experimental design with respect to PAA concentration, time, and temperature. In all the washing runs, the ratio solution volume to weight produce was of 3 L per kilogram of fruit. Subsequently, washed and nonwashed cut strawberries (control) were packaged in polyethylene bags and immediately stored at −80°C until analysis.

23.2.3 Anthocyanin Content

Five grams of crushed fruit was homogenized with 75 mL acetone/water (80:20) and sonicated in an ultrasonic bath for 5 min. The mixture was centrifuged at 12,000 g for 30 min at 4°C with an MSE Mistral 4L refrigerated centrifuge. The supernatant was separated and used for analysis. The total anthocyanin content was determined on the extracts with the pH differential method according to Jin-Heo and Yong-Lee (2005). Absorbance was measured at 510 and 700 nm in a Milton Roy spectrophotometer in buffer solutions at pH 1 and 4.5. Absorbance readings were converted to milligrams of pelargonidin-3-glucoside/100 g fresh fruit (mg P3 G/100 g fresh fruit) using a molar extinction coefficient of 22, 400 L/mol cm, an optical path of 1 cm, and an absorbance of

$$A = \left\lfloor \left(A_{510} - A_{700}\right)_{\text{pH}=1} - \left(A_{510} - A_{700}\right)_{\text{pH}=4.5} \right\rfloor \tag{23.1}$$

All samples were prepared in duplicate.

23.2.4 Color Measurement

Color was determined on samples crushed (about 50 g per sample) and placed in 1 cm deep white plastic cell, using a Minolta 508d spectrophotometer, under the following conditions: illuminant D65, observer angle of 10°, and SCE (specular component excluded), evaluating the CIE system parameters—L^*, a^*, b^*, C^*_{ab}, h_{ab}. The parameter L^* measures the degree of lightness ($L^* = 100$: white, $L^* = 0$: black), a^*, the degree of red or green component ($a^* > 0$: red, $a^* < 0$: green), and b^*, the degree of yellow or blue component ($b^* > 0$: yellow, $b^* < 0$: blue). The $L^*C^*_{ab}h_{ab}$ space uses the same diagram of the $L^*a^*b^*$ space, but in cylindrical coordinates. The brightness is the same in both systems, and for their part, C^*_{ab} (chroma) and h_{ab} (hue angle) are defined from a^* and b^*, where $C^*_{ab} = \left(a^{*2} + b^{*2}\right)^{0.5}$ and $h_{ab} = \text{arctg}\ (b^*/a^*)$ (0°, red; 90°, yellow; 180°, green; 270°, blue). On each plastic cell, five measurements were performed.

23.2.5 Experimental Design

Response surface methodology (RSM) using a Box–Behnken design (three factors at three levels) was used to study the operation of washing-disinfection. It was assumed that there was a mathematical function for each of the responses expressed as a percentage of total anthocyanin retention (TAR, %) and color change (%), according to three independent factors of process:

$$Y = f(C, T, t) \quad (23.2)$$

where
- Y is the response
- C is the PAA concentration (ppm)
- T is the temperature of the washing solution (°C)
- t is the time of treatment(s)

The second-order polynomial equation as a model is proposed:

$$Y_k = \beta_{ko} + \sum_{i=1}^{3} \beta_{ki} X_i + \sum_{i=1}^{3} \beta_{kii} X_i^2 + \sum_{i=1}^{2} \sum_{j=i+1}^{3} \beta_{kij} X_i X_j \quad (23.3)$$

where
- β_{ko}, β_{ki}, and β_{kij} are the coefficients
- X_i's are the independent variables coded

The independent variables, the coded variables, and their levels are presented in Table 23.1. The TAR is expressed as $(TA/TA_o) \times 100$, where TA and TA_o are the total anthocyanin content in washed and nonwashed (control) fresh-cut strawberries after and before washing-disinfection operation, respectively. The color parameter changes (δL^*, δa^*, δb^*, δC^*_{ab},δh_{ab}) are expressed as a percentage of the difference of the washed fresh-cut strawberries' parameter value (L^*, a^*, b^*, C^*_{ab}, and h_{ab}) minus the unwashed cut strawberries' parameter value (control) (L^*_o, a^*_o, b^*_o, $C^*_{ab\,o}$, and $h_{ab\,o}$), divided by the control value, for example, $\delta L^* = [(L^* - L^*_o)/L^*_o] \times 100$.

Table 23.1 Experimental Design of Box–Benhken, Variables, and Coded and Uncoded Levels

	Coded Variable			
		Levels		
Independent Variable	**Symbols**	**−1**	**0**	**1**
PAA concentration (ppm)	X_1	0	50	100
Temperature (°C)	X_2	4	22	40
Time (s)	X_3	10	65	120

23.2.6 Statistical Analysis

Statgraphics Plus 7.1 was used to fit the second-order polynomial equations to the experimental data shown in Table 23.2, to obtain the coefficients of the equations, the regression analyses, and the analyses of variance (ANOVA). The significance of each term of the model was evaluated which referred to the pure error. Tests to verify that the residuals satisfied the assumptions of normality, independence, and randomness were also done.

23.3 Results and Discussion

The nonwashed cut strawberries (control) had a total anthocyanin content of 18.5 ± 0.75 mg P3 G/100 gff, and the CIELAB values and the derivate parameters were $L^*_o = 31.9 \pm 0.6$, $a^*_o = 25.4 \pm 0.7$, $b^*_o = 12.2 \pm 0.6$, $C^*_{abo} = 28.2 \pm 0.9$, and $h_{abo} = 25.7 \pm 0.5$. The responses of the experimental design and the ANOVA of the models are presented in Tables 23.2 and 23.3, respectively. In the case of total anthocyanins, the response surface model described the experimental data adequately ($R^2 = 0.896$), exhibiting no significant lack of fit ($p \geq 0.05$). As it is shown in Figure 23.1a, the TAR (%) decreases significantly as PAA concentration increases. Figure 23.1b shows through the contour lines the different combinations of condition variables (PAA concentration and time at 22°C) which result in different loss of anthocyanin content. Working with low concentration of PAA and short washing time, high retention of anthocyanins can be achieved (95%).

The models for the instrumental color change fitted the experimental data adequately (all the $R^2 > 0.67$), exhibiting no significant lack of fit ($p \geq 0.05$) (Table 23.3). Among the processing parameters, PAA concentration had the greatest effect on the color quality of strawberries. Specifically, the L^* change was only affected by PAA concentration. Positive and negative values in L^* change (δL^*) were obtained indicating that washed fresh-cut strawberries were lighter and darker than untreated cut strawberry samples (control), respectively, depending on washing conditions. Figure 23.2a indicates that lighter fruits were obtained when they were washed with high PAA concentration and long times (>100 s) at 22°C. Özkan et al. (2002) demonstrated that anthocyanin degradation occurs in the presence of hydrogen peroxide in strawberry juice. Taking into account that PAA solution contains hydrogen peroxide, these compounds could be responsible for the oxidation and subsequent degradation of anthocyanins in this case. The C^*_{ab} change was affected by the PAA concentration and time. Positive values in C^*_{ab} change indicate that washed fresh-cut strawberries were more vivid than untreated cut strawberry samples. Figure 23.2b indicates that more vivid fresh-cut fruits were obtained when they were washed with low PAA concentration and long times at 22°C.

The h_{ab} change was affected by the PAA concentration, time, and temperature. Positive and negative values in h_{ab} change indicate that washed fresh-cut

Table 23.2 Experimental Data for Percentage of TAR and Percentage of Color Change (δL^*, δa^*, δb^*, δC^*_{ab}, and δh_{ab}) in the Washing-Disinfection with PAA of Fresh-Cut Strawberries

Run N	Concentration PAAC	Temperature, T	Time, t	TAR (%)	δL^* (%)	δa^* (%)	δb^* (%)	δC^*_{ab} (%)	δh_{ab} (%)
1	50	22	65	81.1	−3.0	13.3	10.8	12.9	−1.9
2	50	22	65	91.4	−7.3	15.4	15.0	15.4	−3.2
3	50	22	65	90.3	−5.8	13.6	12.3	13.4	−1.0
4	0	22	10	97.6	−0.6	16.2	19.6	16.0	3.4
5	0	22	120	88.9	−1.9	15.7	18.5	16.2	2.1
6	100	22	10	77.3	2.2	12.5	18.6	13.7	4.7
7	100	22	120	62.1	6.7	7.6	12.8	7.9	4.9
8	50	4	10	72.3	4.9	2.3	3.9	1.8	2.2
9	50	4	120	72.4	−1.7	7.7	18.5	8.4	10.3
10	50	40	10	97.8	−7.6	7.9	11.8	8.1	3.7
11	50	40	120	66.8	0.3	12.7	33.3	16.9	15.4
12	0	4	65	93.0	−1.2	11.4	19.0	12.9	5.9
13	100	4	65	67.9	−0.3	2.9	9.9	4.2	5.9
14	0	40	65	83.2	7.1	3.6	21.4	7.2	14.5
15	100	40	65	62.0	4.5	3.4	18.6	6.5	12.4

Note: Experimental runs were performed in random order.

Table 23.3 ANOVA of Percentage of TAR and Percentage of Color Change of Fresh-Cut Strawberries

Source	d.f.	Sum of Squares					
		TAR (%)	δL^* (%)	δa^* (%)	δb^* (%)	δC^*_{ab} (%)	δh_{ab} (%)
X_1 (C)	1	1092.1$^+$	11.7	52.1$^+$	43.2	50.0$^+$	0.5
X_2 (T)	1	2.2	0.9	1.3	142.8$^+$	16.2	58.9^{++}
X_3 (t)	1	375.7	2.5	2.9	106.6$^+$	12.0^{++}	43.7^{++}
X_1^2	1	43.9	102.4$^+$	10.7	23.3	2.2	53.1^{++}
X_1X_2	1	3.9	3.2	17.5	939	16.0	1.1
X_1X_3	1	10.4	8.3	4.8	5.5	9.0	0.6
X_2^2	1	214.8	25.7	183.8^{++}	15.0	108.7	231.4^{++}
X_2X_3	1	242.6	52.7	0.1	11.9	1.2	3.2
X_3^2	1	26.0	10.5	1.4	17.3	0.4	15.0
Residual	5						
Lack of fit	3	165.4	91.2	70.3	251.3	96.7	68.4
Pure error	2	64.0	9.4	2.5	9.1	3.5	2.5
R^2		0.896	0.671	0.790	0.588	0.683	0.845

$^+p \leq 0.05$; $^{++}p \leq 0.01$

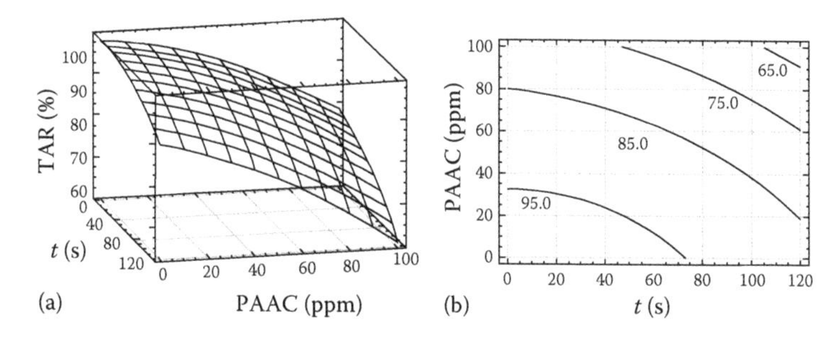

Figure 23.1 (a) Response surface and (b) contour plot of TAR in the washing-disinfection of fresh-cut strawberries at 22°C.

strawberries were more yellow and redder than untreated cut strawberry samples, respectively. A broad experimental area of different times and PAA concentrations demonstrate the possibility of doing washing-disinfection, preserving the initial quality (δh_{ab} equal to 0% of change) or getting redder fresh-cut strawberries (δh_{ab} less than 0%). The increment in red color in this experimental area could be justified because of the low pH of PAA solution which intensifies the red color of anthocyanins with minimum degradation.

These results are in agreement with Reyes et al. (2007) that described a negative change in h_{ab} (strawberries were redder after washing), a slight increment in chroma, that is, strawberries were more vivid ($\delta C^*_{ab} < 2.5\%$) and lighter ($\delta L^* < 5\%$) when they were washed with 100 ppm of PAA for 2 min.

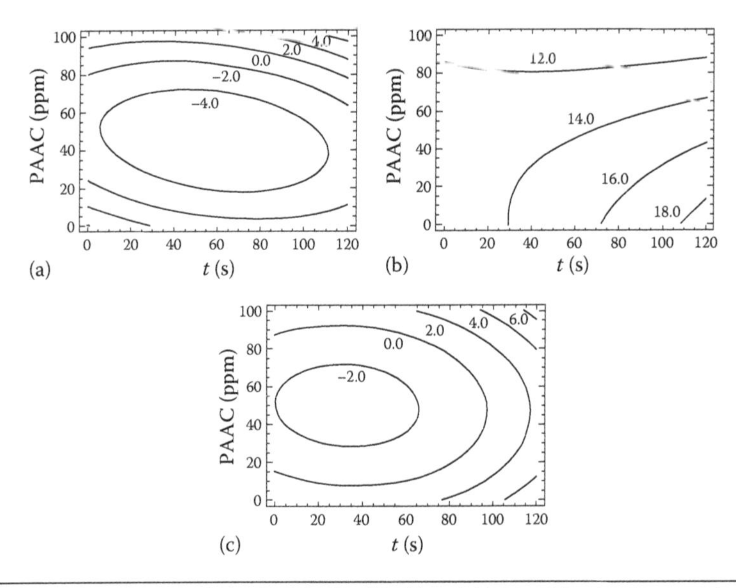

Figure 23.2 Contour plots of (a) δL^* (%), (b) δC^*_{ab} (%), and (c) δh_{ab} (%) color change on fresh-cut strawberries washed at 22°C.

The predictive models obtained here are shown to be adequate. They could be used as a way of improving the washing-disinfection operation by complementing them with studies about microbiological load reduction and preservation of nutritional and bioactive compounds.

Acknowledgments

This study was partly supported by CAI+D of Universidad Nacional del Litoral (Argentina) and ANPCyT (Argentina) through PICTO 35685.

References

Castañeda-Ovando, A., M. Pacheco-Hernández, M. Páez-Hernández, J. Rodríguez, and C. Galán-Vidal. 2009. Chemical studies of anthocyanins: A review. *Food Chemistry* 113: 859–871.

Jin-Heo, H. and C. Yong-Lee. 2005. Strawberry and its anthocyanins reduce oxidative stress-induced apoptosis in PC12 cells. *Journal of Agricultural and Food Chemistry* 53: 1984–1989.

Ölmez, H. and U. Kretzschmar. 2009. Review: Potential alternative disinfection methods for organic fresh-cut industry for minimizing water consumption and environmental impact. *LWT—Food Science and Technology* 42 (3): 686–693.

Özkan, M., A. Yemenicioglu, N. Asefi, and M. Cemeroglu. 2002. Degradation kinetics of anthocyanins from sour cherry, pomegranate, and strawberry juices by hydrogen peroxide. *Journal of Food Science* 67: 527–529.

Pirovani, M., A. Piagentini, D. Güemes, and S. Arkwright. 2004. Reduction of chlorine concentration and microbial load during washing-disinfection of shredded lettuce. *International Journal of Food Science and Technology* 39: 341–347.

Pirovani, M., F. Van de Velde, A. M. Piagentini, and D. R. Güemes. 2010. Desinfección de vegetales frescos cortados, técnicas convencionales y emergentes. In *Actas del V Simposio Internacional de Ciencia y Tecnología de Alimentos Dr. Mario Yannes*, CD-ROM. Villa Hermosa, México, pp. 169–176.

Reyes, M. S., A. M. Piagentini, D. R. Güemes, and M. E. Pirovani. 2007. Procesamiento y calidad de frutillas frescas cortadas. In *Avances tecnológicos en el procesado mínimo hortofrutícola. aspectos sensoriales y nutricionales*, eds. G. González Aguilar and F. Ayala Zavala. México, pp. 71–76.

Silveira, A. C., A. Conesa, E. Aguayo, and F. Artés. 2008. Alternative sanitizers to chlorine for use on fresh-cut "Galia" (*Cucumis melo* var. catalupensis) melon. *Journal of Food Science* 73: 405–411.

Van de Velde, F., A. Tavella, A. Piagentini, D. Güemes, and M. Pirovani. 2010. Retención de compuestos bioactivos en el lavado-desinfección de frutillas mínimamente procesadas con ácido peracético. *Revista Iberoamericana de Tecnología Postcosecha* 11 (2): 162–170.

Vandekinderen I., F. Devlieghere, B. De Meulenaer, P. Ragaert, and J. Van Camp. 2009. Optimization and evaluation of a decontamination step with peroxyacetic acid for fresh-cut produce. *Food Microbiology* 26: 882–888.

CHAPTER 24

Effects of Postharvest Treatments in Ruby Red Grapefruit Quality

ANA MARÍA SANCHO, SILVINA GUIDI, ANDREA BIOLATTO, ADRIANA PAZOS, and GABRIELA MARÍA GRIGIONI

Contents

24.1 Introduction

Consumers of the developed countries are increasingly demanding different aspects of citrus fruits quality. Some intrinsic quality factors, such as color, texture, juice percentage, and maturity index, can be measured by laboratory techniques. Others, like taste, palatability, and ease of peeling, are assessed by sensory analysis. In fruit, appearance affects the consumer decision of purchase. In particular, rind color influences consumer acceptance for fresh consumption (Martínez-Javega 1997, Hwang et al. 2004).

Argentina is one of the most important citrus producers in the southern hemisphere (Federcitrus 2009), and Ruby Red (*Citrus paradisi* Macf.) grapefruit is a variety cultivated in NOA (Argentine Northwest) region. International trade often requires fruit to be stored for long periods at low temperature. This condition could be detrimental to the quality of subtropical fruits such as citrus because they are known to be susceptible to chilling injury development during cold storage (Chalutz et al. 1985). This sensitivity to low temperatures has serious economic implications because cold storage also provides an important quarantine treatment required by many countries to export citrus to Mediterranean fruit fly (*Ceratitis capitata*)–free zone (USDA s.d.). One approach to avoid chilling injury under quarantine treatment is to apply postharvest heat treatments to induce cold tolerance and to reduce the development of chilling injury symptoms (Wang 1994). The postharvest handlings simulate shipment periods to overseas markets such as Japan and United States, marketing conditions, and storage at nonchilling temperature. However, the conditions employed in these postharvest treatments can induce several physiological and biochemical alterations, such as peel injury and changes in color, that greatly reduce fruit marketability. Therefore, the objective of the present research was to study the effect of postharvest treatments in relation to color index (CI), juice percentage (J%), and deformation index (DI).

24.2 Materials and Methods

24.2.1 Plant Material and Treatments Applied

The variety of grapefruit (*Citrus paradisi* Macf.) used was "Ruby Red," selected in the Argentinean northern province of Salta. This grapefruit, subjected to temperature conditioning, was harvested 7 days earlier than fruit for other treatments. Fruits were degreened for 80 h with 3.5 mg/kg ethylene and 1.5 mg/kg CO_2 at of 26°C and 90% relative humidity. Then, all fruit were washed, disinfected with 0.5/100 g sodium orthophenylphenate (SOPP), rinsed, dried, and coated with a polyethylene-based wax (18/100 g solid matter) containing 5000 mg/L thiabendazole (TBZ). After that, all fruits were packed in boxes, which were randomly distributed among different treatments before being transported to the laboratory, located about 1200 km south (36 h trip), in refrigerated transport at 7°C ± 1°C. Six different treatments (T1–T6), simulating different stages of postharvest handling, were applied: normal postharvest treatments (temperature conditioning and cold storage), simulated shipment periods to overseas markets such as Japan and United States, marketing conditions, and storage at nonchilling temperature (control treatments) as described in Table 24.1.

Three boxes with 40 fruits each were used for the analysis of each of the six treatments. Grapefruits were sampled immediately after transportation (S) and marketing conditions (SC). Color measurements (L^*, a^*, and b^* coordinates) were performed in triplicates, and average values were calculated for each fruit and converted into $CI = (1000 \times a^*)/(L^* \times b^*)$ using a Byk Gardner color view model 9000 with D65 illuminant and large vision area. J% was performed on three fruit batches. Each batch was obtained by squeezing six randomly chosen fruit. The DI was determined as the equatorial zone of the

Table 24.1 Scheme of Treatments Applied and Sampling Time

	Days					
	Stage S	**Stage SA**	**Stage SB**		**Stage SC**	
Treatment	**0**	**1–18**	**19–22**	**23–29**	**30–35**	**36–42**
T1	R	Q		Storage		Marketing
T2	R	Q	Storage	Marketing	—	—
T3	C+R	Q		Storage		Marketing
T4	C+R	Q	Storage	Marketing	—	—
T5	R		Storage			Marketing
T6	R	Storage		Marketing		—

R, transport under refrigeration to laboratory; C, conditioned, 7 days at 15°C–16°C in packinghouse; Q, quarantine simulation, 2°C 85% RH; storage, 13°C 85% RH; marketing, simulation of marketing conditions at 20°C.

fruit by means of an Instron test machine. The compression cell used was 50 kg, and the applied force was 2 kg with a loading velocity of 5 cm min^{-1}. DI was calculated as follows:

$$DI = 100 * \varepsilon / d,$$

where

ε is the fruit diameter before the compression cycle
d is the fruit diameter after the compression cycle

24.2.2 Statistical Analysis

The experiment procedure consisted of a randomized design, with three replicates per treatment, with each box being replicate. The data were analyzed using analysis of variance (ANOVA) considering stage and treatment effects and Pearson correlation, using SPSS® (v 12.0, Illinois, United States).

24.3 Results and Discussion

In this study, quality parameters were analyzed at the beginning of the assay (S) and at the end of the marketing simulation period (SC). CI, J%, and DI values did not show significant differences among postharvest treatments at the end of refrigerated transport (S) (Table 24.2).

Table 24.3 shows the results corresponding to the same variables analyzed at stage SC. ANOVA showed significant differences between treatments for CI ($p < 0.005$), J% ($p < 0.01$), and DI ($p < 0.005$).

Fruits submitted to T5 and T6 (just exposed to refrigerated transport to laboratory plus storage and marketing periods to 13°C and 20°C, respectively) showed the highest CI values. The other treatments showed lower CI values in the following order: T3 = T2 = T1 > T4. This result suggests that the exposure

Table 24.2 Means and Standard Deviations of Fruits Treatments in Stage S

S	Treatments	CI	Juice %	DI
	T1	2.8 ± 0.4	47.9 ± 1.7	0.600 ± 0.004
	T2	2.9 ± 0.6	48.9 ± 2.1	0.540 ± 0.007
	T3	2.8 ± 0.4	49.7 ± 2.8	0.570 ± 0.006

No significant differences ($p > 0.05$) were obtained.

Table 24.3 Means and Standard Deviations of CI, J%, and DI at Commercial Stage (SC)

Treatments	CI	J%	DI
T1	2.9 ± 0.5 b	48.9 ± 1.6 a	0.57 ± 0.03 b
T2	3.1 ± 0.7 b	47.9 ± 5.4 a	0.56 ± 0.09 b
T3	2.9 ± 0.9 b	43.2 ± 6.0 b	0.57 ± 0.08 b
T4	2.3 ± 0.7 c	47.5 ± 3.5 a	0.61 ± 0.06 ab
T5	3.5 ± 1.0 a	47.9 ± 3.3 a	0.65 ± 0.08 a
T6	3.7 ± 0.6 a	46.8 ± 1.8 a	0.62 ± 0.05 a

Means with different letters indicated significantly different ($p < 0.05$, LS means).

of fruits to higher temperatures has a positive effect on the color changes. Regarding this effect, it is well known that the color change processes involve both the destruction of chlorophylls already present in the cell and the development of carotenoid pigments responsible for the orange color (Wheaton and Stewart 1973). These enzymatic reactions have different optimal temperatures. For the chlorophyll degradation, the induction and inhibition temperatures are 28°C and 40°C, respectively, while for carotenoids synthesis are 18°C and 30°C, respectively (Cuquerella and Navarro 1997). Therefore, the condition applied in the T5 and T6 treatments would have induced the carotenoids synthesis, rendering a higher CI value.

Fruit submitted to T5 and T6 showed the highest values of DI, indicating that fruit subjected to upper temperatures could be more susceptible to softening during storage period. Instead, the other treatments that involved quarantine showed less DI.

Regarding J%, the lowest value was observed for T3 treatment.

The response of a particular fruit to the heat treatments results from a combination of factors, including physiological age of the commodity, time and temperature of exposure, treatment methods, and storage temperature (Lydakis and Aked 2003).

In general, the longer treatments, including conditioning for 3–7 days and heat treatments for 2–3 days, significantly increased fruit weight loss, enhanced changes in fruit peel color, and decreased juice acid content. On the other hand, Porat et al. (2000) reported that the enhancement of fruit pigmentation by the long heat treatments may be beneficial.

Moreover, in this work, a high correlation coefficient ($r = 0.94$) between the a^* value and the CI value was found. Similar results were obtained by Sepúlveda et al. (2010) who reported that a^* coordinate would be a good indicator for pigmentation in red grapefruit juice. Further studies would be necessary to confirm that this color coordinate is a good indicator for pigmentation in peel and grapefruit juice.

24.4 Conclusion

In this present research, it was found that Ruby Red grapefruit did not develop chilling injury under the postharvest practices applied at the end of simulated marketing period. Cold treatment is an effective quarantine method to kill fly eggs and larvae in grapefruit and is commercially applied by various citrus-importing countries. The application of heat treatments before cold disinfection may overcome the risk of potential fruit damage from chilling injury after quarantine treatment.

On the basis of these results, it can be concluded that cold quarantine treatment and temperature conditioning have an important commercial application for Ruby Red grapefruit (*Citrus paradise* Macf.) without adversely affecting its quality.

Acknowledgments

This work was supported by Instituto Nacional de Tecnología Agropecuaria (INTA), Argentina, and Asociación Fitosanitaria del Noroeste Argentino.

References

Chalutz, E., J. Waks, and M. Schiffmann-Nadel. 1985. Reducing susceptibility of grapefruit to chilling injury during cold treatment. *HortScience* 20: 226–228.

Cuquerella, J. and P. Navarro. 1997. Medidas objetivas de calidad en frutos cítricos con tratamiento de cuarentena por frío. In *Medición de la calidad en frutos tropicales y subtropicales con tratamientos con cuarentena*, eds. V. C. Saucedo and J. M. Martínez. Moncada, Valencia: Instituto Valenciano de Investigaciones Agrarias, pp. 1–20.

Federcitrus (Federación Argentina del Citrus). 2009. La actividad citrícola argentina. http://www.federcitrus.org/actividad-citricola.

Hwang, A. S., K. L. Huang, and S. H. Hsu. 2004. Effect of bagging with black paper on coloration and fruit quality of 'Ruby' grapefruit. *J. Agric. Res. China* 53: 229–238.

Lydakis, D. and J. Aked. 2003. Vapor heat treatment of Sultanina table grapes. II Effects on postharvest quality. *Postharvest Biol. Technol.* 27: 117–126.

Martínez-Javega, J. M. 1997. La frigoconservación en naranjas y mandarinas. *Phytoma* 90: 136–140.

Porat, R., D. Pavoncello, J. Peretz, S. Ben-Yehoshua, and S. Lurie. 2000. Effects of various heat treatments on the induction of cold tolerance and on the postharvest qualities of 'Star Ruby' grapefruit. *Postharvest Biol. Technol.* 18: 159–165.

Sepúlveda, E., C. Sáenz, A. Peña, P. Robert, B. Bartolomé, and C. Gómez-Cordovés. 2010. Color is one of the most important parameters when making a sensorial evaluation of food quality. Influence of the genotype on the anthocyanin composition, antioxidant capacity and color of Chilean pomegranate (*Punica granatum L.*) juices. *Chilean J. Agric. Res.* 70 (1): 50–57.

USDA (United States Department of Agriculture). s.d. Animal and Plant Health Inspection Service. Plant protection and quarantine. Treatment manual, chapter 5: Treatment schedules. T100—Schedules for fruit, nuts, and vegetables. T107—Cold treatment. http://www.aphis.usda.gov/import_export/plants/manuals. Accessed May 08, 2011.

Wang, C. Y. 1994. Chilling injury of tropical horticultural commodities. *HortScience* 29: 986–994.

Wheaton, T. A. and I. Stewart. 1973. Optimum temperature and ethylene concentrations for postharvest development of carotenoids pigments in citrus. *J. Am. Soc. Hort. Sci.* 98: 337–340.

CHAPTER 25

Color Variation in Nut Kernels during Storage under Different Drying Methods

LEONOR PILATTI, ANA MARÍA SANCHO, MARTÍN IRURUETA, and GABRIELA MARÍA GRIGIONI

Contents

25.1 Introduction

Color, flavor, and texture are considered to be the major attributes that contribute to the overall quality of food products. In this context, assessment of the color has become an important part of quality product and process management (Grigioni et al. 2010).

Color intensity and its progression with time depend on the composition of the product, processing technology, and preservation conditions (Buera et al. 1986).

In nuts, darkening is a phenomenon that depends on the variety and on the harvest and postharvest conditions. Furthermore, it is an important element in commercial practice since dark colors are often associated with poor quality and are related to the exposure of nuts to adverse conditions (Senter et al. 1984, López et al. 1995).

In dry nuts, the conditions of drying processes should ensure the preservation of its original properties and constitute a key element in the final quality of the product (Prunet and Herman 1995).

Drying temperatures are determined by the type and variety of nuts, since each one has a characteristic size, structure, and sensitivity to darkening (Muncharaz Pou 2001). It is generally recognized that exposure of the grains to temperatures above 38°C for prolonged periods causes deterioration in their color.

Also, exposure to temperatures higher than 43°C can accelerate autooxidation and lead to undesirable changes in color and flavor (Forbus et al. 1980, Senter et al. 1984, Erickson et al. 1994, Muncharaz Pou 2001).

Only adequate and controlled storage conditions can delay the natural tendency of the grain to get dark. Nevertheless, the proportion of dark walnut is one of the indicators that they are damaged or aged (Aleta 1999).

The aim of this study was to evaluate the effect of drying process and storage conditions of three walnut cultivars (*Juglans regia* L.) through color determination by instrumental method.

25.2 Materials and Methods

25.2.1 Sample Preparation and Storage

The varieties of nuts used in this study—Criolla, California, and Chandler—were harvested between February and March 2007 in a commercial plantation in the town of Mutquín, Pomán Department, Catamarca Province, Argentina.

Harvesting was performed after physiological maturity; harvesting was indicated when the hull begins to split from the nuts.

The method used for harvesting was shaking, which is a traditional manual method that consists of hitting the tree branches with a cane. Then, the collection was performed to prevent insect damage and darkening of the fruits. Subsequently, the nuts were subjected to two drying treatments. First, the natural treatment (T1), in which nuts were placed on trays in the sun for 7 days and exposed to temperatures in the range of 35°C–37°C during the day and 15°C–20°C at night, with 30% relative humidity (RH); and second, the treatment in which the nuts are dried in an oven (T2). In this case, the nuts were placed in an oven for 48 h with a steady flow of hot air at 38°C ± 2°C until the moisture content of nuts was in 4%–6% range. The samples, three replicates for each variety and treatment, were stored in perforated polyethylene bags for 225 days at an average temperature of 21.5°C ± 0.5°C (maximum of 23.1°C and minimum of 19.0°C).

25.2.2 Color Assessment

Color measurements were carried out using a reflectance spectrophotometer (Byk Gardner Color View model 9000, United States) according to CIELAB scale, with a 5 cm port area and D65 illuminant.

CIELAB color space is a three-dimensional spherical system defined by three colorimetric coordinates. The coordinate L^* is called the lightness. The coordinate a^* defines the deviation from the achromatic point corresponding to lightness, to red when it is positive and toward the green if negative. Similarly, the coordinate b^* defines the turning to yellow if positive and to blue if negative (Pérez Alvarez 2006).

Nut color was measured every 45 days along a period of 225 days according to the methodology described by Forbus et al. (1980). At each sampling time, an aliquot of 20 ± 1 g of kernel with teguments intact, for each treatment and variety, was selected and measured in an optical cylindrical cell (2.4 cm of internal diameter, CC-6136, Byk Gardner, United States). Three replicate measurements were made for each sample.

Index of browning (BR) was calculated as described by Buera et al. (1986) and Oro et al. (2008):

$$BR = 100 \times \frac{x - 0.31}{0.172} \tag{25.1}$$

where x was determined from the conversion of the parameters L^*, a^*, and b^*. The x value was calculated using the Color Converter online software (http://www.yellowpipe.com/yis/tools/hex-to-rgb/color-converter.php). BR can be explained by nonenzymatic darkening produced in the nuts due to time and storage temperature; higher BR corresponds to a greater extent of darkening.

25.2.3 Statistical Analysis

Data color was examined by applying a factorial design with two factors: treatment and time, with two (T1 and T2) and six (T0, T45, T90, T135, T180, and T225) levels, respectively. The homogeneity of variances was checked by means of Levene's test, establishing a significance level of 0.05. The statistical analysis was done using SPSS® software (version 12, Illinois, United States).

The model was as follows:

$$y_{ijk} = \mu + \alpha_i + \tau_j + (\alpha\tau)_{ij} + e_{ijk}$$

where

y_{ijk} is the dependent variable
α_i is the effect of treatment ($i = 1, 2$)
τ_j is the effect of time ($j = 1$–6)
$(\alpha\tau)_{ij}$ is the interaction between ith effect of treatment and jth effect of time
e_{ijk} is the residual error

25.3 Results and Discussion

Color parameter results at the beginning of the storage are presented in Table 25.1. For the parameter L^*, statistically significant differences were observed among the three varieties. Nuts of the Chandler variety (54.0) were the lightest, followed by California (49.0) and Criolla (42.0). Coordinates a^* and b^* showed significant differences in Criolla and in other varieties.

Regarding drying treatments, significant differences were observed in the three varieties for the parameter a^*. Nuts dried in the oven were redder than those dried at natural conditions.

During storage, a significant increase was observed for parameter a^* in Criolla variety. On the other hand, no changes were detected for the parameters L^* and b^*. In relation to drying methods, the L^* and b^* values for the nuts subjected to oven-drying treatment were lower than the values for the nuts dried under traditional conditions (Table 25.2).

For the Chandler variety, statistically significant differences were observed during storage in the three parameters that define the color. There was a marked decrease in L^*, ranging from 53.74 at the beginning of storage to a value of 47.56 at 225 days. Parameter a^* increased along time, while b^* decreased over 180 days of storage. In this variety, there is a significant difference between drying treatments; values obtained for T1 were greater than those for T2 for L^* and b^*, while the parameter a^* yields the highest values for T2 (Table 25.3).

For the California nuts, the parameters L^* and b^* decreased at 135 and 180 days, respectively. Nevertheless, an increase of parameter a^* was noted at 225 days. Parameters a^* and b^* reflected differences between drying treatments. The nuts subjected to T1 were more yellow and less red than those corresponding to T2 (Table 25.4).

Table 25.1 L^*, a^*, and b^* Values in Criolla, Chandler, and California Nuts Subjected to Natural (T1) and Oven (T2) Drying at the Beginning of Storage

	L^*		a^*		b^*	
Variety	**T1**	**T2**	**T1**	**T2**	**T1**	**T2**
Criolla	42.22 ± 3.50 C	41.25 ± 0.74 C	9.73 ± 0.97 A b	10.56 ± 0.55 A a	29.62 ± 1.77 A	29.25 ± 1.20 A
California	49.28 ± 0.66 B	48.33 ± 2.74 B	7.06 ± 0.28 B b	7.69 ± 0.24 B a	27.34 ± 1.14 B	26.95 ± 0.74 B
Chandler	53.74 ± 1.41 A	50.72 ± 0.86 A	6.67 ± 0.30 B b	7.20 ± 0.64 B a	29.11 ± 0.38 AB	27.74 ± 0.40 AB

T1, Natural drying; T2, oven drying.

Capital letters indicate significant differences ($p < 0.05$) between varieties by Tukey test.

Lowercase letters indicate significant differences ($p < 0.05$) between treatments by Student test.

Table 25.2 L^*, a^*, and b^* Parameters for Criolla Nuts Subjected to Natural (T1) and Oven (T2) Drying along 8 Months of Storage

	L^*		a^*		b^*	
Time (Days)	T1	T2	T1	T2	T1	T2
0	42.22 ± 3.50 a	41.25 ± 0.74 b	9.73 ± 0.97 AB	10.56 ± 0.56 AB	29.62 ± 1.77 a	29.25 ± 1.20 b
45	42.23 ± 1.13 a	37.78 ± 0.88 b	9.98 ± 0.11 AB	10.39 ± 0.61 AB	28.26 ± 0.59 a	26.33 ± 0.43 b
90	43.76 ± 1.60 a	38.70 ± 3.38 b	10.73 ± 0.39 AB	10.28 ± 0.95 AB	30.35 ± 1.33 a	26.19 ± 2.52 b
135	42.93 ± 1.99 a	39.75 ± 2.63 b	10.00 ± 0.14 B	10.17 ± 0.75 B	28.06 ± 1.54 a	24.64 ± 2.99 b
180	41.10 ± 2.95 a	38.10 ± 2.62 b	10.64 ± 0.22 AB	10.19 ± 0.47 AB	29.09 ± 1.16 a	25.94 ± 1.39 b
225	40.11 ± 0.83 a	39.19 ± 2.55 b	11.38 ± 0.14 A	10.86 ± 0.49 A	27.49 ± 0.69 a	27.04 ± 1.76 b

T1, Natural-drying secado; T2, oven drying.
Capital letters indicate significant differences ($p < 0.05$) in the effect of time.
Lowercase letters indicate significant differences ($p < 0.05$) between treatments by Student test.

Table 25.3 L^*, a^*, and b^* Parameters for Chandler Nuts Subjected to Natural (T1) and Oven (T2) Drying along 8 Months of Storage

	L^*		a^*		b^*	
Time (Days)	T1	T2	T1	T2	T1	T2
0	53.74 ± 1.41 a A	50.72 ± 0.86 b A	6.67 ± 0.30 b A	7.20 ± 0.64 a A	29.11 ± 0.38 a A	27.73 ± 0.40 b A
45	51.48 ± 1.38 a AB	50.47 ± 1.86 b AB	7.49 ± 0.30 b B	8.09 ± 0.17 a B	28.51 ± 0.19 a A	28.60 ± 0.98 b A
90	51.40 ± 0.53 a AB	49.04 ± 0.60 b AB	7.62 ± 0.26 b B	7.90 ± 0.18 a B	28.55 ± 0.91 a A	27.50 ± 0.75 b A
135	52.07 ± 1.87 a AB	50.16 ± 0.65 b AB	7.88 ± 0.25 b B	8.44 ± 0.16 a B	28.08 ± 0.87 a A	27.76 ± 0.57 b A
180	48.64 ± 1.88 a B	50.51 ± 0.87 b B	8.46 ± 0.36 b B	8.20 ± 0.25 a B	25.89 ± 0.50 a B	26.17 ± 0.30 b B
225	47.56 ± 1.44 a C	45.71 ± 1.30 b C	10.04 ± 0.58 b C	10.32 ± 0.32 a C	27.16 ± 0.23 a B	26.48 ± 0.25 b B

T1, Natural-drying secado; T2, oven drying.
Capital letters indicate significant differences ($p < 0.05$) in the effect of time.
Lowercase letters indicate significant differences ($p < 0.05$) between treatments by Student test.

Table 25.4 *L**, *a**, and *b** Parameters for California Nuts Subjected to Natural (T1) and Oven (T2) Drying along 8 Months of Storage

Time (Days)	*L**		*a**		*b**	
	T1	T2	T1	T2	T1	T2
0	49.27 ± 0.65 AB	48.32 ± 2.74 ABC	7.06 ± 0.28 b C	7.68 ± 0.24 a C	27.33 ± 1.14 a BC	26.94 ± 0.74 b BC
45	47.70 ± 1.52 ABC	48.14 ± 0.65 ABC	8.14 ± 0.85 b B	8.30 ± 0.47 a B	28.79 ± 0.72 a A	29.41 ± 1.17 b A
90	48.81 ± 0.96 AB	46.21 ± 1.95 BC	7.67 ± 0.39 b B	8.68 ± 0.68 a B	28.68 ± 0.42 a AB	28.08 ± 1.35 b AB
135	50.85 ± 1.31 A	44.40 ± 1.02 C	8.19 ± 0.17 b B	9.37 ± 0.18 a B	30.09 ± 0.29 a A	27.74 ± 0.49 b A
180	47.28 ± 1.93 ABC	44.41 ± 0.83 C	8.83 ± 0.17 b AB	10.06 ± 0.41 a AB	28.84 ± 1.31 a AB	27.44 ± 0.44 b AB
225	45.13 ± 0.44 BC	44.51 ± 1.26 C	9.69 ± 0.55 b A	10.05 ± 0.35 a A	26.46 ± 1.26 a C	26.42 ± 1.18 b C

T1, Natural-drying secado; T2, oven drying.
Capital letters indicate significant differences ($p < 0.05$) in the effect of time.
Lowercase letters indicate significant differences ($p < 0.05$) between treatments by Student test.

Color results obtained in this study are consistent with those reported by other authors for nut storage under diverse conditions.

Forbus et al. (1980) analyzed *L*, *a*, and *b* parameters, with a HunterLab D25D, of Stuart pecans over 12 weeks at ambient temperature. Samples were placed into perforated polyethylene bags and stored in darkness under the accelerated conditions of 21°C and 65% RH.

The authors reported a significant variation of the parameters during storage, showing a linear decrease of the coordinates *L* and *b*. The slopes for the regression equations show that the rate of change was greater for *L* values (34–29) than for b values (16–13).

Erickson et al. (1994) conducted a similar study using pecans; the fruits were subjected to a 241 day storage period at 24°C at two RH conditions (55% and 65% RH). Tristimulus color was measured using a Pacific Scientific XL 800 Series Colorimeter. Authors reported a linear decrease of L and b, and an increase in parameters. The changes noted in this study occurred primarily during the first half of storage.

Descalzo et al. (1999) conducted a trial using pecans at two storage temperatures, 2°C and 20°C. Parameters *L**, *a**, and *b** were measured with a reflectance spectrophotometer Byk Gardner Color View. In the fruit stored at 20°C, the value of *L** declined over time due to the browning of fruit.

Oro et al. (2008) investigated the changes in the color of nuts stored for 150 days in two types of packaging (nylon-polyethylene vacuum and polypropylene). Color parameters were determined every 30 days using a colorimeter Minolta chroma meter CR 400. The values of L^* showed a significant decrease for both containers with the consequent darkening of nuts during the entire period. A similar behavior was noticed in b^* parameter, while no significant changes were seen for a^* values.

Mexis et al. (2009) studied changes in almonds in two packaging conditions for 12 months of storage. They showed that in both treatments, there was a decrease in the L^* values and a parallel increase in the a^* values.

Table 25.5 presents the results obtained for the BR for the three varieties and both treatments throughout storage.

An increase of BR in the California variety throughout the storage for both treatments, being statistically significant at 180 days for California and 225 days for Chandler, can be observed. No significant differences were observed for Criolla variety.

The data obtained for the BR of the nuts studied agree with those obtained by Oro et al. (2008); they studied the behavior of this parameter in pecans peeled, packed in two packaging conditions (polypropylene and nylon-polyethylene vacuum), and subjected to 150 days of storage. The authors found significant increases in BR, varying from 82 to 97 between the beginning and end of storage for nuts packed in polypropylene and from 75 to 87 for those packaged in nylon-polyethylene vacuum.

The differences obtained for BR values by these authors were higher over a shorter storage time than those obtained in this experimental work, this

Table 25.5 BR for Criolla, Chandler, and California Nuts Subjected to Natural (T1) and Oven (T2) Drying along 8 Months of Storage

	Index of Browning					
	Criolla		**Chandler**		**California**	
Time (Days)	**T1**	**T2**	**T1**	**T2**	**T1**	**T2**
0	73.72 ± 5.14	74.4 ± 2.11	56.74 ± 1.84 A	58.24 ± 2.92 B	58.02 ± 1.66 B	59.13 ± 1.79 B
45	71.61 ± 1.79	73.55 ± 3.07	58.99 ± 2.23 AB	61.30 ± 1.79 AB	61.61 ± 0.7 AB	64.32 ± 1.92 AB
90	72.85 ± 1.84	73.08 ± 3.55	59.18 ± 2.21 AB	59.49 ± 0.44 B	61.12 ± 1.23 AB	63.02 ± 3.95 AB
135	70.85 ± 4.54	66.2 ± 4.99	58.37 ± 0.06 AB	60.17 ± 1.16 B	62.23 ± 0.84 AB	65.66 ± 2.15 A
180	74.77 ± 2.68	71.4 ± 1.8	58.89 ± 2.75 AB	56.84 ± 1.38 B	64.71 ± 0.70 A	68.06 ± 0.94 A
225	74.67 ± 1.33	73.9 ± 3.56	63.70 ± 2.98 A	64.69 ± 0.98 A	63.33 ± 3.28 A	66.08 ± 2.16 A

T1, Natural-drying secado; T2, Oven drying.
Capital letters indicate significant differences ($p < 0.05$) in the effect of time.

could be attributed to the fact that nuts without shell are more susceptible to darkening.

25.4 Conclusion

Color results support the conclusion that postharvest drying treatments and storage time have an effect on the characteristic color of grains. The behavior is variety dependent. Nuts from Chandler variety showed a decrease in lightness of 11.5% as an average, while this amount was 8.5% for California nuts. No significant changes in lightness were observed for Criolla nuts, being these nuts darker than the other varieties considered in the assay. Comparing oven- and natural-drying methods, it was observed that kernels from the first method were less bright, redder, and less yellow than those dried by the natural method. Instrumental color determination became a useful tool to characterize the darkening of nuts throughout storage.

Abbreviations

a^* CIELAB red-green parameter
a HunterLab red-green parameter
b^* CIELAB blue-yellow parameter
b HunterLab blue-yellow parameter
L^* CIELAB lightness-darkness parameter
L HunterLab lightness-darkness parameter

Acknowledgments

We thank Mónica Pecile and Luis Sanow for their skillful assistance. This work was supported by INTA, AETA Project number 2681.

References

Aleta, N. 1999. Conservación, normativa y tipificación de nuez. In *Congreso Internacional de Nogalicultura 1999, Proceedings*. Catamarca, Argentina: PROSER.

Buera, M. P., R. Lozano, and C. Petriella. 1986. Definition of color in the non-enzymatic browning process. *Die Farbe* 32/33: 316–326.

Descalzo, A. M., A. Biolatto, G. Grigioni, L. Rossetti, and F. Carduza. 1999. Estabilidad oxidativa y su relación con la vida útil, valor nutricional y aspectos sensoriales de nuez pecán. In *Producción de pecán en Argentina*. Buenos Aires, Argentina: INTA.

Erickson, M. C., C. R. Santerre, and M. E. Malingre. 1994. Oxidative stability in raw and roasted pecans: Chemical, physical and sensory measurements. *Journal of Food Science* 59 (6): 1234–1238.

Forbus, W. R., S. D. Senter, B. G. Lyon, and H. P. Dupuy. 1980. Correlation of objective and subjective measurements of pecan kernel quality. *Journal of Food Science* 45: 1376–1379.

Grigioni, G., A. Biolatto, L. Langman, A. Descalzo, M. Irurueta, R. Páez, and M. Taverna. 2010. Colour and pigments. Milk and dairy products. In *Practical food research*, ed. Rui Cruz. New York: Nova Science Publisher, Chapter XI.

López, A., M. T. Pique, A. Romero, and N. Aleta. 1995. Influence of cold-storage conditions on the quality of unshelled walnuts. *International Journal of Refrigeration* 18 (8): 544–549.

Mexis, S. F., A. V. Badeka, and M. G. Kontominas. 2009. Quality evaluation of raw ground almond kernels *(Prunus dulcis):* Effect of active and modified atmosphere packaging, container oxygen barrier and storage conditions. *Innovative Food Science and Emerging Technologies*. DOI: 10.1016/j.ifset.2009.05.002.

Muncharaz Pou, M. 2001. Postrecoleccion. In *El Nogal*. Madrid, Spain: Mundi Prensa, Chapter XVI.

Oro, T., P. Ogliari, R. Dias de Mello, C. Amboni, D. Barrera-Arellano, and J. M Block. 2008. Evaluación de la calidad durante el almacenamiento de nueces Pecán [Carya illinoinensis (Wangenh.) C. Koch] acondicionadas en diferentes envases. *Grasas y Aceites* 59 (2): 132–138.

Pérez Alvarez, J. 2006. Color. In *Ciencia y tecnología de carnes*, eds. Y. In Hui, I. Guerrero, and M. Rosmini. Mexico: Limusa, pp. 199–228.

Prunet, J. P. and I. Herman. 1995. Noyer: déperissement lié aux excès d'eau et au Phytophthora. *Infos-Ctifl* 113: 42–44.

Senter, S. D., W. R. Forbus, S. O. Nelson, and R. J. Horvat. 1984. Effects of dielectric and steam heating treatments on the pre-storage and storage color characteristics of pecan kernels. *Journal of Food Science* 49: 1532–1534.

CHAPTER 26

Kinetics of Melanosis in Shrimp

Effect of Pretreatment Using Chemical Additives

MARÍA ANA LOUBES, CARLOS ALBERTO ALMADA, and MARCELA PATRICIA TOLABA

Contents

26.1 Introduction

The Argentine fishing industry is basically oriented to exportation, and within that sector, marketing of crustaceans is the activity of high economic value. Melanosis or blackening of the shrimps can reduce their market value between 10% and 25%, or it can even mean its rejection (SAGPyA 2007).

The polyphenol oxidase (PPO) is the enzyme responsible for the phenomenon of melanosis, and it is present in an inactive form in the hemolymph of crustaceans. After their death, different physiological processes involve its activation and production of colored pigments. These pigments are not harmful and do not affect flavor or aroma, but consumers reject them for their appearance (Díaz López et al. 2003).

Crustaceans are marketed with sulfiting agents in order to prevent and control melanosis since the PPO remains active during refrigeration. Sodium metabisulfite is most often used for its effectiveness and low cost (Lucien-Brun 2006). However, numerous publications about possible adverse effects and cases of allergies (e.g., urticaria, angioedema, bronchial constriction, pruritus, contact eczema, rhinitis, anaphylactic shock) in sensitive or asthmatic individuals have led to the search for alternative treatments in the prevention of melanosis (Brasó Aznar 2003).

L-ascorbic acid has been considered the safest food additive to prevent melanosis. It has the advantage of being very soluble in water, but it is oxidized irreversibly to dehydroascorbic acid; therefore, it only has a temporary effect. Moreover, its oxidation products may give rise to off-flavors and present undesirable yellow coloration in the exoskeleton.

Cysteine is an inhibitor of the enzyme melanosis, and although it has been used commercially as an antioxidant, the necessary concentration to inhibit melanosis in an acceptable way has a negative effect on the shellfish taste.

Several studies demonstrate the efficacy of 4-hexylresorcinol (4-HR) in the control of melanosis in Mediterranean shrimp. In addition, it has been shown that its use do not affect the taste, texture, or color of the product. 4-HR is a water-soluble and chemically stable compound. It has been used since 1920 in human and veterinary medicine as an antiparasitic, oral antiseptic, dilutant in medicines for colds, and as an antibrowning agent in food. There have been studies on toxicity, mutagenicity, carcinogenicity, and allergic potential which show that 4-HR does not represent risk to consumer health at the levels used to inhibit melanosis. Based on the results of these studies, the FDA of the United States called the 4-HR as GRAS (generally recognized as safe) additive, which means that it is not necessary to control its residual amount.

Moreover, research by Taoukis et al. (1990), using five different proteases, showed that only ficin had melanosis inhibitor activity, inhibiting the formation of dark spots on shrimp (*Penaeus duorarum*) for 4 days under refrigeration.

The aim of this chapter is to study the effects of several chemical additives to prevent melanosis of Argentine red shrimps (*Pleoticus muelleri*). Three chemical additives—sodium metabisulfite, 4-HR, and ficin—were used. An untreated sample was adopted as control. The monitoring of the progression of melanosis during 96 h was made through digital images of the samples, captured with an experimental device designed for that purpose. Digital images were analyzed using a computer program that enabled calculation of the percentage of darkened area (PDA) for each treatment. The kinetics of shrimp blackening was modeled by means of zero-order reaction.

26.2 Materials and Methods

26.2.1 Material and Chemical Treatments

Atlantic white shrimps (*Pleoticus muelleri*) with the size of 31–70 g were purchased from the dock in Rawson, Chubut, Argentina. The shrimps were washed and stored for 2 h in ice until used. Chemical treatments were performed using the following aqueous solutions. Sodium metabisulfite (0.6%; 1.25%, g/100 mL of solution) was purchased from Química Oeste SA (Buenos Aires, Argentina), 4-HR (0.0025%, 0.005%, 0.01%, g/100 mL of solution) was obtained from Aldrich (St. Louis, Missouri), and ficin (0.5%, 2%, g/100 mL of solution) was procured from Enzyme Development Corporation (New York).

The shrimps were immersed in each solution at a shrimp/solution ratio of 400 g/1 L at 20°C for 1 min (McEvily et al. 1991). A sample soaked in pure water was adopted as control. The soaked shrimps were drained and stored in sealed bags at −20°C for a month until melanosis was determined.

26.2.2 Color Measurements

Quantification of the color change (darkening) of fresh samples during iced storage (96 h at 4°C) was based on the measurement of PDA. Darkened area was defined selecting "shadows" option from palette function of Adobe Photoshop CS3 Extended Version 10.0 software. The lighting conditions in the measuring chamber and the location of the samples were standardized. An experimental device (Figure 26.1), where the sample was illuminated with daylight fluorescent lamp, was designed to capture digital images of the samples at predetermined times of storage.

For each treatment, the PDA was calculated from the total area and the darkened area by counting the pixels of the image:

$$\text{PDA} = \frac{\text{Darkened Pixels}}{\text{Total Pixels}} \times 100 \tag{26.1}$$

A digital camera SONY Cyber-shot (DSC-W80, Sony Corp.) was used to capture an 18.7 cm × 14.5 cm objective area. All measurements were carried out by duplicate. Digital images in JPEG format (3072 × 2304 pixels) were analyzed by means of Adobe Photoshop CS3 Extended Version 10.0 software.

26.3 Results and Discussion

The effect of melanosis in shrimps can be recognized in Figure 26.1, where the fresh material was shown together with untreated specimen and optimal specimen treated by chemical additive. Melanosis was characterized by the appearance of melanin black spots in the exoskeleton of shrimp; it started on

Figure 26.1 Experimental device used to capture shrimp images.

the last pair of appendages (paddles), gills, carapace intersegmental area, in the coxae of the legs, and the bottom of the cephalothorax.

Figure 26.2 shows the best results obtained with each of the additives used. Although, in all cases, the inhibitory effect is observed, it is possible to notice some differences of tone and luminosity between the treatments (see also Figure 26.3).

Digital images of control and treated samples were obtained and analyzed according to the procedure described. PDA was calculated by means of Equation 26.1. Experimental data of PDA obtained during ice storage of shrimps are shown in Table 26.1 being the relative experimental error of 2%. From results of Duncan's multiple range test, it can be appreciated that time and chemicals significantly affected the average values of darkened area. A significant reduction of darkness (60%) was observed at 24 h for pretreatment samples in comparison with control sample (PDA: 30.32%). PDA mean values were higher for control and ficin treatments in comparison with MBS and 4-HR treatments independent of storage time investigated.

In all treatments, it can be observed that darkening was significantly ($p < 0.05$) increasing during ice storage. This fact is evidenced by comparison

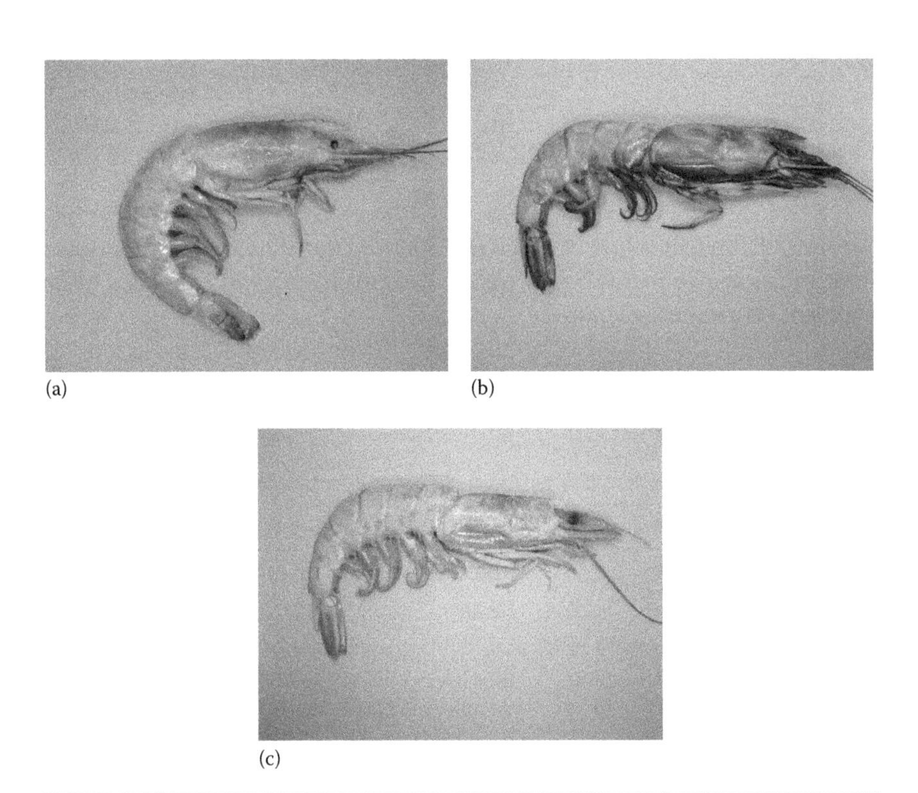

Figure 26.2 (a) Fresh specimen, (b) untreated specimen at 96 h, and (c) specimen treated by 4-HR (0.0025%) at 96 h.

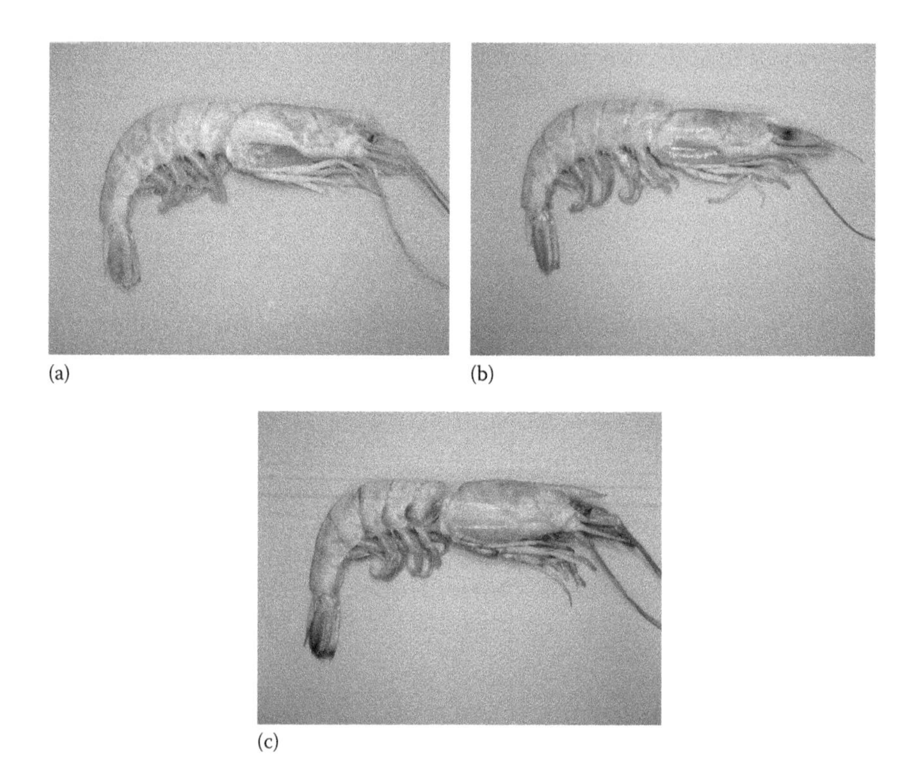

Figure 26.3 After 96 h of treatment of the specimen treated by (a) sodium metabisulfite (0.6%), (b) 4-HR (0.0025%), and (c) ficin (2%).

between PDA mean values of initial (15.19%) and final (25.14%) storage times. PDA values shown in Table 26.1 were satisfactorily ($r^2 > 0.85$) modeled by an apparent zero-order reaction:

$$\frac{d(\text{PDA})}{dt} = k \rightarrow \text{PDA} = \text{PDA}_0 + k \times t \qquad (26.2)$$

Table 26.1 PDA Obtained Using Different Chemical Treatments to Avoid Melanosis in Shrimps

Time (h)	Control	MBS 0.6%	MBS 1.25%	4-HR 0.0025%	4-HR 0.005%	4-HR 0.01%	Ficin 0.5%	Ficin 2%
24	$30.32^{a,2}$	$8.67^{c,3}$	$11.02^{c,3}$	$8.86^{c,3}$	$8.55^{c,3}$	$8.91^{c,3}$	$14.52^{b,3}$	$12.93^{c,3}$
48	$37.17^{a,1,2}$	$10.6^{c,2,3}$	$11.99^{c,2,3}$	$9.55^{c,2,3}$	$10.31^{c,2,3}$	$11.23^{c,2,3}$	$16.47^{b,2,3}$	$15.28^{c,2,3}$
72	$41.42^{a,1}$	$12.79^{c,1,2}$	$14.82^{c,1,2}$	$10.83^{c,1,2}$	$12.82^{c,1,2}$	$12.74^{c,1,2}$	$31.18^{b,1,2}$	$16.12^{c,1,2}$
96	$44.05^{a,1}$	$14.2^{c,1}$	$19.12^{c,1}$	$12.42^{c,1}$	$15.31^{c,1}$	$16.06^{c,1}$	$31.47^{b,1}$	$18.09^{c,1}$

a, b, c Data in the same row with the same letter do not differ significantly at the significance level of 0.05.

1, 2, 3 Data in the same column with the same number in the same kind of sample do not differ significantly at the significance level of 0.05.

where

k is the melanosis rate constant

t is the time

Kinetics constants, obtained by linear regression, are shown in Table 26.2. It can be appreciated that PDA_0 values of treated samples were smaller than the control, which did not have the protective effect of chemical additive after thawing.

4-HR (0.0025%) showed a k value of 4.98×10^{-2} h^{-1}, which was 73.65% lower than the control ($k = 18.9 \times 10^{-2}$ h^{-1}); in return, ficin (0.5%) had a k value of 27.3×10^{-2} h^{-1}, being 44.44% higher than control. The results indicate that 4-HR at a very low concentration significantly retards melanosis.

26.4 Conclusion

The use of 4-HR, additive considered GRAS, at very low doses (0.0025%) prevented melanosis during frozen storage and thawing. In addition, there were no side effects caused by sulfites, which indicate a very convenient alternative. Moreover, the maximum ficin concentration tested (2%) showed an efficacy of less than 4-HR.

The image analysis, based on the darkening of the samples, has been useful to follow the kinetics of shrimp deterioration. However, this is a simplified approach because the original appearance and color of the product undergo changes during spoilage beyond mere darkening. New techniques will be required to identify the color palette representative of the product and to quantify changes in the distribution of colors. Surely the sensory analysis will be a valuable reference when a more complex instrumental technique is applied.

Table 26.2 Kinetics Constants of Melanosis in Shrimps

Treatment	PDA_0 (%)	$k \times 10^2$ (h^{-1})	r^2
MBS 0.6%	6.87 ± 0.30	7.82 ± 0.46[a]	0.9931
MBS 1.25%	7.46 ± 1.44	11.3 ± 2.2[b]	0.9298
4-HR 0.0025%	7.43 ± 0.39	4.98 ± 0.59[c]	0.9719
4-HR 0.005%	6.05 ± 0.35	9.5 ± 0.53[b]	0.9938
4-HR 0.01%	6.49 ± 0.67	9.57 ± 1.01[b]	0.9780
Ficin 0.5%	7.02 ± 5.31	27.3 ± 8.08[d]	0.8510
Ficin 2%	11.53 ± 0.54	6.8 ± 0.82[a]	0.9719
Control	26.88 ± 1.84	18.9 ± 2.8[d]	0.9582

[a, b, c, d] Data in column followed by the same letter do not differ significantly at the significance level of 0.05.

Acknowledgments

The authors acknowledge the financial assistance of Universidad Nacional de Luján and Secretaría de Ciencia y Técnica de la Universidad de Buenos Aires (FCEN-UBA), Argentina.

References

Brasó Aznar, J. V. 2003. *Manual de alergia clínica*. Barcelona, Spain: Masson SA.

Díaz López, M., I. Martínez Díaz, T. Martínez Moya, M. Montero García, M. Gómez Guillén, M. Zamorano Rodríguez, and O. Martínez Álvarez. 2003. *Estudios de los agentes conservantes e inhibidores de la melanosis en crustáceos*. Andalucía, Spain: Junta de Andalucía Consejería de Agricultura y Pesca.

Lucien-Brun, H. 2006. Melanosis, black spot and sodium metabisulfite. *Panorama Acuícola Magazine* 4: 30–38.

McEvily, A., R. Iyengar, and W. Otwell. 1991. Sulfite alternative prevents shrimp melanosis. *Food Technology* 45 (9): 80–86.

SAGPyA (Secretaría de Agricultura, Ganadería, Pesca y Alimentos de Argentina). 2007. Pesquerías de calamar y langostino. Situación actual. http://www.minagri.gov.ar/SAGPyA/pesca/pesca_maritima.pdf. Accessed November 17, 2008.

Taoukis, P. S., T. P. Labuza, J. H. Lillemo, and S. W. Lin. 1990. Inhibition of shrimp melanosis (black spot) by ficin. *Lebensmittel Wissenschaft und Technologie* 23: 52–54.

CHAPTER 27

Color of Dried Pears as Affected by Prior Blanching and Sugar Infusion

SILVIA B. MAIDANA, MABEL B. VULLIOUD, and DANIELA M. SALVATORI

Contents

27.1 Introduction

Color changes in fruits involve several mechanisms and different properties, which are dependent on the state of the product and its specific composition. Pear fruit is a potential substrate for the majority of the mechanisms of color deterioration, which is related to the amount of phenolic compounds in the tissue and also to the high level of reducing sugars (Kadam et al. 1995, Khalloufi and Ratti 2003).

It is well known that, among the dehydration techniques, air drying causes browning discoloration in foods. Browning could be provoked by enzyme action, taking place in early stages of processing, prior to polyphenol oxidase inactivation, or by nonenzymatic reactions during drying and later storage

(Acevedo et al. 2008). The use of osmotic dehydration—also known as *sugar infusion* (SI)—as a predrying step has received increasing attention in the field of fruit preservation processes to improve quality of fruit products. The main advantages are the inhibition of the enzymatic browning, the retention of volatile compounds, the partial dehydration of the food, and therefore the reduction of energy consumption during further drying (Krokida et al. 2001, Torreggiani and Bertolo 2002). The addition of sugars during the osmotic step combined with other factors, such as a slight thermal treatment, the addition of antifungal and antimicrobial agents, and texture preservatives before drying, could extend shelf life and permit the development of products with fresh quality attributes (Alzamora and Salvatori 2006). Blanching is usually performed prior to drying, not only to inactivate enzymes responsible for various undesirable enzymatic reactions, but also to increase the drying rate, hence reducing the drying time (Chiewchan et al. 2010, González-Fesler et al. 2008). Pear fruits have not been traditionally used as matrices for these technologies because they have a well-defined target toward fresh consumption and especially toward international market. In addition, pear is very sensitive to color deterioration induced by temperature and heating during drying (Khalloufi and Ratti 2003), which suggests that a diversification of available technologies is needed to obtain novel dried products which preserve the characteristic of fresh fruits.

The objective of this chapter was to investigate the effect of blanching and/or SI prior to the drying process on surface color of dried fruits obtained from pears (*Packham's var.*) produced in Argentine Patagonia.

27.2 Materials and Methods

27.2.1 Sample Preparation and Pretreatments

Fresh pears (*Packham's var.*) from the Upper Valley zone of Río Negro (Argentina) were selected. Fruits (85.1% water content, 0.97 water activity (a_w), 14° Brix) were washed, peeled, and cut into disks (3 cm diameter × 0.6 cm thickness). The following treatments were carried out prior to drying process:

1. Blanching (B): Using water vapor at 100°C for 1.5 min and then cooling with cold water at 4°C for 1.5 min. This was used to inactivate enzymes responsible for enzymatic browning.
2. Dry SI: Osmotic dehydration processes were performed by immersing the fruits into a mixture of humectants (sucrose or glucose) and antioxidant and antimicrobial preservatives (potassium sorbate and sodium sulfite). The amount of sugars and chemical agents were determined according to the weight of the fruit and the final levels required after equilibration of the product (a_w = 0.83 and a_w = 0.96). Final system pH was adjusted to 3.5 with citric acid.

Pretreatments were combined and applied to the cut material as follows: (a) SI1: SI with sucrose ($a_w = 0.83$); (b) SI2: SI with sucrose ($a_w = 0.96$); (c) B + SI1: blanching + SI with sucrose ($a_w = 0.83$); (d) B + SI2: blanching + SI with sucrose ($a_w = 0.96$); (e) GI2: SI with glucose ($a_w = 0.96$); (f) B + GI2: blanching + SI with glucose ($a_w = 0.96$); and (g) C: control corresponding to pears without pretreatments.

27.2.2 Drying Process

Pears, with and without pretreatment, were subjected to convective dehydration in a fluid bed dryer operated at 65°C, 4 m/s air velocity, and 6% relative humidity for 4 h. After drying, samples were analyzed for water and soluble solid content, water activity, and surface color.

27.2.3 Color Measurement

Color of dehydrated pears was measured by a photocolorimeter Minolta CR400 (2° observer, illuminant C, CIELAB color space) in the central point of sample surface. Measurements were performed on 10 samples at two positions on the top and bottom surfaces. A white background of reflectance provided by the manufacturer was used. Color changes of pears were evaluated through L^*, a^*, b^* components where L^* indicates lightness, a^* indicates chromaticity on a green (–) to red (+) axis, and b^* chromaticity on a blue (–) to yellow (+) axis. These numerical values were converted into global color change ($\Box E^*_{ab}$) and browning index (BI) functions in order to analyze the color evolution caused by treatments:

$$\Delta E^*_{ab} = \sqrt{(\Delta L^*)^2 + (\Delta a^*)^2 + (\Delta b^*)^2} \tag{27.1}$$

where $\Delta L^* = \left(L^* - \overline{L^*_0}\right)$, $\Delta a^* = \left(a^* - \overline{a^*_0}\right)$, $\Delta b^* = \left(b^* - \overline{b^*_0}\right)$, and $\overline{L^*_0}$, $\overline{a^*_0}$, $\overline{b^*_0}$ are the average tristimulus results for fresh fruit.

$$\text{BI} = [100\ (x - 0.31)]/0.172 \tag{27.2}$$

where $x = X/(X + Y + Z)$.

With the sole purpose of evaluating translucency or opacity characteristics, L^*_{ab} values of slices were measured by placing them over both white and black backgrounds.

27.2.4 Statistical Analysis

All statistical analyses were carried out using the Statgraphics Plus package. Analysis of variance (ANOVA) was done to establish significant differences in color parameters after infusion and after drying processes. Significance level was set at $p < 0.05$, and multiple comparisons were performed using the Tukey test.

27.3 Results and Discussion

Fresh fruit color measurements were $L_0^* = 80.76 \pm 1.34$, $a_0^* = -2.91 \pm 0.74$, and $b_0^* = 14.21 \pm 2.08$. The development of translucency was confirmed in pear samples after infusion, which changed from a cream-white opaque color to a translucent yellow color. This behavior can be explained, at least partially, by the replacement of gas by external solution in the intercellular spaces during infusion, resulting in a more homogeneous refractive index in the tissue and a consequent reduction in light scattering. A similar process was described by Talens et al. (2002) to explain the development of translucency in processed kiwi and by Lana et al. (2006) in tomato after storage. After the drying process, pear samples recovered opacity due to solid concentration and structure compaction.

Many authors have reported that a decrease in L^* value and an increase in a^* value are indicative of darkening in apples and pears (Gómez et al. 2010, Sapers and Douglas 1987, Taiwo et al. 2001). As can be observed in Figure 27.1, the simultaneous change in both values, L^* and a^*, is a useful indicator of browning, probably resulting from enzymatic and nonenzymatic reactions during processing. After both the infusion and drying steps, pear surfaces were darker (lower L^* values) and less green (higher a^* value) compared to fresh-cut samples. These changes were quite pronounced after the first step, especially in L_{ab}^* values (Figure 27.1A). It seems clear that the greater values registered for this parameter in only infused samples were due to translucency development, rather than tissue browning. As it was expected, an increase in a^* values was observed in dried samples when compared with only infused samples (Figure 27.1B).

During infusion pretreatments, samples are partially dehydrated. Nevertheless, due to the open structure of the tissue, diffusion of sugars in the intercellular spaces and cut external cells also takes place. High sugar concentration in SI pears treated at $0.83a_w$ caused smaller alterations in color after infusions than lower concentrations (SI at $0.96a_w$), resulting in L^* and a^* values more similar to those obtained in fresh fruit. Some authors have observed a protective role of sugars on some plant pigments and a reduction of activity of polyphenol oxidase responsible for enzymatic browning (Chiralt and Talens 2005). Also, the higher water loss achieved in SI1 samples could promote the formation of a superficial sugar layer, contributing to the color preservation.

Blanching significantly increased the L^* parameter in infused samples (B + SI1, B + SI2, and B + GI2). This may be due to the fact that, during heating, air in the plant tissue is replaced by water, reducing light scattering and contributing to translucency development. On the other hand, a^* values showed a certain stability compared with infused pears without previous heat treatment, which experienced a greater darkening.

During drying, changes of color coordinates of material are affected by water removal, shrinkage, and alterations of the structure of the material surface.

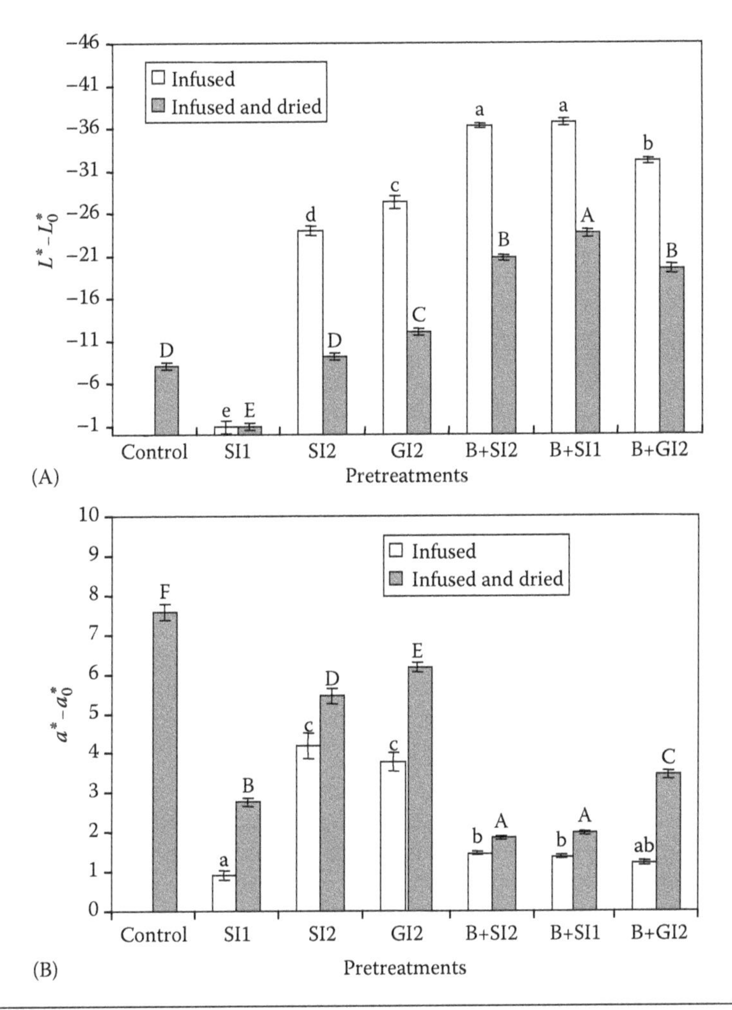

Figure 27.1 Color values obtained after infusion and after drying compared to the fresh sample: (A) lightness change ($L^* - L_0^*$) and (B) redness change ($a^* - a_0^*$). Bars with the same letter were not significantly different ($p < 0.05$).

Water is replaced by air in porous materials, and they appear as pale (Lewicki 2004). Decreased luminance and more saturated color of the material during final stages of drying are caused by browning reactions as well as by the effect of sample concentration. Our results showed that sugar penetration in the fruits causes a relative stability after drying, since variation of a^* parameter was less drastic than in control samples (Figure 27.1). Although the same tendency was observed in color changes when compared with only infused samples, dried pears exhibited an increase in a^* values. All dried samples (with and without pretreatments) showed a decrease in L^* values, except for SI pears pretreated with SI1, which again maintained a lightness similar to

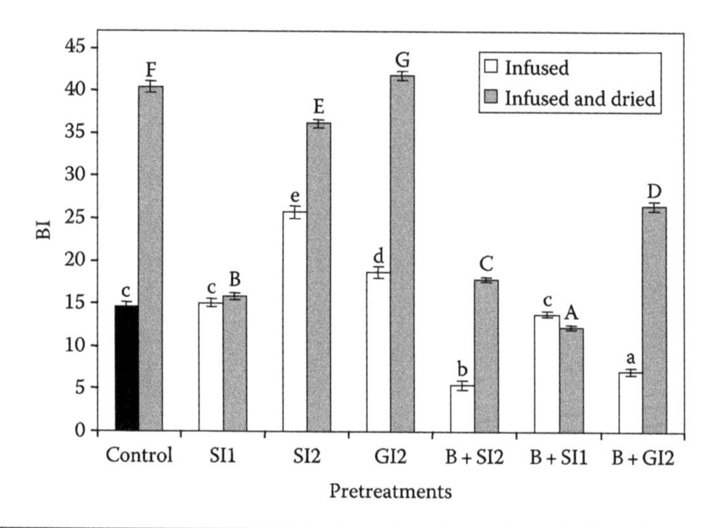

Figure 27.2 BI of pears after infusion and after drying. Bars with the same letter were not significantly different ($p < 0.05$).

that of fresh fruit. The highest variation was observed for samples pretreated with GI2 (Figure 27.1B). The nature of the humectants seemed to affect color behavior in dried samples since significant differences were obtained between sucrose (SI2) and glucose (GI2) pretreatments.

According to the BI (Figure 27.2), dried samples pretreated with less humectant ($a_w = 0.97$) developed higher browning than those with infusion at $0.87a_w$, which retained a light color due to the presence of sugars on tissue surface. Samples infused in glucose at $0.97a_w$ (GI2) showed minor color retention during drying than those pretreated in sucrose at the same a_w. Although sodium sulfite had been added to the osmotic mediums in order to inhibit nonenzymatic browning, the uptake of a reducing sugar during GI2 infusion could have resulted in increased Maillard reaction during further drying stage.

Although the more important changes in global color occurred in blanched samples (Figure 27.3), this thermal treatment promoted enzymatic inactivation and led to less discolored pears in all cases when compared with the corresponding dried pears not previously blanched (Figures 27.2 and 27.4). The higher ΔE^*_{ab} values observed in samples with previous blanching could be due to the significant changes in lightness developed.

27.4 Conclusions

The different behaviors obtained in samples after drying show that the type of pretreatment significantly affects color changes. Since pear slices pretreated with sucrose infusion at $0.87a_w$ before drying appeared with minor browning, the protective effect of sugar concentration was confirmed. These samples

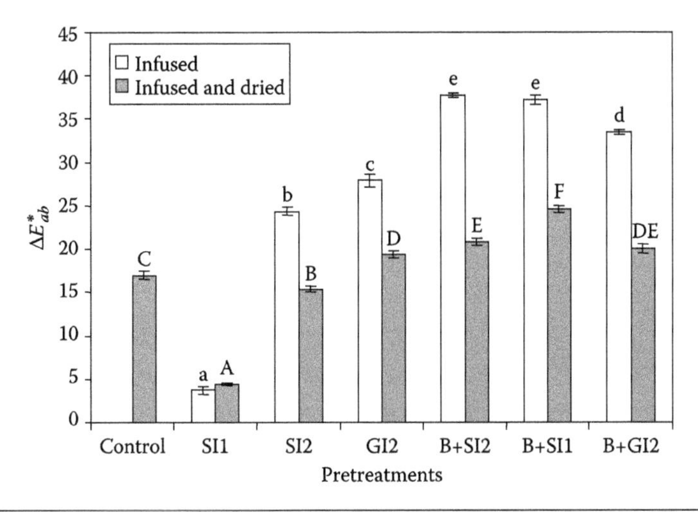

Figure 27.3 Total color difference □ E^*_{ab} of pears after infusion and after drying. Bars with the same letter were not significantly different ($p<0.05$).

Figure 27.4 Photographs taken from fresh and dried pears with and without pretreatments. (A) fresh, (B) dried control, (C) SI1, (D) SI2, (E) GI2, (F) B+GI2, (G) B+SI1, and (H) B+SI2.

showed no changes in lightness and a small increase of redness without being affected by a previous blanching, which indicates that color preservation occurred. Blanching reduced browning during drying, except in glucose pretreated samples. Also, transparency changes during infusion had an impact on color attributes, especially on lightness.

According to the results, an adequate control of blanching and infusion treatments prior to drying of pears seems to be useful as a tool for improving product color.

Acknowledgments

The authors acknowledge the financial support from Universidad Nacional del Comahue, Conicet, and Agencia Nacional de Promoción Científica y Tecnológica of Argentina.

References

Acevedo, N. C., V. Briones, P. Buera, and J. M. Aguilera. 2008. Microstructure affects the rate of chemical, physical and color changes during storage of dried apple discs. *Journal of Food Engineering* 85: 222–231.

Alzamora, S. M. and D. M. Salvatori. 2006. Minimal processing foods. In *Handbook of Food Technology and Food Engineering*, ed. Y. H. Hui. Boca Raton, FL: CRC Press, Taylor & Francis Books, vol. III, Chapter 118, pp. 1–16.

Chiewchan, N., C. Praphraiphetch, and S. Devahastin. 2010. Effect of pretreatment on surface topographical features of vegetables during drying. *Journal of Food Engineering* 101: 41–48.

Chiralt, A. and P. Talens. 2005. Physical and chemical changes induced by osmotic dehydration in plant tissues. *Journal of Food Engineering* 67: 167–177.

Gómez, P., S. Alzamora, M. Castro, and D. Salvatori. 2010. Effect of ultraviolet-C light dose and storage on color changes of fresh-cut apple. *Journal of Food Engineering* 98 (1): 60–70.

González-Fesler, M., D. M. Salvatori, P. Gómez, and S. M. Alzamora. 2008. Convective air drying of apples as affected by blanching and calcium impregnation. *Journal of Food Engineering* 87: 323–332.

Kadam, P., S. A. Dhumal, and N. N. Shinde. 1995. Pear. In *Fruit Science and Technology. Production, Composition, Storage and Processing*, eds. D. K. Salunkhe and S. S. Kadam. New York: Marcel Dekker, pp. 183–202.

Khalloufi, S. and C. Ratti. 2003. Quality deterioration of freeze-dried foods as explained by their glass transition temperature and internal structure. *Journal of Food Science* 68: 892–903.

Krokida, M., Z. Maroulis, and G. Saravacos. 2001. The effect of the method of drying on the colour of dehydrated products. *International Journal of Food Science & Technology* 36: 53–59.

Lana, M. M., L. M. Tijskens, A. de Theije, M. Hogenkamp, and O. van Kooten. 2006. Assessment of changes in optical properties of fresh-cut tomato using video image analysis. *Postharvest Biology and Technology* 41: 296–306.

Lewicki, P. 2004. Water as the determinant of food engineering properties. A review. *Journal of Food Engineering* 61: 483–495.

Sapers, G. M. and F. W. Douglas. 1987. Measurement of enzymatic browning at cut surfaces and in juice of raw apple and pear fruit. *Journal of Food Science* 52: 1258–1262.

Taiwo, K. A., A. Angersbach, B. I. O. Ade-Omowaye, and D. Knorr. 2001. Effects of pre-treatments on the diffusion kinetics and some quality parameters of osmotically dehydrated apple slices. *Journal of Agriculture and Food Chemistry* 49: 2804–2811.

Talens, P., N. Martínez-Navarrete, P. Fito, and A. Chiralt. 2002. Changes in optical and mechanical properties during osmodehydrofreezing of kiwi fruit. *Innovative Food Science and Emerging Technologies* 3: 191–199.

Torreggiani, D. and G. Bertolo. 2002. The role of an osmotic step: Combined processes to improve quality and control functional properties in fruit and vegetables. In *Engineering and Food for the 21st Century*, eds. J. Welti Chanes, G. V. Barbosa Cánovas, and J. M. Aguilera. Boca Raton, FL: CRC Press, pp. 651–670.

CHAPTER 28

Color Changes in Fresh-Cut Fruits as Affected by Cultivar, Chemical Treatment, and Storage Time and Temperature

ANDREA PIAGENTINI, LORENA MARTÍN, CECILIA BERNARDI, DANIEL GÜEMES, and MARÍA PIROVANI

Contents

28.1 Introduction

Visual appearance of a fresh-cut fruit or vegetable is the attribute most immediately obvious to the consumer and which strongly affects its commercial shelf life. Many unrelated factors influence appearance, from wound-related effects to drying or to microbial colonization (Toivonen and Brummell 2008). Appearance, which is significantly impacted by color, is one of the first attributes used by consumers in evaluating food quality. Color is also critically important in the many dimensions of food choice and influences the perception of other sensory characteristics by the consumers. Color is a result of a variety of factors both endogenous and exogenous to the food that may be affected by genetics and pre- and postharvest treatments (Clydesdale 1998). Color in fruits may be influenced by naturally occurring pigments such as chlorophylls, carotenoids, and anthocyanins in fruits, or by pigments resulting from browning reactions. Browning of fruits and fruit products is one of the major problems in the food industry and is believed to be probably the first cause of quality loss during postharvest handling, processing, and storage. Browning can also adversely affect flavor and nutritional value (Lozano 2006).

Although it is well known that minimal processing accelerates the end of shelf life due to a reduction in visual quality, further studies are needed to evaluate the effect of postcutting treatments, including modified atmosphere packaging (MAP) and chemical dips on delaying softening and browning. Wounding of fruit tissues induces a number of physiological disorders that need to be minimized to obtain freshlike quality products. This is one of the reasons that the control of discoloration (pinking, reddening, or blackening) or browning at cut surfaces is an important issue in fresh-cut fruit processing (Gil et al. 2006).

Cut-edge browning is a particular problem in fruit with white flesh such as apples and pears. Oxidative browning is usually caused by the polyphenol oxidase (PPO) enzyme which, in the presence of O_2, converts phenolic compounds in fruits and vegetables into dark-colored pigments. Most raw fruits contain polyphenols and PPO, located in different compartments in the cell structure. Enzymatic browning occurs in fruits after bruising, cutting, or during storage, when substrates and oxygen come into contact with each other, and a lot of reactions start that finally lead to the formation of insoluble brown pigments (melanins). Browning reactions in fruits are complex because of the large number of secondary reactions that may occur. This is reflected in the range of color produced even in the same product (Toivonen and Brummell 2008).

The enzymatic browning needs to be controlled during the processing of fruits. PPO activity, as for most of enzymes, may be minimized by reducing agents, heat inactivation, lowering the pH of the fruit product, and the presence of enzyme inhibitors, among other techniques (Lozano 2006).

The occurrence of this defect is mainly affected by cultivar, processing techniques, packaging, and temperature management. Different strategies may be used to reduce PPO-mediated cut surface discoloration: refrigerated storage, reducing the amount of O_2 in the package, heat treatment, dipping fruits in mildly acidic food grade solution, or reducing agents. Nevertheless, enzymatic browning still represents a major challenge with fresh-cut fruit (Beaulieu and Gorny 2002).

To effectively inhibit or control the enzymatic browning in fresh-cut fruits, an accurate determination of the kinetics of deterioration can be followed through color measurements, which is a simple and effective way for studying the phenomenon. Although relatively few of the phenolic compounds in fruits serve as substrates for PPO, the stoichiometry of complex reaction like enzymatic browning in fruits is practically unknown. Therefore, instead of determining the consumption of reactive (phenols), or formation of products (melanins), the kinetics of color development is commonly used for studying the browning reactions (Lozano 2006, Piagentini et al. 2008).

Therefore, the objective of this work was to evaluate the color changes of fresh-cut apples and peaches from different cultivars during storage, studying the effect of an antibrowning solution or different storage temperatures.

28.2 Materials and Methods

28.2.1 Plant Material and Sample Preparation

Fruits from two apple cultivars (*Granny Smith* and *Princesa*) and four peach cultivars (yellow pulp: *Early Grande*, *Flordaking*, *Hermosillo*; white pulp: *Tropic Snow*) were studied. *Granny Smith* is a well-known high-cold-requirement apple variety from the Rio Negro province (Argentina), and *Princesa* is a low-cold-requirement apple variety cultivated in the central-east region of Santa Fe (Argentina). Peach cultivars, adapted to the central-east region of Santa Fe province, came from an experimental production.

Apples and peaches were prepared as fresh-cut fruits. They were washed with a 100 mg/L solution of sodium hypochlorite (NaClO) for 2 min, peeled, cored, and cut in eight wedges with a sharp stainless steel knife. Then, fruit wedges were washed in water containing 30 mg/L available chlorine as NaClO and pH 7, and a water to produce ratio of 3 L/kg. Fruit wedges were drained by gravity and then over blotting paper. Half of apple samples were treated with an antioxidant solution (1% ascorbic acid plus 1% citric acid) for 3 min and drained. Both treated (TQ) and untreated (STQ) fresh-cut apples were packaged in polyethylene terephthalate (PET) clamshell trays (80–100 g) and stored at 2.5°C for 14–15 days. Fresh-cut peaches were also packed (80–100 g) in PET clamshell trays and stored at 1.5°C, 5°C, 10°C, and 15°C for 10–11 days.

28.2.2 Color Determination

Color changes of fresh-cut apples and peaches were measured on fruit wedges from two packages by sample and storage time. Surface color of three to four wedges from each replicate of each treatment was measured. Measurements were made at the middle point of the two cut surfaces of each fruit wedges.

Color (CIELAB and *XYZ* values) was measured using a Minolta spectrophotometer (model CM-508d/8, Minolta, Tokyo, Japan), calibrated using the standard white tile. D65/10° was used as the illuminant/viewing geometry and specular component excluded (SCE).

Chroma value $\left[C^*_{ab} = \left(a^{*2} + b^{*2}\right)^{0.5}\right]$, hue angle [$h_{ab}$ = arctangent (b^*/a^*)], and total color difference $\left[\Delta E^*_{ab} = ((\Delta L^*)^2 + (\Delta a^*)^2 + (\Delta b^*)^2)^{0.5}\right]$ were also determined. □E^*_{ab} was calculated for each sample at each storage time with respect to its initial value ($t = 0$ day). Finally, the browning index (BR) was determined [$BR = 100(x - 0.31)/0.172$, where $x = X/(X + Y + Z)$], as proposed by Matiacevich and Buera (2006).

28.2.3 Statistical Analysis

Analysis of variance (ANOVA) was used to determine significant differences among sample color parameters in response to fruit variety, chemical treatment, and storage time for fresh-cut apples, and in response to fruit variety, and storage time and temperature for fresh-cut peaches. Duncan's multiple range tests were used to determine significant differences among treatment means. All statistical analyses were performed using Statgraphics Plus 7.1.

28.3 Results and Discussion

28.3.1 Color Changes of Fresh-Cut Granny Smith and Princesa Apples

Chemical antibrowning agents have been commonly used to prevent browning of fruits and fruit products. In this work, a solution with antioxidant and acid-chelating compounds were used. Antibrowning agents are compounds that either act primarily on the enzyme or react with the substrates and/or products of enzymatic catalysis in a manner that inhibits colored product formation. The enzyme PPO can be inhibited by acids like citric acid and chelating agents and reducing agents like ascorbic acid; both acid compounds are used here to treat fresh-cut apples. Ascorbic acid does not inhibit PPO directly but acts as a reducing compound and reduces the orthoquinones to dehydroxyphenols. As the concentration of ascorbic acid decreases, the quinone concentration increases and causes the formation of the brown pigments (Lozano 2006).

As can be seen in Figures 28.1 and 28.2, the decrease of L^* values (darker), and increase of a^* and BR values during storage, clearly represented browning development in fresh-cut apples. Piagentini et al. (2008) also found that browning development in fresh-cut apples from five different cultivars was represented by L^* and h_{ab} decrease, and a^*, b^*, and C^*_{ab} increase. Browning development occurred in fresh-cut apple samples during refrigerated storage, but it was significantly lower for samples treated with ascorbic plus citric acid. Treated samples (TQ) showed higher L^* and lower a^* and BR values than nontreated samples (STQ) for both apple cultivars (Figures 28.1 and 28.2).

By comparing both apple cultivars, *Granny Smith* samples showed lower a^* and BR values, and higher L^* values than *Princesa* samples. Other authors also found differences in susceptibility to browning development among

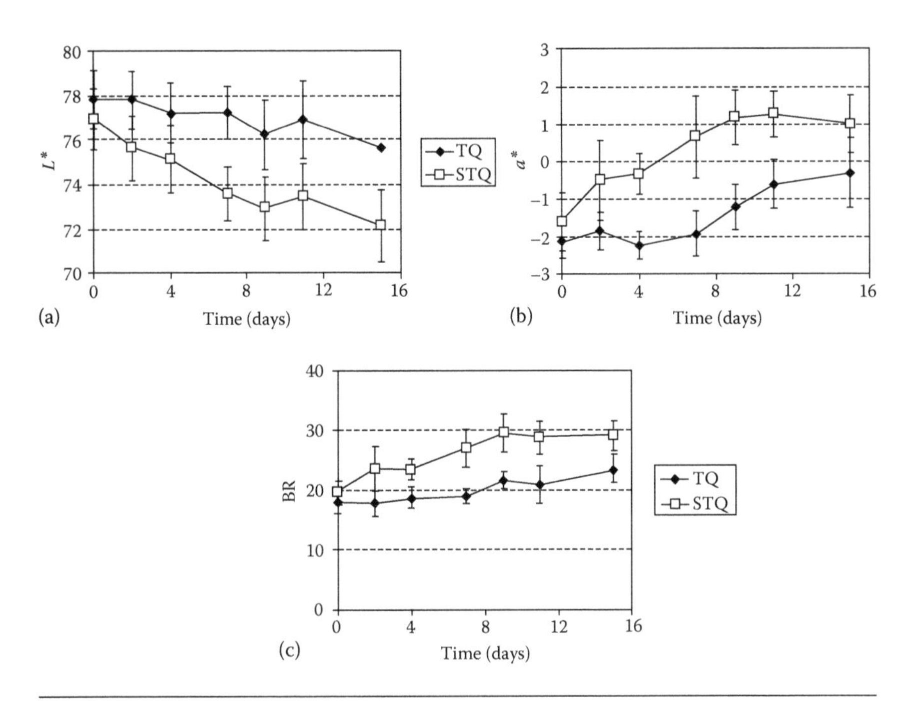

Figure 28.1 Color changes of fresh-cut *Granny Smith* apples: (a) L^*, (b) a^*, and (c) BR.

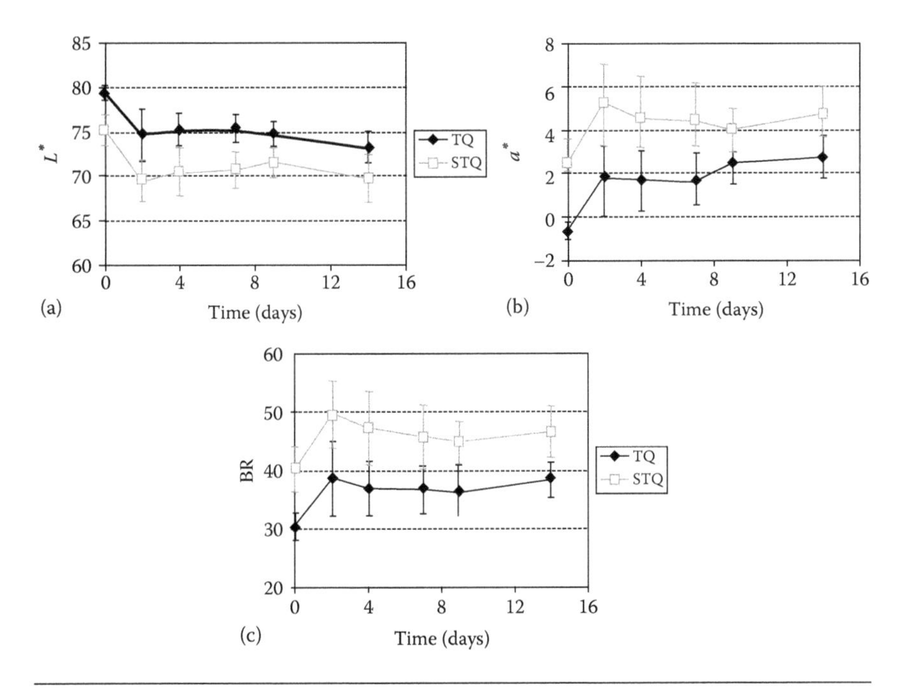

Figure 28.2 Color changes of fresh-cut *Princesa* apples: (a) L^*, (b) a^*, and (c) BR.

different apple cultivars. For example, it was reported that *Eva*, *Granny Smith*, and *Princesa* varieties presented higher values of L^* and h_{ab}, and lower a^*, b^*, and C^*_{ab} values than *Caricia* and *Red Delicious* cultivars 90 min after being cut (Piagentini et al. 2008).

The lowest changes in L^*, a^*, and BR during storage were found for treated fresh-cut *Granny Smith* apples (Figure 28.1). After 14–15 days of refrigerated storage, $\square E^*_{ab}$ was about 3.5 and 8 for TQ and STQ *Granny Smith* fresh-cut apples, respectively, showing the efficacy of the chemical treatment used as Limbo and Piergiovanni (2006) indicated that a value of $6 < \Delta E^*_{ab} < 12$ showed strong color differences. Sapers and Zoilkowski (1987) also found that ascorbate was effective in preventing surface browning in *Winesap* and *Red Delicious* apple plugs stored 24 h, and citric acid enhanced their effectiveness. It was also found that *Fuji* apple slices treated with 2% ascorbate had no browning or loss of visual quality for up to 15 days when stored at 10°C (Gil et al. 1998).

28.3.2 Color Changes of Fresh-Cut Peaches from Different Cultivars

Cultivar selection is probably the most important consideration in fresh-cut fruit processing because cultivars can vary greatly in characteristics such as flesh texture, skin color, and browning potential (Gorny et al. 1999, Gil et al. 2006, Piagentini et al. 2008).

On the other hand, it is known that refrigeration retards browning. In a broad sense, for every 10°C reduction in temperature, the rate of enzyme-catalyzed reactions decreases by half or a third. This effect was attributable to a decrease in both mobility and "effective collisions" necessary for the formation of enzyme-substrate complexes and their products. Cold preservation and storage during distribution and retailing are essential for the prevention of browning in cut fruit, since refrigerated temperatures are effective in lowering PPO activity (Beaulieu and Gorny 2002, Lozano 2006).

In the case of fresh-cut peaches, all color parameters changed significantly with cultivar, and storage time and temperature. After 4 days of storage at 15°C, fresh-cut *Early Grande* peaches showed the greatest color changes ($\Delta E^*_{ab} = 32$), followed by *Flordaking* and *Hermosillo* samples ($\Delta E^*_{ab} = 10$), and finally by *Tropic Snow* samples ($\Delta E^*_{ab} = 5$) (Figure 28.3). Total color differences were mainly due to decrease in L^*, a^*, and b^* for *Early Grande* samples, but for the other yellow pulp cultivars (*Flordaking* and *Hermosillo*), color differences were found mainly due to decrease in L^* and increase in a^*. For the white pulp cultivar, $\square E^*_{ab}$ was mainly due to increase in a^*. No color differences were found between samples stored at 1.5°C and 5°C for all peach cultivars. However, L^* and h_{ab} values of *Early Grande*, *Flordaking*, and *Hermosillo* samples decreased with higher storage temperatures (Figure 28.4). Fresh-cut *Tropic Snow* peaches, the white pulp cultivar, showed an increase in a^* values at the higher temperatures.

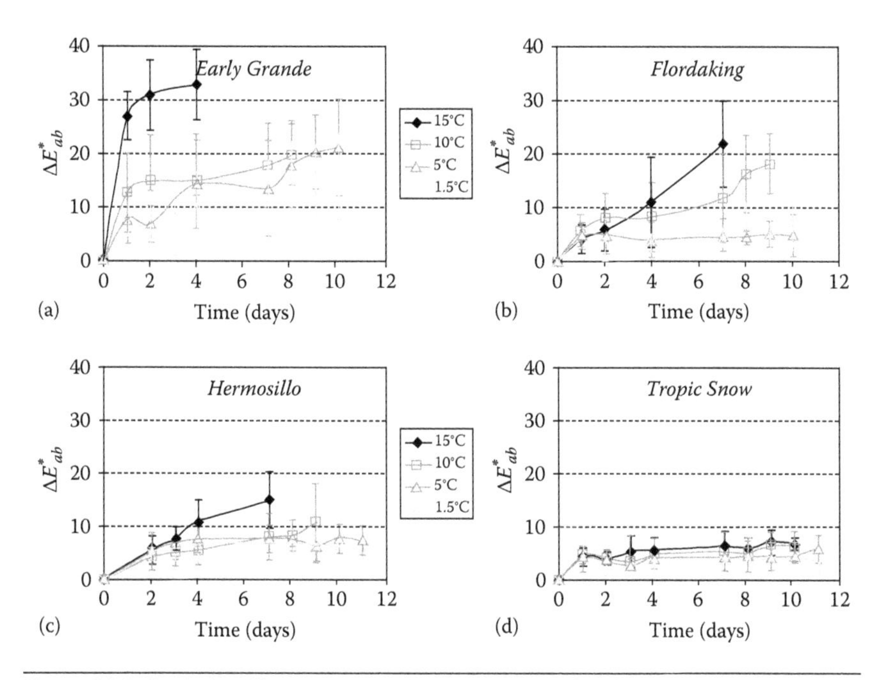

Figure 28.3 Total color difference changes of fresh-cut peaches from four cultivars: (a) *Early Grande*, (b) *Flordaking*, (c) *Hermosillo*, and (d) *Tropic Snow*.

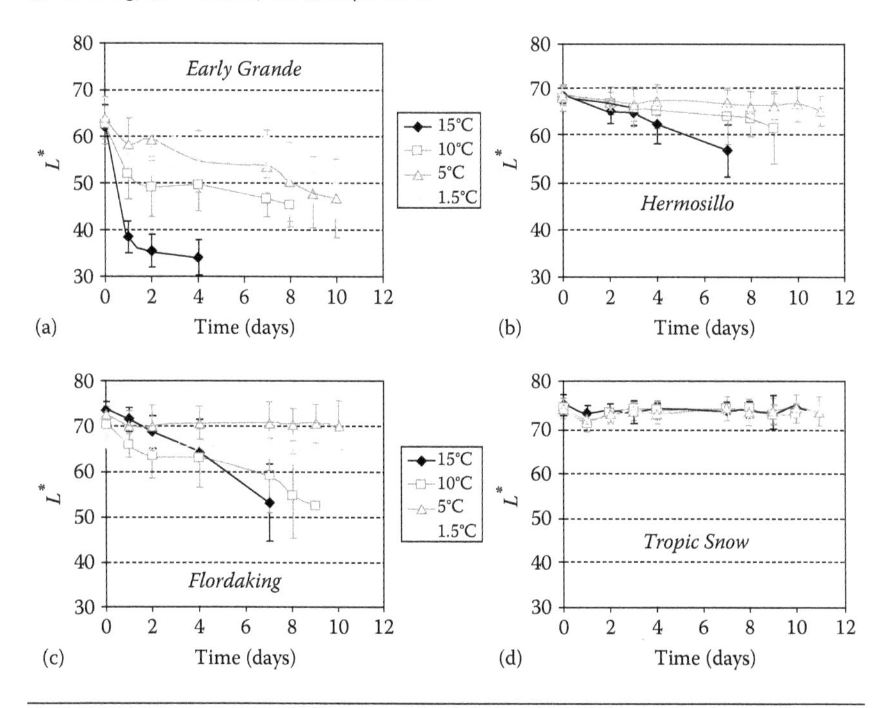

Figure 28.4 L^* value changes of fresh-cut peaches from four cultivars: (a) *Early Grande*, (b) *Hermosillo*, (c) *Flordaking*, and (d) *Tropic Snow*.

Finally, it was found that *Hermosillo* (yellow pulp cultivar) and *Tropic Snow* (white pulp cultivar) fresh-cut peaches stored at 1.5°C or 5°C showed the lowest browning development, in agreement with Gorny et al. (1999). They determined that white-fleshed peaches had a comparable shelf life to yellow-fleshed cultivars, with similar browning characteristics at 0°C. These authors reported that selection of appropriate cultivars and proper storage temperature were two of the most important factors that determine the shelf life of fresh-cut peach slices.

Acknowledgments

This study was partly supported by CAI+D of Universidad Nacional del Litoral (Argentina). We gratefully acknowledge the technical assistance of Maillén Seipel and Cecilia Depetris.

References

Beaulieu, J. C. and J. R. Gorny. 2002. Fresh-cut fruits. In *The Commercial Storage of Fruits, Vegetables, and Florist and Nursery Stocks. USDA Agricultural Handbook, Number 66*, eds. K. Gross, C. Y. Wang, and M. E. Saltveit. http://www.ba.ars.usda.gov/hb66/146freshcutfruits.pdf

Clydesdale, F. M. 1998. Color: Origin, stability, measurement, and quality. In *Food Storage Stability*, eds. I. Taub and R. P. Singh. Boca Raton, FL: CRC Press, pp. 175–190.

Gil, M. I., E. Aguayo, and A. A. Kader. 2006. Quality changes and nutrient retention in fresh-cut versus whole fruits during storage. *Journal of Agricultural and Food Chemistry* 54: 4284–4296.

Gil, M. I., J. R. Gorny, and A. A. Kader. 1998. Responses of *Fuji* apple slices to ascorbic acid treatments and low-oxygen atmospheres. *HortScience* 33: 305–309.

Gorny J. R., B. Hess-Pierce, and A. A. Kader. 1999. Quality changes in fresh-cut peach and nectarine slices as affected by cultivar, storage atmosphere and chemical treatments. *Journal of Food Science* 65: 541–544.

Limbo, S. and L. Piergiovanni. 2006. Shelf life of minimally processed potatoes. Part 1. Effects of high oxygen partial pressures in combination with ascorbic and citric acids on enzymatic browning. *Postharvest Biology and Technology* 39: 254–264.

Lozano, J. E. 2006. *Fruit Manufacturing. Scientific Basis, Engineering Properties, and Deteriorative Reactions of Technological Importance*. New York: Springer.

Matiacevich, S. B. and M. P. Buera. 2006. A critical evaluation of fluorescence as a potential marker for the Maillard reaction. *Food Chemistry* 95: 423–430.

Piagentini, A. M., M. E. Pirovani, and D. R. Güemes. 2008. Evaluación del color para determinar la susceptibilidad al pardeamiento enzimático de diferentes variedades de manzanas. In *ArgenColor 2008, Noveno Congreso Argentino del Color*. Santa Fe.

Sapers, G. M. and M. A. Zoilkowski. 1987. Comparison of erythorbic and ascorbic acids as inhibitors of enzymatic browning in apple. *Journal of Agricultural and Food Chemistry* 52: 1732–1747.

Toivonen, P. M. A. and D. A. Brummell. 2008. Biochemical bases of appearance and texture changes in fresh-cut fruit and vegetables. *Postharvest Biology and Technology* 48: 1–14.

CHAPTER 29

Cross-Sectional Color Evaluation in Borage Stems

GUILLERMO ALCUSÓN, ANA MARÍA RUIZ DE CASTRO, MARÍA CONCEPCIÓN URZOLA, ROSA ORIA, and ÁNGEL I. NEGUERUELA

Contents

29.1 Introduction

Measuring the physical properties of fruit and food has always been crucial and increasingly concerns the food industry. Food technologists as well as agricultural engineers are interested in physical properties of food materials in order to determinate how foods or fruits will be handled during processing. To get an indication of the products' quality and to understand the consumers' preference for certain foods, color is an important quality factor, and it has been widely studied (Francis and Clydesdale 1975).

Color is used in the agricultural industry: first, color is an indicator of the product quality; second, color measurement is used as an indicator in the development of storage techniques; and finally, the most important use of color is to provide a buying criterion in the purchase of raw materials and/or products (Abdullah et al. 2004). For consumers, the color and general appearance are the most important attributes when they are selecting a product since the first evaluation is visual. Color is, therefore, of great commercial importance, and its study is very important to evaluate the quality of the products.

In recent years, consumption of fresh-cut vegetable products has been increasing in developed countries because their appearance and nutritional value are very similar to the fresh product (Rico et al. 2007). Damage and alteration of the color are some crucial factors in the evaluation of the quality of fruit that have been studied along time for various fruits, like apples. The first method of study to evaluate the damage and defects was based on the use of interferential filters, then color estimation and more recently the use of techniques that combine infrared and visible spectrum, and the incorporation of hyperspectral images.

One of the main causes of disruption of minimally processed fruits and vegetables is the enzymatic browning, which modifies the color on the cut surface.

Thus, the conservation of a product color is of primary concern as it is not only associated with its quality but also with its correct technological processing. It should also be pointed out that color is used as an indicator of defects in the product. These defects are mainly caused by polyphenol oxidase (PPO) that is found in most fruits and vegetables (Galeazzi and Sgarbieri 1981, Cano et al. 1990, López et al. 1994, Martínez and Whitaker 1995, Whitaker 1995, Aydemir 2004). Its reactions can change flavor, texture, and nutritional value besides color, one of the sensory characteristics most valued by consumers. PPO catalyzes the hydroxylation of monophenols and oxidation of *o*-diphenols to quinones, which, in the presence of oxygen, are polymerizable to form melanins (brown compounds) (Espín et al. 1998).

Owing to its gastronomic and nutritional qualities, borage (*Borago officinalis*) is a very popular vegetable in the Ebro Valley (Spain), through its market is not as widespread as should be expected. Borage is a vegetable that is mainly commercialized fresh but may also be presented in "fourth range" or fresh cut, cut and packaged in expanded polystyrene trays covered with stretch PVC films. However, in the stages of cutting and washing during the process of borage stems, physical damage at the surface takes place that allows the reactions of enzymatic browning.

The process of cutting produces cell rupture allowing the polyphenols to come into contact with enzymes (mainly PPO), which leads to the browning. This browning modifies the color of the borage, thus affecting its quality and, therefore, its shelf life. This enzymatic browning activity can be reduced by the application of cold, by modified atmosphere conservation, and by the use of antioxidant substances, as ascorbic acid (AA), EDTA, citric acid, cysteine, and their combination.

In this chapter, we have studied the use of two knives (stainless-steel knife for vegetables and ceramic knife) and the AA application in the ceramic knife to retard the enzymatic browning on the cut of borage surface stems. Besides, we present the application of technologies of scanning and digital image processing in obtaining information regarding color changes in all the causes that have been studied.

29.2 Material and Methods

The borage used in this study was harvested in optimum condition in greenhouses in Zaragoza (in the Ebro Valley, Spain). It was transported and stored in refrigeration until it was analyzed.

29.2.1 Enzyme Assays

Prior to the analysis of the cut, we have studied the enzyme activity of PPO from borage stems and the effect of AA applied at different percentages to test their effectiveness.

Extraction of PPO was determined according to the modified method of Galeazzi and Sgarbieri (1981). Borage stems (20 g) were homogenized in sodium phosphate buffer at physiological product pH (6.4), containing polyvinylpyrrolidone and Triton X-100. The homogenate was centrifuged at 4000 rpm for 30 min at 4°C. The supernatant was filtered with Watman no. 4, obtaining the enzyme solution. AA concentrations were added to the enzyme extract from a 40% solution (w:v).

The PPO activity was tested according to the method of López et al. (1994). The assay medium consisted of sodium phosphate buffer of pH 6.4 containing 40 mM pyrocatechol. The reaction was started by adding to 900 μL of the assay mixture and 100 μL of the enzyme sample. Its activity was determined by spectrophotometry at 400 nm at 25°C. One unit of enzymatic activity is defined as an absorption increase of 0.1 min^{-1} (Flurkey and Jen 1978).

29.2.2 Study of Borage Stem Cuttings

To perform the study of color evolution, a series of stalks were selected as homogeneous as possible. These were cut in 1 cm thick pieces. The knives used were a Granton® stainless-steel knife for vegetables and a Kyocera® ceramic knife. Twenty pieces were cut with each of the knives, and the browning evolution of each cut was observed for 8 h. All the tests were performed in triplicate.

For image acquisition, an HP G4010 Scanjet scanner (1200 ppi resolution, reproduction in millions of colors) was used, and the images were saved in TIFF format. Digital image processing was performed using Matrox Inspector 8.0® software.

In order to obtain the CIELAB coordinates, a color calibration was performed of the images obtained with the HP scanner/Matrox combination using the 300 NCS color samples which make up the UNE 48-103-94 Spanish color norm. The means of the R, G, and B coordinates were obtained for each color sample, and a square multiple regression was applied between the L^*, a^*, and b^* (CIELAB) coordinates of the UNE norm and the corresponding R, G, and B values and their products R^2, G^2, B^2, RB, RG, and GB of the measured color samples so as to obtain the corresponding transformation functions (Martínez-Verdú 2001).

In order to study the evolution of browning of each of the borage stalk cuts, 3×3 pixels ROIs (region of interests) were used in the three most affected areas. The mean pixel of each ROI and the corresponding R, G, and B coordinates were obtained followed by the means and the standard deviations. Applying the transformation functions obtained to these values, the corresponding L^*, a^*, b^*, C^*_{ab}, and h_{ab} coordinates were calculated.

To quantify the browning area, the obtained images were previously processed in Photoshop 6.0®, eliminating the characteristic hair of the borage from each image as this provokes errors in the analysis. In addition, the background color of the image was changed to black ($R=0$, $G=0$, $B=0$) so that there would be no interference. Once the image had been modified, the remaining digital image processing was performed using Matrox Inspector 8.0® software. First, the image was calibrated to express the results in millimeters, and the cut area was delimited as an ROI for which the total area was calculated. From this image, eliminating the green band, another image was obtained, corresponding to the nongreen (browning) zone, whose area was calculated.

29.3 Results

29.3.1 Enzyme Assays

AA prevents enzymatic browning by chemically reducing the *o*-quinones to colorless diphenols (McEvily et al. 1992). In this chapter, the enzyme activity is reduced, but it was not statistically significant for the use of percentages of AA less than 2% (Table 29.1). For 2%, reduction in enzyme activity was significant, but at higher concentrations, results are not statistically significant compared with the results obtained for 2%.

In other work on fruits and vegetables, the use of AA has been found to be effective in reducing the PPO activity in stored artichokes (Lattanzio et al. 1989), sugarcane juice (Mao et al. 2007), peach slices (Li-Qin et al. 2009), and fresh-cut apple (Janga and Moon 2011).

29.3.2 Study of Borage Stem Cuttings

The evolution of color over time (in the selected ROIs) may be seen in the variation of the h_{ab} coordinate (Figure 29.1). It can be observed that the cut

Table 29.1 PPO Activity (UA/g) with Different Percentages of AA

% AA	0%	0.5%	1%	1.5%	2%	3%	4%
PPO activity (UA/g)	57.7 ± 13.6	40.9 ± 14.6	35.6 ± 7.0	36.9 ± 7.6	28.2 ± 10.1	28.8 ± 3.9	24.9 ± 8.9

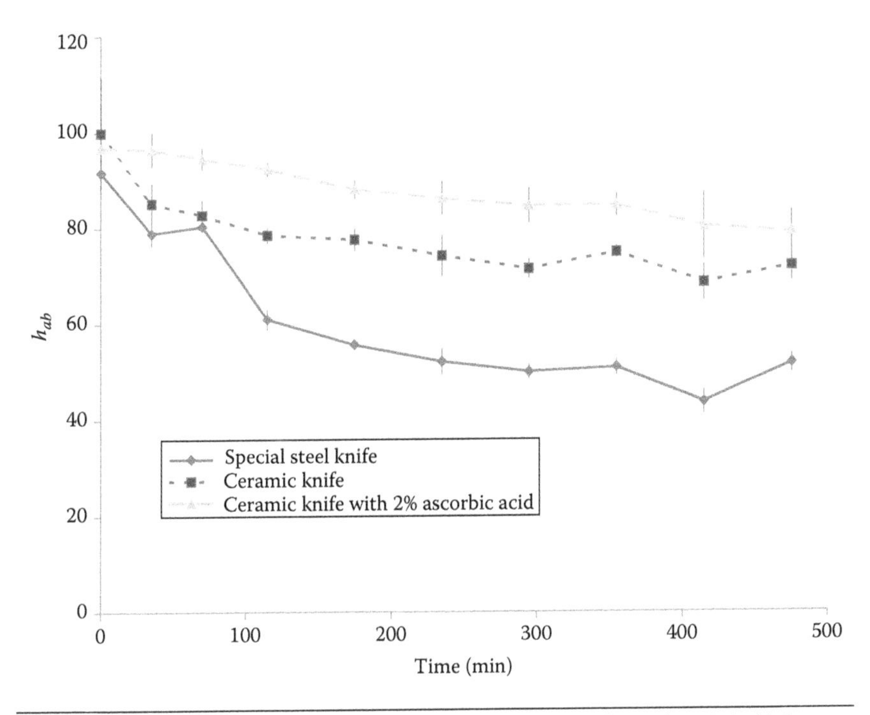

Figure 29.1 Evolution of the h_{ab} coordinate in selected areas of cuts of ceramic and steel knives.

made with the ceramic knife and the cut made with ceramic knife with 2% AA maintain the yellow-green color typical of borage better than the stainless-steel knife that produces a color evolution toward more orange hues (lower h_{ab} values).

In Figures 29.2 and 29.3, we show some examples of the evolution, over the time of the test, of a stalk cut with each knife. First, the image presents small pink areas, which correspond to the browning areas. In time, the pink area increases. It should be pointed out that the borage stalks lost water over the time of test that caused a surface area decrease.

As it can be seen in the images, the browning area of the borage cut with the ceramic knife (Figure 29.2) is smaller at the end of the test than that corresponding to the special stainless-steel knife for vegetables (Figure 29.3).

In Figure 29.4, we can see the evolution of the percentage of the browning area with regard to the total area of the cut, as well as the standard deviation for each moment of the test, which is lower for the ceramic knife. One explanation for this may be that the ceramic knife cut is finer and causes less damage on the cut area than the stainless-steel knife. Besides, results obtained by the use of ceramic knife with 2% AA are better than the results obtained up to 3 h after cutting borage stems. For later times, all of its mean values are lower than the other knives, but the standard deviation is higher. This is because the

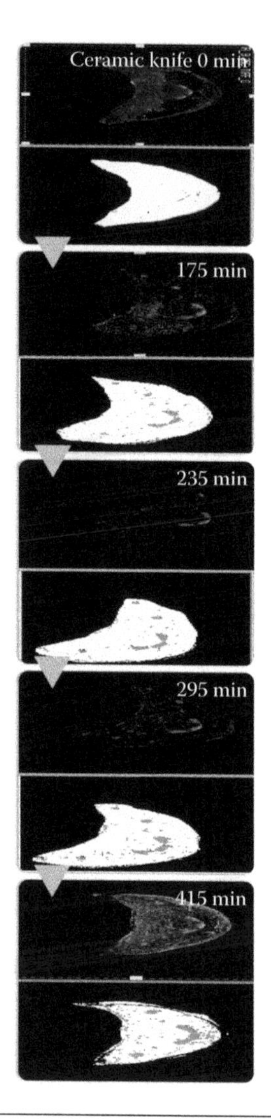

Figure 29.2 Borage stems' image cuts with ceramic knife.

AA is consumed and the browning enzymatic reaction is active again generating pigmented compounds.

29.4 Conclusion

The use of a scanner and a digital image-processing program permits the study of the browning evolution of borage and may be extended to similar vegetable products.

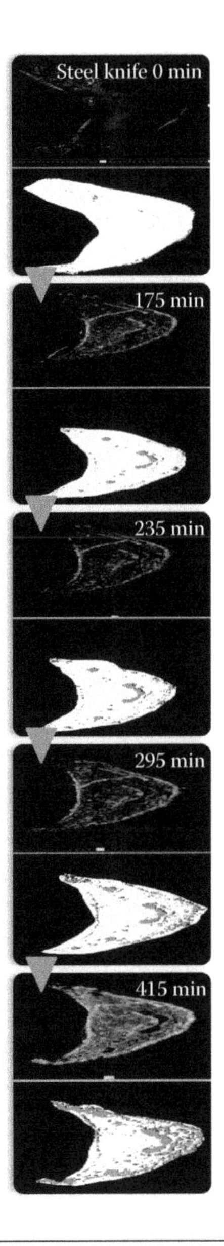

Figure 29.3 Borage stems' image cuts with steel knife.

The study of the browning of the cut shows that both the color change (measured as the h_{ab} coordinate variation) and the affected area depend on the type of knife chosen for cutting the borage. From this, we may conclude that for vegetables, the ceramic knife presents advantages when cutting the stalks since it causes less browning.

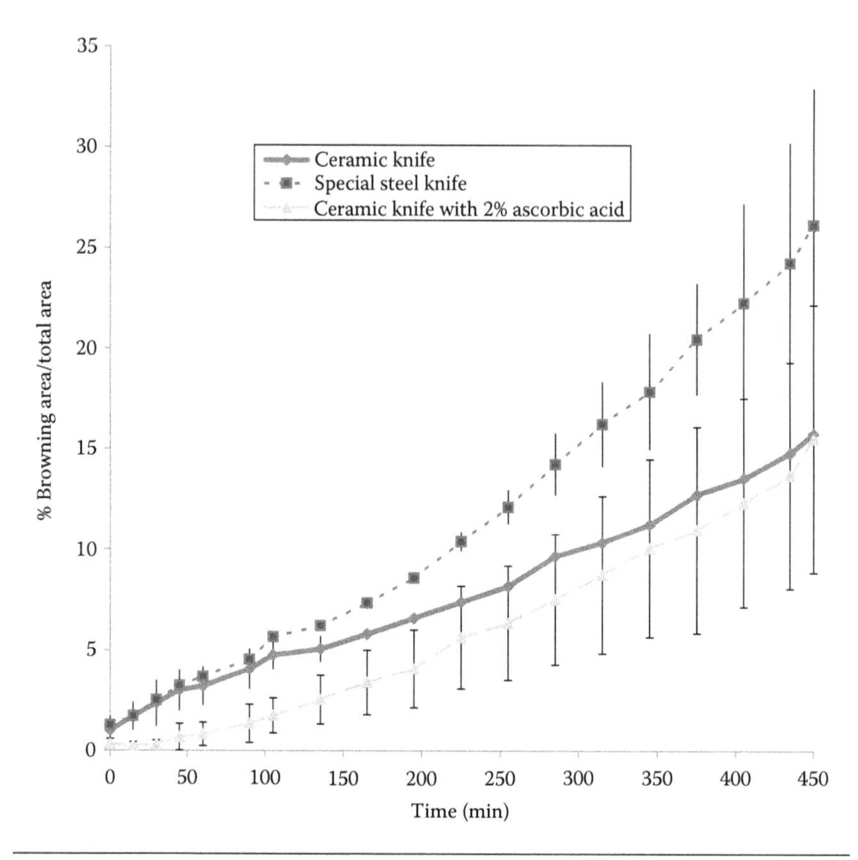

Figure 29.4 Evolution of % browning area of cuts of steel knife, ceramic knife, and ceramic knife with 2% AA.

Furthermore, the use of AA reduces the enzymatic browning and can maintain the overall quality and color for 3 h after cutting borage stems. This time is sufficient for modified atmosphere packaging to be completed.

References

Abdullah, M. Z., L. C. Guan, K. C. Lim, and A. A. Karim. 2004. The applications of computer vision system and tomographic radar imaging for assessing physical properties of food. *Journal of Food Engineering* 61: 125–135.

Aydemir, T. 2004. Partial purification and characterization of polyphenol oxidase from artichoke. *Food Chemistry* 87: 59–67.

Cano, M. P., M. A. Marín, and C. Fúster. 1990. Effects of some treatments on polyphenoloxidase and peroxidase activity of banana. *Journal of Agricultural and Food Chemistry* 51: 223–231.

Espín, J., P. García-Ruiz, and J. Tudela. 1998. Monophenolase and diphenolase reaction mechanisms of apple and pear polyphenol oxidases. *Journal of Agricultural and Food Chemistry* 46: 2968–2975.

Flurkey, W. H. and J. J. Jen. 1978. Peroxidase and polyphenol oxidase activities in developing peaches. *Journal of Food Science* 43: 1826–1831.
Francis, F. J. and F. M. Clydesdale. 1975. *Food Colorimetry: Theory and Applications.* Westport, CN: Avi Publishing.
Galeazzi, M. and V. C. Sgarbieri. 1981. Substrate specificity and inhibition of polyphenoloxidase from a dwarf variety of banana. *Journal of Food Science* 46: 1404–1406.
Janga, J. H. and K. D. Moon. 2011. Inhibition of polyphenol oxidase and peroxidase activities on fresh-cut apple by simultaneous treatment of ultrasound and ascorbic acid. *Food Chemistry* 124 (2): 444–449.
Lattanzio, V., V. Linsalata, S. Palmieri, and C. F. Van Sumere. 1989. The beneficial effect of citric and ascorbic acid on the phenolic browning reaction in stored artichoke (*Cynara scolymus* L.) heads. *Food Chemistry* 33 (2): 93–106.
Li-Qin, Z., Z. Jie, Z. Shu-Hua, and G. Lai-Hui. 2009. Inhibition of browning on the surface of peach slices by short-term exposure to nitric oxide and ascorbic acid. *Food Chemistry* 14 (1): 174–179.
López, P., F. J. Sala, J. L. De la Fuente, S. Condón, J. Raso, and J. Burgos. 1994. Inactivation of peroxidase, lipoxygenase and polyphenol oxidase by manothermosonication. *Journal of Agricultural and Food Chemistry* 42: 252–256.
Mao, L. C., Y. Q. Xua, and F. Quea. 2007. Maintaining the quality of sugarcane juice with blanching and ascorbic acid. *Food Chemistry* 104 (2): 740–745.
Martínez, V. and J. R. Whitaker. 1995. The biochemistry and control of enzymatic browning. *Trends in Food Science & Technology* 6: 195–200.
Martínez-Verdú, F. 2001. Chromatic subjects about image capture (II). *Optica Pura y Aplicada* 34: 1–16.
McEvily, A., R. Iyengar, and S. Otwell. 1992. Inhibition of enzymatic browning in foods and beverages, critical review. *Food Science and Nutrition* 32 (3): 253–273.
Rico, D., A. Marindiana, J. Barat, and C. Barry Ryan. 2007. Extending and measuring the quality of fresh-cut fruit and vegetables: A review. *Trends in Food Science & Technology* 18 (7): 373–386.
Whitaker, J. R. 1995. Polyphenol oxidase. In *Food Enzymes—Structure and Mechanism*, ed. D. W. S. Wong. New York: Chapman & Hall, pp. 271–307.

PART IV

Color as an Index of Food Composition and Properties

CHAPTER 30

Spectral Signatures

A Way to Identify Species and Conditions of Vegetables

JOSÉ D. SANDOVAL, SERGIO R. GOR, JACQUELINE RAMALLO, ANA SFER, ELISA COLOMBO, MERITXELL VILASECA, and JAUME PUJOL

Contents

30.1 Introduction

When optical radiation reaches the surface of any of the numerous components of the environment, it is subject to one or more of the following processes: it can be reflected, transmitted, or absorbed, according to the energy conservation laws. The characteristics and intensity of this behavior depend on the material and the quality of the surface that the radiation is impinging on. The particular combination of elements making up the material stuff, their proportions, quantity, size, and form will determine the characteristics of the interaction, setting which aspects of the incident radiation will be modified and in what extent. The energy of the electromagnetic wave is related to its wavelength in such a way that the shorter the wavelength, the more energy a given wave contains. When this energy reaches the surface of a body, it is either reflected from, absorbed by, or transmitted by it, as stated earlier. The degree and intensity of each process is determined by the wavelength and the physical and chemical properties of that body (Scotford and Miller 2005). In turn, the spectral reflectance in plants is influenced, besides the absorption of their elements, by the structure of the surface and the cells in the leaves (Zwiggelaar 1998). Leaf optical properties are a function of leaf components and structure, water content, and the concentration of biochemicals (Asner 1998). According to numerous research studies and scientists, the spectral reflectance in the visible (VIS) and near-infrared (NIR) regions is a powerful and useful tool to evaluate properties and situations of plants and crops. Consequently, most agricultural studies use measurements in the VIS (400–700 nm) and NIR (700–2500 nm) regions of the electromagnetic spectrum. Many studies performed in the last few decades sustain that optical properties in these regions can potentially detect physiological and biological functions of plants and crops, offering potential information for applications in agriculture (Scotford and Miller 2005). Some researchers proposed a set of wavelengths in which the reflectance values are capable to offer much valuable information about the status and functionality of the plant (Gausman and Allen 1973). Particularly, they pointed out to 550 nm (green reflectance peak), 650 nm (chlorophyll absorption band), 850 nm (on infrared reflectance plateau), 1450 nm (water absorption band), 1650 nm (reflectance peak following water absorption band at 1450 nm), 1950 nm (water absorption band), and 2200 nm (reflectance peak following water absorption band at 1950 nm). All of those research works suggest that if the body under consideration is a plant and the characteristics of one or more of these processes can be measured, from the data we could infer useful information about conditions and functionality of the plant. To start with this proposal, we try to discriminate and recognize different plant species by means of the analysis of their reflectance or absorption functions. Among the characteristics of the interaction determined by the matter structure, we are particularly concerned in reflection and absorption. Those, expressed by means of spectral reflectance or absorption functions of materials, especially of vegetables and named here as "spectral signatures," allow us to obtain information about the constitution and condition of the material analyzed: measuring the spectral signature with enough precision will allow, under specific conditions and by means of an adequate treatment of data, identifying not only the species to which the signature corresponds to but also its phenology and nutritional condition as well as the presence or absence of diseases, affections, and scarcities of the plant from which the sample comes from.

30.2 Materials and Methods

In order to compare and identify the species a leaf belongs to, several samples of leaves of different plants were collected, and their spectral reflectance and absorption (spectral signatures) were measured. The species considered in this study were two of the most common varieties of sugarcane cultivated in Tucuman, Argentine, named TUCCP 77-42 and LCP 85-384 (labeled "742" or "S" and "384" or "T," respectively), and four types of citrus: tangerine, grapefruit, orange, and lemon.

In the case of sugarcane, about 240 samples were measured, half of each variety, including two forms to prepare the samples to be measured: finely minced and coarsely chopped. In the case of citrus, about 160 samples were measured, both sides of leaves (front and back) being measured. Sample subsets were collected every fortnight during 4 months and measured within the following 24 h. Spectral signatures were measured between 400 and 2500 nm, at 2 nm intervals, by means of a scanning near-infrared spectrophotometer (Foss NIRSystems 6500) with 0°/45° measuring geometry (Foss NIRSystems, Silver Spring, MD). The spectral data obtained were processed and statistically analyzed with STATA 11.0, applying principal component analysis (PCA) to suitably grouped subsets of data.

30.3 Results

30.3.1 Measurements

Some spectral signatures for sugarcane samples, in this case in terms of spectral absorption (strictly speaking, log $1/R$), are shown in Figure 30.1. They are plotted between 400 and 2498 nm for the two varieties of sugarcane analyzed.

Visual inspection of the figure indicates that it is fairly hard to discriminate if one particular sample belongs to one or the other of the varieties considered.

30.3.2 Analysis

As visual inspection of Figure 30.1 indicates, it is fairly hard to discriminate if one particular sample belongs to one or another of the considered varieties. Thus, spectral signatures should be statistically analyzed. For this purpose, PCA was used. This is a technique extremely useful to "summarize" all the information contained in the original data matrix (X-matrix) in an easily understandable form. PCA works by decomposing the X-matrix as the product of two smaller matrices, which are called loading and score matrices. The loading matrix contains information about the variables: it is composed of a few vectors (principal components, PCs) which are linear combinations of the original X-variables. The score matrix contains information about the samples. Each sample is described in terms of its projections onto the PCs instead of the original variables (Volsurf—statistic s.d.). PCA not only permits reducing the dimensionality of the problem but also provides a way to reveal the sometimes

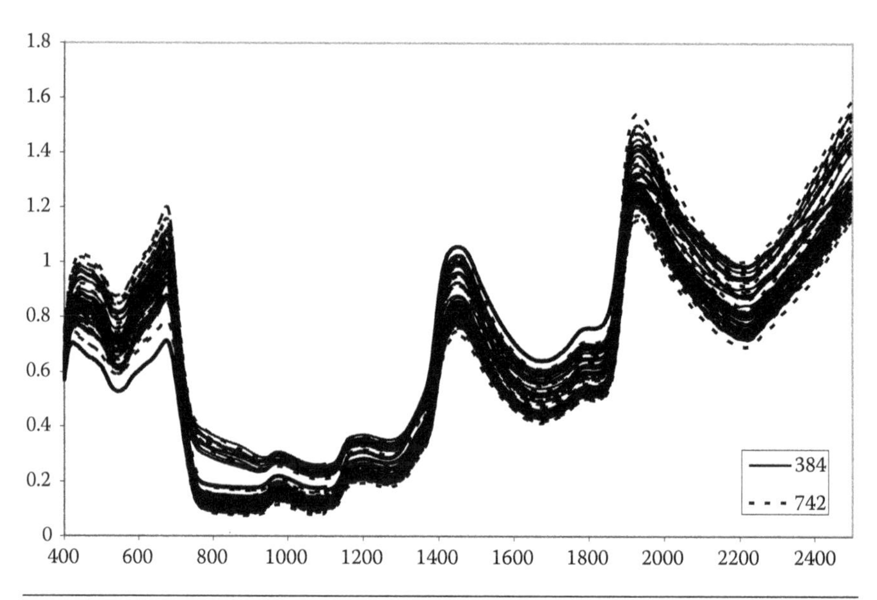

Figure 30.1 Spectral signatures (in terms of absorption) for sugarcane samples of varieties "384" and "742."

hidden structures that often underlie it, in our case, the variety to which the analyzed sample belongs to (Pla 1986, Shlens 2009). To perform this analysis, we initially applied a PCA process to the whole set of measured data. Figure 30.2 shows PCA scores for the three main principal components, corresponding to 121 measurements of "384" and 122 of "742" spectral signatures, considered between 400 and 820 nm (the range chosen to test the ability of the method is due to the limited capacity of the spreadsheet to manage large data matrices). The samples considered include sugarcane leaves, minced ("m") and chopped, the last ones measured at the face ("a") and the reverse ("b") of the leaf.

From Figure 30.2, we cannot make a decision on the group that represents each variety. In the graph, three clouds can be seen, but only the small cloud on the upper left corner is clearly differentiated from the rest of the clouds, which represents the scores for the measurements of a white reference disk included in the data analyzed. In this representation, different types of treated samples are included together: minced and chopped leaves, front and back sides. The chopped leaves are grouped in the central cloud, while the minced ones are gathered in the upper right cloud. The situation does not change substantially if we only analyze data corresponding to minced samples, as shown in Figure 30.3.

If we now work on the same type of samples (minced), but considering separately 70 measurements with data covering a spectral band from 400 to 820 nm, the results from the PCA are rather different (see Figure 30.4), being the points corresponding to the 742 variety located mainly at the right side of the graph and the 384 variety on the left. It should be noted that points

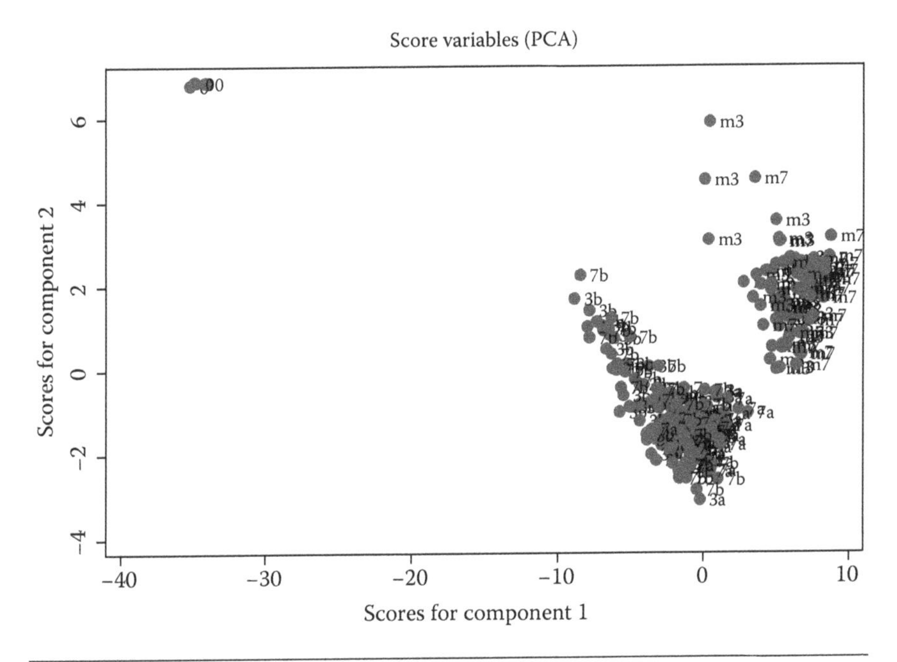

Figure 30.2 Scores for samples of two sugarcane varieties labeled 3 (from 384) and 7 (from 742) respectively; m stands for "minced," a for "face," and b for "reverse."

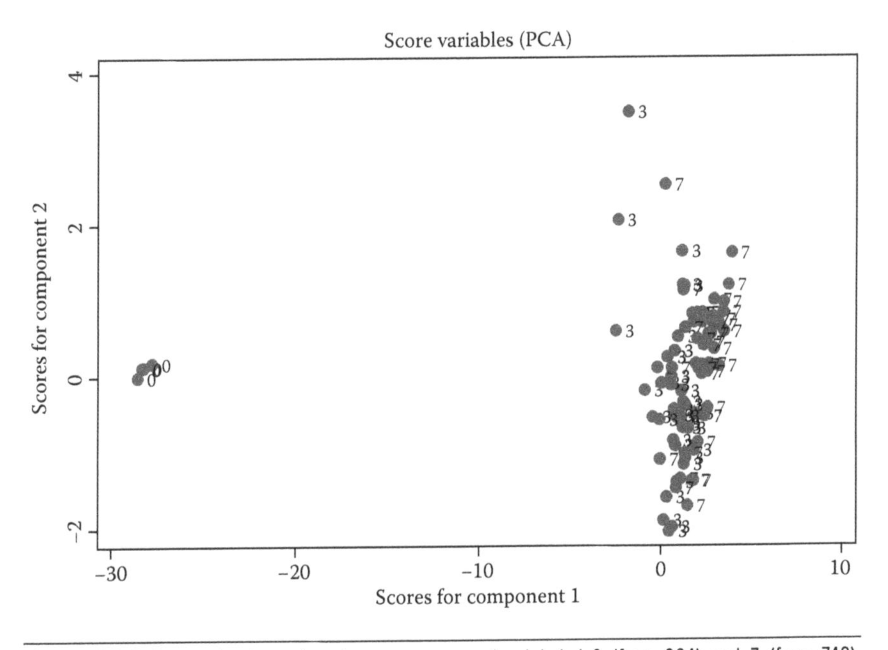

Figure 30.3 Scores for two minced sugarcane samples labeled 3 (from 384) and 7 (from 742), respectively.

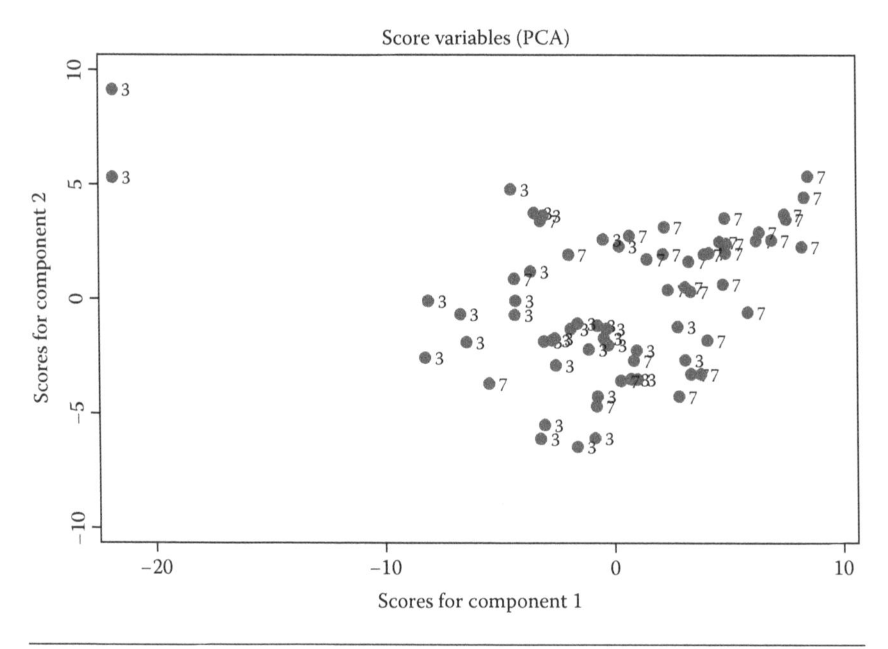

Figure 30.4 Scores for two minced sugarcane samples labeled 3 (from 384) and 7 (from 742) respectively, considering the spectral range from 400 to 820 nm.

corresponding to each variety are not completely separated, which tells us about the importance of the conditioning (characteristics, spectral range, quantity) of the samples to be treated.

We applied the method to a new set of data, checking if it was capable of identifying which variety a given sample came from. The new set is composed of 21 spectral signatures of minced sugarcane, 10 of each variety ("742" and "384") plus an additional "unknown" sample. The spectral signatures were measured from 400 to 1100 nm. In this case, the additional sample corresponded to a variety named "742." Figure 30.5 displays the resulting PCA scores for this dataset. The "unknown" sample is easily identified closest to the component 1 axis, on the left side of the graph: the side corresponding to variety "742."

An analogous procedure was applied to the comparison of four citrus samples, considered between 400 and 1100 nm. Figure 30.6 shows scores for all citrus samples without any special preparation. At the left side of Figure 30.6, the PCA scores for all the data in the complete range analyzed can be seen. From this, it is rather hard to separate and identify the type of citrus one given sample comes from. On the right side of the figure, the PCA scores for the same citrus samples are presented, but this time, the NIR range (800–1100 nm) was the spectral range analyzed. For vegetables, this range gives much information about their structure and conditions, allowing us to grouping by pairs, making it possible to separate samples of orange and tangerine (N and M) from lemon and grapefruit (L and P).

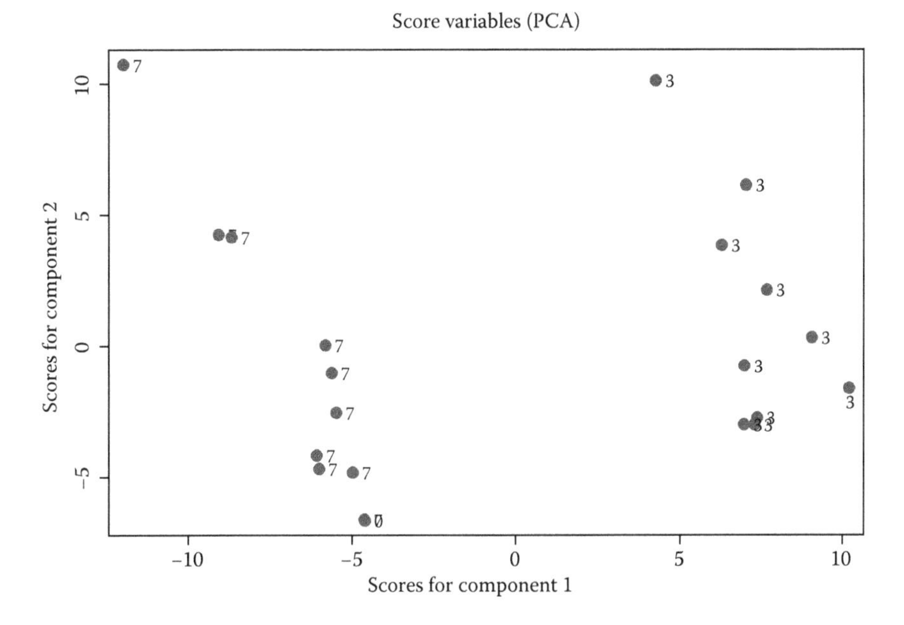

Figure 30.5 Scores for minced sugarcane samples labeled 3 (from 384) and 7 (from 742), respectively, considering the spectral range from 400 to 1100 nm.

When we applied the method to data corresponding to leaves of lemon (L) and orange (N) drawn from the previous dataset, the identification of each type was very evident, as shown in Figure 30.7.

30.3.3 Refinement

The reflectance from plants is caused by scattering from discontinuities in the refractive index of leaves. The spectral reflectance depends on the cell structure (number and orientation of air/water/cell–wall interfaces). The spectral reflectance is closely related to the spectral absorption of pigments present in the leaf, too. The most important absorption pigments and their peak wavelengths are listed in Table 30.1 (Zwiggelaar 1998).

Therefore, it would be expected for different plants (species and/or varieties) having different pigment content and/or cell structure to present differences in their spectral reflectance curves, particularly in some of those noticeable wavelengths (Kokaly et al. 2009, Sims and Gamon 2002, Zwiggelaar 1998). This suggests that the analysis could be successfully performed for a selected subset of wavelengths instead of using the complete range or a partial subset as it was previously done. To verify this hypothesis, we worked on a new set of data, now keeping only the reflectance values of the samples at certain specific wavelengths: 400, 420, 424, 434, 440, 444, 450, 470, 474, 480, 500, 528, 530, 536, 550, 570, 598, 600, 646, 650, 660, 662, 670, 672, 674, 676, 678, 680, 698, 700, 704, 720, 722, 740, 750, 800, 900, 970, and 1050 nm. These wavelengths

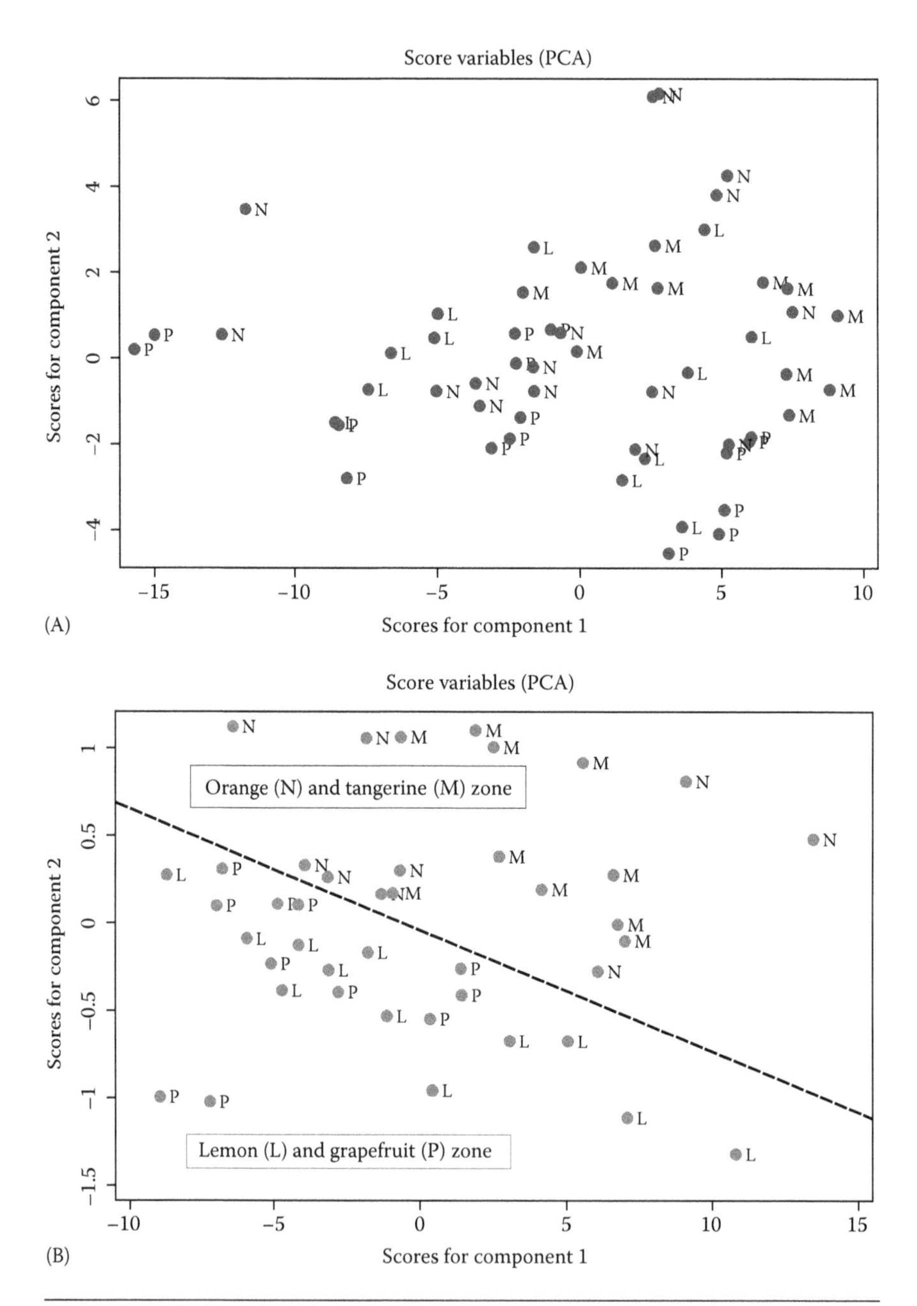

Figure 30.6 (A) Scores for four citrus samples labeled N (orange), M (tangerine), L (lemon), and P (grapefruit), respectively. Data analyzed between 400 and 1100 nm. (B) Same data considered in the NIR (800–1100 nm) range.

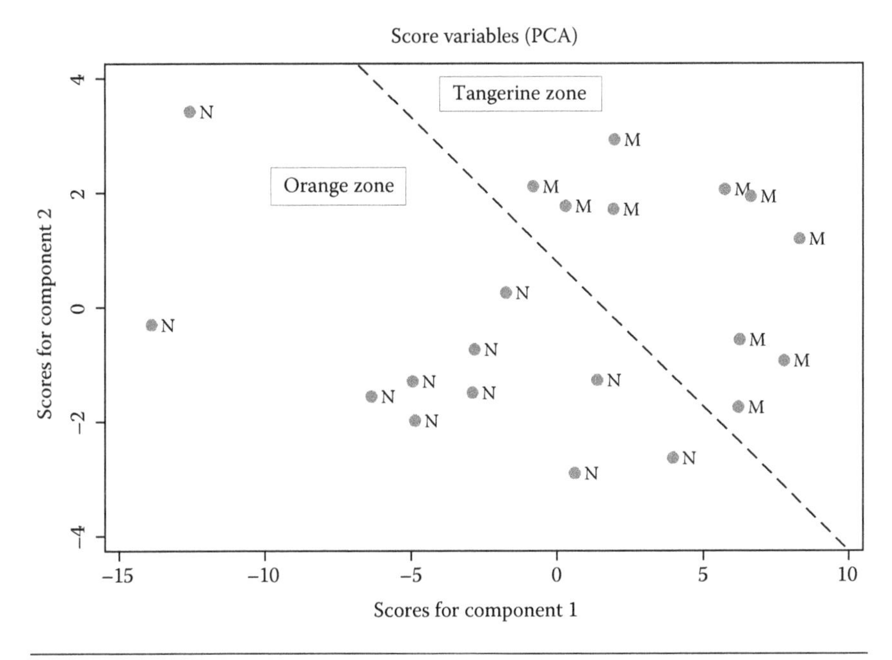

Figure 30.7 Scores for two citrus samples labeled N (orange) and M (tangerine), respectively.

Table 30.1 Plant Pigments and Characteristic Absorption Wavelengths

Plant Pigment	Important Spectral Bands and Peak Wavelengths (nm)
Chlorophyll a	435, 670–680, 740
Chlorophyll b	480, 650
α-Carotenoid	420, 440, 470
β-Carotenoid	425, 445, 475
Anthocyanin	400–550
Lutein	425, 445, 475
Water	970, 1450, 1944

were selected in order to take into account several phenomena of absorption produced by pigments, water, etc., that occur within the leaves, as explained earlier.

The PCA analysis performed on a set of data corresponding to 36 spectral reflectances of minced sugarcane samples, 19 of variety "742" (labeled S) and 17 of "384" (labeled T), gave the results showed in Figure 30.8.

It can be clearly seen that the analysis allows us separating the varieties of sugarcane by taking into account the sign of the principal component 1, which is positive for the S-labeled variety ("742") and negative for the T-labeled

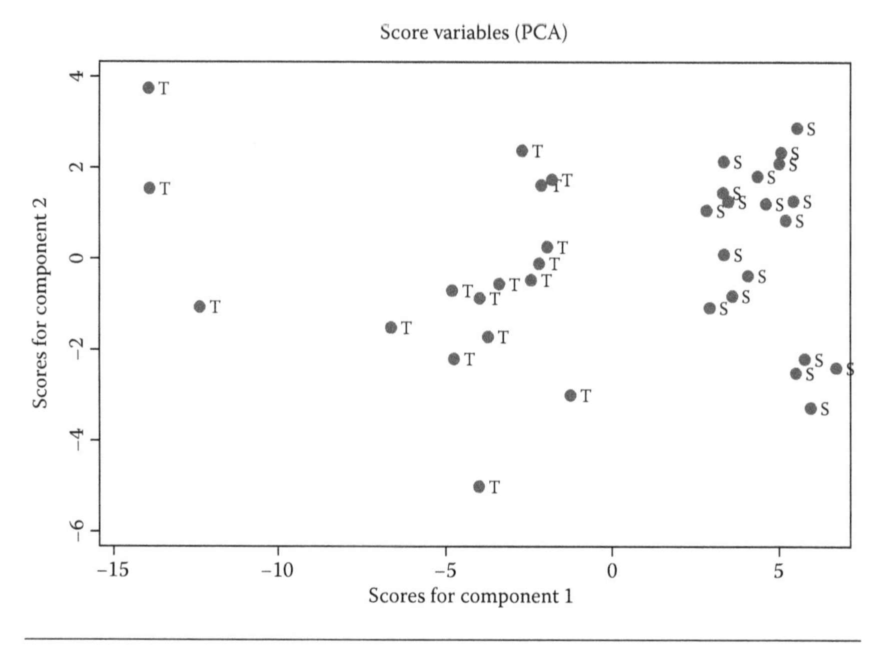

Figure 30.8 Scores for two varieties of sugarcane labeled S (from 742) and T (from 384), respectively.

variety ("384"). To improve the separation, we can apply orthogonal Varimax rotation to the multidimensional system. A Varimax rotation is a change of coordinates used in PCA that maximizes the sum of the variances of the squared loadings. This rotation is an integral part of component analysis whose objective is making the rotated components as simple as possible to interpret. A principal component is a linear function of all the p original variables. If the coefficients or loadings for a principal component are all of a similar size, or if a few are large and the remainder small, the component looks easy to interpret. If there are intermediate loadings, as well as large and small ones, the component can be more difficult to interpret. In this case, one might decide that the first m components account for most of the variation in a p-dimensional dataset. Then, it is more important to interpret simply the m-dimensional space defined by these m components than it is to interpret each individual component. To reach this objective, the axes can be rotated within this m-dimensional space in a way that simplifies the interpretation of the axes as much as possible (Jolliffe 2002). The loading plot (column vectors) gives us a summary of the variable properties, while the score plot (row vectors) gives us a summary of the relationship among the observations (or samples). Loading plot is a means to interpret the patterns seen in the score plot. Both are complementary and superimposable (Digital library s.d.) which authorize us to continue using score plots after the rotation. With the application of this rotation, the interpretation of the results of the PCA will be simplified, as can be seen in Figure 30.9 for our specific case.

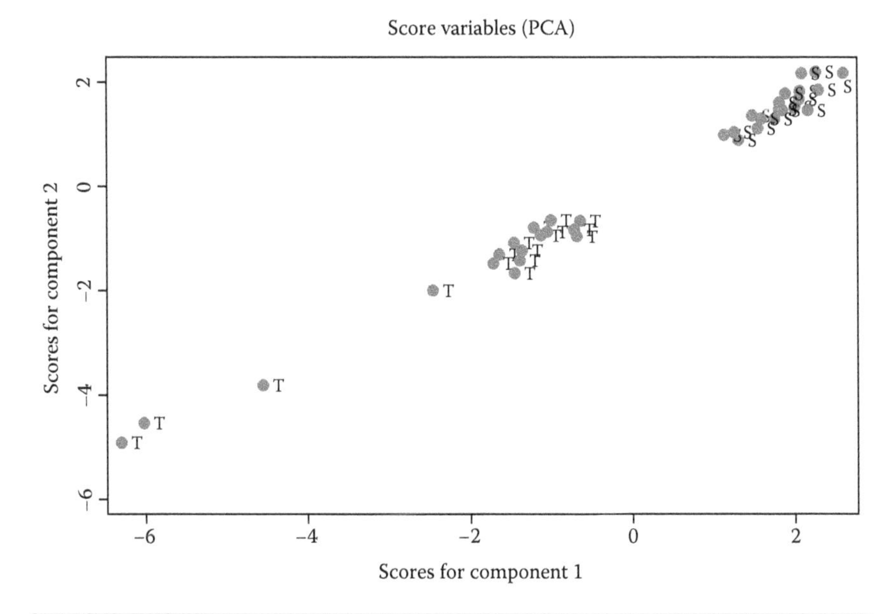

Figure 30.9 Scores for two varieties of sugarcane labeled S (from 742) and T (from 384), respectively, when the rotation is applied. Rotation: Orthogonal Varimax.

Figure 30.9 permits an easy identification of the variety each sample comes from: values of component 1 and component 2 are positive for "742" (labeled S in the graph) and negative for "384" (labeled T). Each species has different location in the component space, making the identification evident. To test the ability of the method to reliably identify the type of plant (species and/or varieties) a given leaf comes from, we consider the last set of data used as an unlabeled reference, adding two spectral reflectance curves named Xs and Xt from the "742" and "384" varieties, respectively. Applying PCA, we obtained the results shown in Figure 30.10. When the rotation is applied in this case (right plot), the two added samples can be immediately and clearly identified: Xs is within the cloud that corresponds to S-labeled samples, that is, the "742" variety, and Xt is within the T samples side, that is, it corresponds to the "384" variety.

30.4 Conclusion

Confirming the statement from Shlens (2009), that is, that the goal of PCA is to identify the most meaningful basis to reexpress a dataset, the results obtained in this work show that the careful selection and suitable preparation of samples together with the precise collection of spectral signature data and the application of an adequate statistical analysis like PCA conform a powerful and reliable technique to recognize and classify plants, allowing us to identify the origin of a given vegetable sample. That technique could be

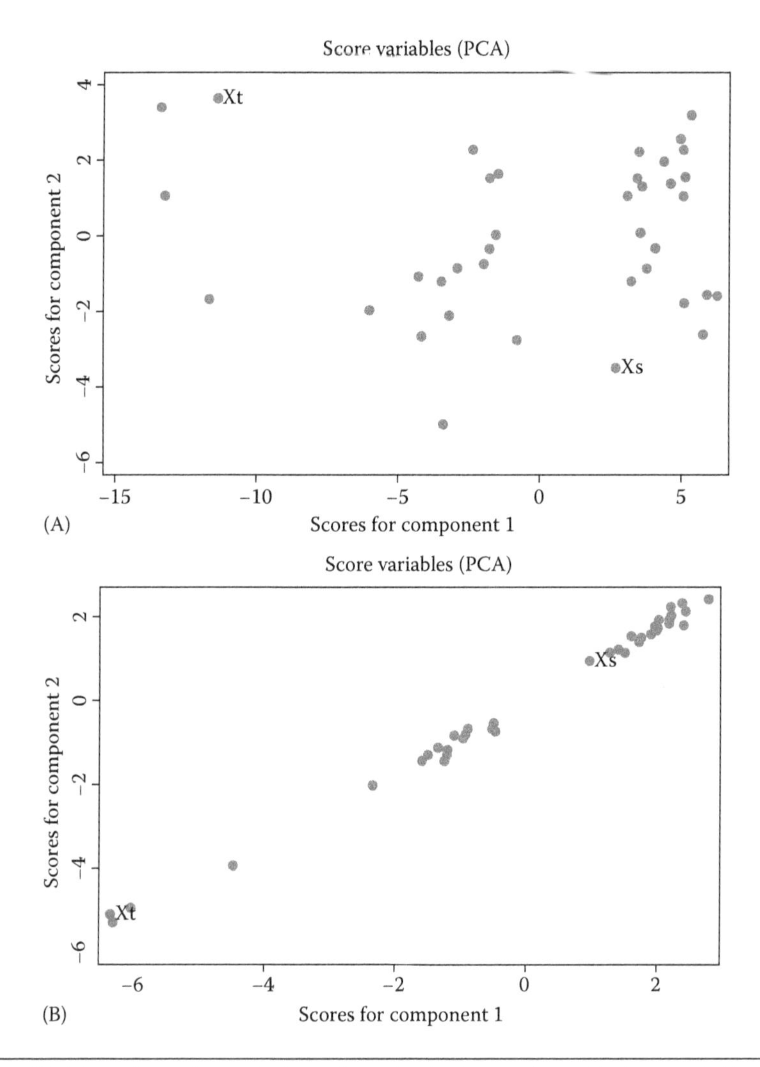

Figure 30.10 (A) Scores for two "unknown" samples. Each one is located in the correct side of the components space: Xt with Ts and Xs with Ss. (B) Rotation applied. Rotation: Orthogonal Varimax.

considerably improved by developing a database of standardized spectral signatures of the main crops in each stage and status. This could be the basis for higher level of plant and crop analysis, allowing us the prediction, diagnosis, and solution of different health and phenologic affections of plants.

Acknowledgments

The authors would like to thank the Research Council of the National University of Tucuman and the Spanish Agency for International Cooperation under grant D/030286/10 for supporting this work.

References

Asner, G. P. 1998. Biophysical and biochemical sources of variability in canopy reflectance. *Remote Sensing of the Environment* 64: 234–253.

Digital library. s.d. Interpretation of PCA results: Through the scores and loadings in interesting projections. http://cosmic.mse.iastate.edu/b-Digital%20library/Mathematical%20Details%20pdfs/pcainterpretation.pdf. Accessed June 1, 2011.

Gausman, H. W. and W. A. Allen. 1973. Optical parameters of leaves of 30 plant species. *Plant Physiology* 52: 57–62.

Jolliffe, I. T. 2002. *Principal Component Analysis*, 2nd edn. New York: Springer.

Kokaly, R. F., G. P. Asner, S. V. Ollinger, M. E. Martin, and C. A. Wessman. 2009. Characterizing canopy biochemistry from imaging spectroscopy and its application to ecosystem studies. *Remote Sensing of Environment* 113: S78–S91.

Pla, L. E. 1986. *Análisis multivariado: método de componentes principales*. Washington, DC: Programa Regional de Desarrollo Científico y Tecnológico, Departamento de Asuntos Científicos, Secretaría General de la Organización de los Estados Americanos, Monograph 27. http://www.elibros.cl/ficha_libro.php?id=29. Accessed February 20, 2011.

Scotford, I. M. and P. C. H. Miller. 2005. Applications of spectral reflectance techniques in Northern European cereal production: A review. *Biosystems Engineering* 90 (3): 235–250.

Shlens, J. 2009. *A Tutorial on Principal Component Analysis*. New York: Center for Neural Sciences, New York University. http://www.snl.salk.edu/~shlens/pca.pdf. Accessed February 20, 2011.

Sims, D. A. and J. A. Gamon. 2002. Relationships between leaf pigment content and spectral reflectance across a wide range of species, leaf structures and developmental stages. *Remote Sensing of Environment* 81: 337–354.

Volsurf—statistic. s.d. http://www.moldiscovery.com/docs/volsurf/statistic.html. Accessed June 1, 2011.

Zwiggelaar, R. 1998. A review of spectral properties of plants and their potential use for crop/weed discrimination in row-crops. *Crop Protection* 17 (3): 189–206.

CHAPTER 31

Control of Animal Stress and Welfare with Measurements of Skin Color Variation

A New Field of Applications of Colorimetry in Applied Psychology

PAULO FELIX MARCELINO CONCEIÇÃO

Contents

31.1 Introduction: Stress and Animal Welfare: From the Farm to the Table

Animals are live and sensitive creatures; although their general appearance reflects their health and welfare, it is not so easy to define this concept in such a way that it can be scientifically assessed. The measurements of skin color variation in stressful situations would be an indicator because the color of the skin may oscillate and become pale yellow in the living of stressful situations, a condition generated by psychological reactions, neurohumoral mechanisms, and physiological responses to symptoms of stress such as fear and fright. Stress has some common symptoms in humans and animals (Conceição 2008, 2010).

Dr. Temple Grandin* wrote her PhD thesis on the effect of the environment on animal behavior and neural development. She explained that fear is a universal emotion that motivates animals to avoid predation—nevertheless, fear and fright, an animal's reactions to handling, are difficult to predict because they depend on how the animal perceives the situation. Two types of animals may have different behaviors and physiological reactions to the same procedure; the nature of early experiences and genetic factors are determiners of how an animal will behave in a fear-provoking situation. Animals with high-strung temperament are more fearful and form stronger fear memories than those with a more placid temperament. The exposure to stress early in life will have effects on its physiological response to stressors later—memories of stress cannot be easily erased (Grandin 1997, Sacks 2006).

Mounier et al. (2006), Buckham Sporer et al. (2008), and Gruber et al. (2010) make clear that prolonged stress has real adverse consequences on animal welfare, on health, and on meat quality, affecting its tenderness, color, pH, and taste.

Buckham Sporer et al. (2008) observe that the transport of young beef bulls causes stress and thus alters physiological variables, with negative impact on animal production. One possible result of transportation stress is severe respiratory diseases, to which animals often succumb. Buckham Sporer et al. (2008) collected blood samples (10 mL each at –24, 0, 4.5, 9.75, 14.25, 24, and 48 h—relative to the beginning of transfer to slaughter) and found that after 9 h of truck transport, there was drastic increase in cortisone levels, significant decrease in protein metabolism, and reduction in testosterone. The study concluded that researches of physiological response to stress may aid future detection of disease-susceptible cattle after transport and suggest the need of further researches to confirm the potential of blood plasma biological biomarkers.

Studies to quantify stress response in farm animals during handling and transport have highly variable results and are difficult to interpret from the animal's standpoint, since stress in farm animal routines depends on how different individuals perceive potential stressor agents according to genetic variability; hereditary dispositions; temperament; physiological stressors such as hunger, thirst, fatigue, injury, and exposure to thermal extremes; and psychological factors perceived on the conflict-handling behavior.

The control of stress levels in animal is already challenging, as the biochemical diagnosis of stress involves the classical determination of cortical levels, in invasive procedures outside the laboratory setting (Conceição 2010).

---

* Oliver Sacks, an expert in neurology and psychology, explains in the narrative of the book *An anthropologist on Mars* Dr. Temple Grandin's personal ability to perceive the animal sensitivity and its differences in behavior.

31.2 Origins of the Paradigmatic View on Stress Studies

From the beginning of the systematic research of stress, Walter B. Cannon devised a series of studies and experiments exposing animals to stress agents with two focuses: revealing the somatic basis of symptoms (stress was seen as a clinical/popular parameter) and the differentiation of the concept of biological balance in the notion of homeostasis—a term of Greek etymology which means homo (same) and stasis (state) or "stable conditions." Cannon proposes the concept of psychosomatic stress due to environment and social circumstances, a process that results from the organism reactions to nonspecific stressors (Nitsch et al. 1981).

The mechanism that regulates the vital conditions tends to keep up a stable equilibrium on the feedback interaction (internal/external) of essential conditions to life. The metabolism of nutrients (proteins, vitamins, water, among others) produces energy with the amounts of nutrients from the ambient, connected with threat perception and opportunities to life sustenance, operating as an open/closed system.

Hans Selye is the founder of the modern stress method of study, the control of stressors with the levels of specific hormones. Selye (cited by Nitsch et al. 1981) carried out experiments by extracting and injecting hormones in animals while simultaneously exposing them to a range of stressful situations. In one of his experiments with hormone injection in rats, Selye discovered a hormone (epinephrine) which, in higher levels, causes several injuries, like the same provoked by constant exposure to stressful situation, and he explained the "triad of stress" as the effects of long exposure to stressful situations and continuous activation of hypothalamic-pituitary-adrenal (HPA) axis for energy redirection in response to nonspecific stressors—characterized by the growth of the cortex of adrenal gland due to excessive production of epinephrine, morphological degeneration of the thymus (gland largely involved in the immune system response), ulcerations of the stomach and duodenum, and bleedings, subsequently inferring the general adaptation syndrome (GAS), explained in three stages:

Alarm reaction (identification of stressful stimuli): Something is identified as a threat and activates the HPA axis to produce energy for defense. It is the condition necessary to fight-or-flight response, characterized by an increase in respiratory frequency, increase in cardiac frequency, increase in arterial pressure, pupil dilation, sweating, anxiety, agitation, involuntary contraction of sphincters, and vasoconstriction*—a potent defense mechanism against stress in

* Vasoconstriction is a potent reaction: there are records of people who have sustained injuries in combats but do not bleed. The effects of internalization of the blood flow on preslaughter stress may confer cattle beef its dark red color.

threatening situations, largely responsible for the skin's pale yellow appearance in the eminence that something unpleasant or violent will happen.

Resistance (the stress agent persists): The organism starts to use higher energy level, as the continuous activation of the sympathetic nervous system, but neither can animals nor machines, however, work indefinitely at a higher energy level or far from regular limits. It becomes necessary trying to develop coping strategies to neutralize or substantially reduce the harmful effects of the stressors on the organism. When there is a lack of strategies to stress coping, this stage results in the increase of the adrenal cortex, ulcerations in the digestive system, irritability, and reduction of sexual activity.

Exhaustion (the last stage): The stress agent still persists, and there is an eventual return to the stress's first stage of alarm reaction in trying to produce even more energy, which results in the classical symptoms of stress: inhibition of the immune system, emergence of inflammations, diseases, fatigue, depression, and even death.

Selye understood the HPA axis as the decisive basis of endocrine response to stress. In this sense, most of the researches of stress in animals are now focused on physiological parameters, such as heart rate, blood pressure, weight, rectal temperature, hormone levels, and others parameters, in the search for new diagnostic markers and biomarkers of stress.

In this context, the measure of skin color oscillation based on reflectance colorimetry appears as a new tool of stress diagnosis and disease prevention (Conceição 2010).

31.3 Skin and Its Color

The skin is the largest and most visible organ of vertebrates, a barrier between the internal and external environments; protects the body against injury, invasion of pathogens, and harmful substances; and allows heat exchange with the environment and energy storage. The epithelial tissue, the nervous system, and the sense organs have the same genetic origin in the primordial layers of the gastrula ectoderm, and the epidermal Langerhans cell has an essential role in the primary immune responses (U.S. National Institute of Health 2011).

The color of the skin is due to three pigments: melanin, the primary determinant of skin color variability, deposited in the upper layers of the epithelial tissue; hemoglobin, a complex molecule responsible for the red color of the blood; and carotene, the least common skin pigment, which gives the skin its yellowish aspect (Bindon 2009).

The visual identification of skin color oscillations depends on the incident light on epithelial tissue in mechanisms of absorption and reflection of the light through skin pigments.

Sharpe et al. (2002) explain the physiological process of color vision as trichromatic but mediated by the interactions among four types of retinal photoreceptors. The rods which contain the photopigment rhodopsin are the most prevalent (more than 95% of all photoreceptors), allowing vision under twilight.

Color vision is due to the remaining photoreceptors: the cones—specialized neurons with specific pigments sensible to specific wavelengths: S-cone blue, short-wave sensitive (around 420 nm); M-cone green, medium-wave sensitive (around 530 nm); and L-cone red, long-wave sensitive (around 560 nm).

The visual discrimination of skin color intensity changes to pale yellow is difficult because it depends on the ambient light sources, and there is a real decrement of human vision in response to yellow (S-cone blue). According to Calkins (2002), the neural locus of the cone antagonism additive light syntheses, blue-ON/yellow-OFF, is critical to yellow as stimulus of presynaptic circuit. There is a loss of the S-cone function to light around 420 nm, as they are relatively sparse, branching in small bistratified cells or the tritanopia.

The International Commission on Illumination (CIE 2009) sees the same reasons for some of the problems with visual differentiation on yellowish variations as the effect of small field, since the (blue) S-cones are rare in the retina and the spatial discrimination in the yellow-blue direction is anatomically limited, identified as small-field bistratified ganglion cells.

31.4 Tristimulus Colorimetric Measurements of Skin Color

There are studies of skin color measure in such fields as cosmetics and prevention programs for skin cancer control with the method established by the CIE expressed in a three-dimensional color space, with the coordinates L^* (lightness), a^* (green-red), and b^* (yellow-blue). Serup et al. (2006) suggest the tristimulus colorimeter Chromameter CR-200 of Minolta for skin color measurements, as it was used in many studies on skin cancer. Nevertheless, they call attention to errors caused by sequential measures in different areas of skin, oiliness, moisture, blood vessels, moles, scars, folds, malformation, and variation of thickness.

31.5 Skin Color Oscillation and Symptoms of Stress

Conceição (2008) evaluated with the tristimulus colorimeter Minolta Chromameter CR 410, aperture 8 mm, the skin color oscillations of athletes ($n = 32$), mean of three sequential valid measures in the inner upper arm of Olympics athletes—before (standard measure) and after low-intensity aerobic exercise session of 30 min.

The color differences between the values obtained in pre- and posttraining measurements, ΔL^* (lightness difference: lighter if positive, darker if negative), Δa^* (green-red difference: redder if positive, greener if negative),

Δb^* (yellow-blue difference: yellower if positive, bluer if negative), and ΔE^* (total color difference), correlated with the list of stress symptoms LSS-VAS. Statistical analyses were performed using SPSS for Windows, version 10, with the nonparametric Spearman correlation coefficient.

There was a significant association between the color direction Δb^* (mean −0.24; min. −1.51; max. 1.14) and the frequency of the feeling of fear* ($p = 0.000$) and fright† ($p = 0.001$) evaluated with a list of symptoms of stress LSS-VAS (Vasconcellos 1985).

There was no statistical association between the color direction Δa^* (green-red) in the inner upper arm measurement and the frequency of the feeling of fear and fright.

There was a no significant statistical association with total color differences expressed in the CIELAB equations ΔE^*_{ab} (mean 1.46; min. 0.53; max. 3.42) and the equation CIEDE2000 (mean 1.26; min. 0.45; max. 3.16), maybe because the principles stated in these equations fail to describe yellow color differences to encompass the human deficit of visual acuity to differentiation of yellowness.

31.6 Animal Stress Skin Colorimetry: Future Perspectives

The preliminary result of the study with athletes suggests new directions for colorimetry and stress diagnosis, since reactions of fear and fright have the same phylogenetic origins in the evolution of vertebrates, for example, the food animal meat industry in the control of animal stress and welfare—meat quality: from the farm to the table (Conceição 2010).

Measures of skin color oscillations during handling, transportation, and slaughtering would be less complicated than invasive technique of analysis of blood and tissue biomarkers of stress.

The measure of skin color variations is a new noninvasive procedure to the control of stress in animals, possibly enabling to prevent the harmful effects of stress on health.

The use of a portable handheld tristimulus reflectance colorimeter to evaluate the skin color oscillations of the live animals may constitute a new approach to the study of stress and animal welfare.

The harmful effects of stress could be minimized with the measure of skin color alterations during breeding, which is also associated with the reduction of diseases and reproductive problems (Buckham Sporer et al. 2008).

The yellowness of the skin reveals symptoms of psychological and physiological responses to stress, which has common implications in human and

---

* The original item-question of the list of symptoms of stress LSS-VAS is *Tenho medo* (I'm scared).

† The original item-question of the list of symptoms of stress LSS-VAS is *Qualquer coisa me apavora* (I am frightened of any sort of thing).

animal health—the same stressor stimulus may affect one individual but not the other.

It is necessary to develop new colorimetric equations to encompass/distinguish subtle color differences which would be useful in many areas of science such as health, food, and marine navigation.

Progress in the control of stress in animal handling and slaughter may help bridge nontariff barriers in the international trade of food and meat products.

References

Bindon, J. R. 2009. Skin color human pigmentation and adaptation. The University of Alabama, Tuscaloosa, AL. http://www.as.ua.edu/ant/bindon/ant570/topics/Skincolor.pdf. Accessed May 14, 2011.

Buckham Sporer, K. R., P. S. D. Weber, J. L. Burton, B. Earley, and M. A. Crowe. 2008. Transportation of young beef bulls alters circulating physiological parameters that may be effective biomarkers of stress. *Journal of Animal Science* 86: 1325–1334.

Calkins, D. J. 2002. Synaptic organization of cone pathway in the primate retina. In *Color Vision: From Genes to Perception*, eds. K. R. Gegenfurtner and L. T. Sharpe. Cambridge, U.K.: Cambridge University Press.

CIE (Commission Internationale de L'Eclairage). 2009. CIE TC1-63. Activity report. Validity of the range of CIEDE2000. http://www.cie.org.au/divisions.html#D1. Accessed May 14, 2011.

Conceição, P. F. M. 2008. "Amarelão" no esporte: das alterações da cor da pele ao coping do estresse por crenças religiosas e o lócus de controle de atletas de handebol, ginástica artística e voleibol. Doctoral thesis in Physical Education. Advisor Antonio Carlos Simões. São Paulo: Universidade de São Paulo, EEFE.

Conceição, P. F. M. 2010. The control of animal stress and welfare with measurements of skin color variations: A new field of applications of colorimetry in applied psychology. In *AIC 2010 Color and Food, Proceedings*, eds. J. Caivano and M. López. Buenos Aires, Argentina: Grupo Argentino del Color, pp. 31–36.

Grandin, T. 1997. Assessment of stress handling and transport. *Journal of Animal Science* 75: 249–257.

Gruber, S. L., J. D. Tatum, T. E. Engle, P. L. Chapman, K. E. Belk, and G. C. Smith. 2010. Relationship of behavioral and physiological symptoms of preslaughter stress to beef longissimus muscle tenderness. *Journal of Animal Science* 88: 1148–1149.

Mounier, L., H. Dubroeucq, S. Andanson, and I. Veissier. 2006. Variations in meat pH of beef bulls in relation to conditions of transfer to slaughter and previous history of the animals. *Journal of Animal Science* 84: 1567–1576.

Nitsch, J. R., H. Allmer, D. Hackfort, J. E. Lazarus, J. E. McGrath, H. Selye, and I. Udris. 1981. *Stress: theorien unteruchugen und massnahmaen*. Stuttgart, Germany: Hans Huber.

Sacks, O. 2006. *Um antropólogo em Marte*. São Paulo, Brazil. Companhia das Letras.

Sharpe, L. T., A. Stokman, H. Jägle, and J. Nathans. 2002. Opsin genes, cone photopigments, color vision and color blindness. In *Color Vision: From Genes to Perception*, eds. K. R. Gegenfurtner and L. T. Sharpe. Cambridge, U.K.: Cambridge University Press.

Serup, J., B. E. J. Gregor, and L. G. Gary. 2006. *Handbook of Non-Invasive Methods and the Skin*, 2nd edn. New York: Taylor & Francis.

U.S. National Institute of Health. 2011. Layers of the skin. SEER Training Modules.

Vasconcellos, E. G. 1985. Stress coping ans sozialize kompetens bei kardiovaskularen erkrankungen. Munich, Germany: Ludwig Maximilians Universität, Doctoral dissertation in Psychology.

CHAPTER 32

Relationship between Mineral Content and Color in Honeys from Two Ecological Regions in Argentina

GERMÁN P. BALBARREY, ANA ANDRADA, JUAN ECHAZARRETA, DIEGO IACONIS, and LILIANA M. GALLEZ

Contents

32.1 Introduction

Honey is the sweet food produced by honeybees and resulting from the nectar of flowers or secretions of living parts of plants or excretions of sucking insects remaining on the surface of the plants that the bees collect, transform, combine with specific substances of their own, store, and leave to mature in honeycombs (Crane 1990). Honey contains nutrients such as carbohydrates, proteins, amino acids, vitamins, and minerals.

Nowadays, there is a tendency toward honey's typification by its geographical and/or botanical origin. Some methods have been developed for this purpose as melissopalynological techniques, sensory characterization, and different analytical studies. Anklam (1998) studied the influence of some single parameters on the authenticity of honeys and found that the analysis of minerals and trace elements, aliphatic organic acids, amino acids, aroma compounds, aromatic carbonyl compounds, flavonoids, oligosaccharides, phenolic acids and esters, proteins, and specific stable isotopic ratios could be suitable for the determination of botanical origin. The content of total polyphenol compounds (e.g., flavonoids and phenolic acids) in honey is stated to be influenced by botanical and geographical origin as well as by climate characteristic of the site (Pyrzynska and Biesaga 2009).

As mineral content varies widely depending mainly on pedoclimatic conditions and environmental or seasonal factors, it is stated that minerals and trace elements found in honeys are considered as indicators of geographical origin as well as environment pollution markers (Anklam 1998, González Paramás et al. 2000). The dominant mineral element in honeys is potassium, followed by chlorine, sulfur, sodium, phosphorus, magnesium, silicon, iron, and copper (Hernández et al. 2005, La Serna Ramos et al. 1999). Nanda et al. (2003) reported that potassium is the most abundant of the elements while copper is the least present element in the Northern Indian honeys. Latorre et al. (1999) determined 11 metals in honeys from Galicia and could differentiate between natural Galicia honeys and processed non-Galician ones.

The mineral content varies from about 0.04% in light honeys to 0.2% in some dark honey samples although floral honeys typically have 0.1%–0.2% of mineral constituents and honeydew honeys have 1% or higher (Anklam 1998, Bianchi 1980).

Raw honey color, according to Crane (1990), is related to mineral content and many other components which are largely unknown. The presence of polyphenolic compounds and a high content of amino acids are also reported in dark honeys (Crane 1990, Bertoncelj et al. 2007). The international market classifies honey according to the Pfund color scale; light honeys is more valued than dark ones (Bogdanov et al. 2004). Pfund scale ranges from 0 to 140 mm, showing seven ranges with different color designations named water white, extra white, white, extra light amber, light amber, amber, and dark. Characterization of honey according to its botanical or geographical origin takes into account the Pfund ranges (Piazza et al. 1991, Mateo Castro et al. 1992, Devillers et al. 2003, Gallez et al. 2009).

Electrical conductivity depends on the ash and the acid content and is frequently used to estimate honey mineral content (Bianchi 1980, Bogdanov et al. 2004). Minerals are transported by the sap to the nectaries. As a result, the mineral composition of blossom honey depends on botanical and geographical origin. Some trace elements could come from vegetation sources and environmental pollution (Bogdanov et al. 2004, 2007, Terrab et al. 2004).

Argentina is one of the world's main suppliers of honey. Two nearby ecological regions in Central Argentina, the Espinal and the Pampean phytogeographic provinces, contribute to the national honey output as apiculture is an important activity in the "Pampean Austral" and the south of the "Caldén" districts (Figure 32.1). These two areas show differences in native vegetation,

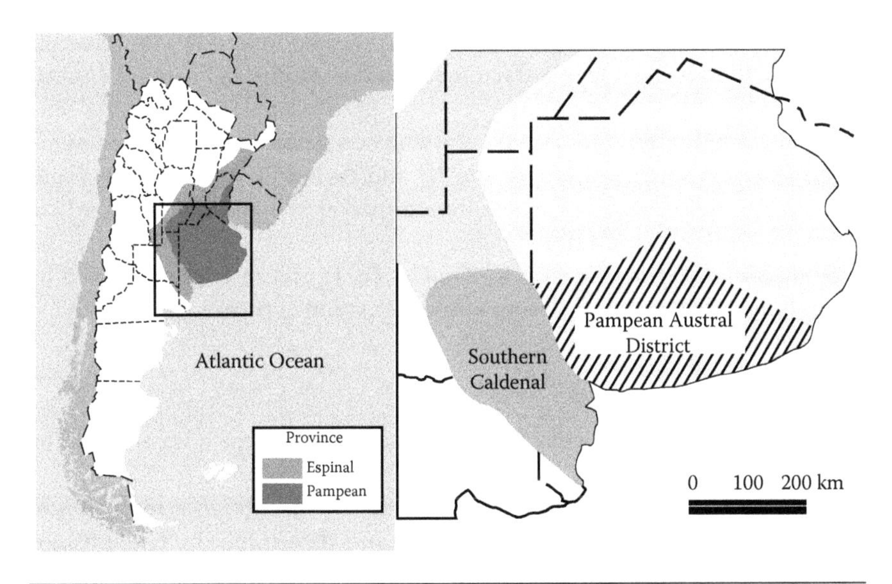

Figure 32.1 Geographical location of the study area in the ecoregions Pampean and Espinal in Argentina.

soil characteristics, as well as in agricultural activities (Cabrera 1976) that should be reflected in certain honey attributes.

The aim of this chapter is to study the relationship between the mineral composition and the physicochemical characteristics of honeys from the "Pampean Austral" and the south of "Caldén" districts, with special emphasis on color.

32.2 Materials and Methods

32.2.1 Honey Samples

A set of 30 representative honey samples, 15 from the "Pampean Austral" district and 15 from the south of "Caldén" district (Argentina), were provided by professional beekeepers from different apiaries during 1997–1999. These samples were obtained by centrifugation.

The botanical origin of honeys was confirmed by pollen analysis, according to Louveaux et al. (1978). Pollen grains were identified using the pollen reference collection from LabEA (Laboratory of Apicultural Studies, Departamento de Agronomía, Universidad Nacional del Sur) (Andrada et al. 2008), as well as specialized literature (Erdtman 1966, Kremp 1968, Heusser 1971, Markgraf and D'Anthoni 1978, Reille 1992).

32.2.2 Physicochemical Parameters

The color of honey samples was assessed by a Pfund grader; the electric conductivity was measured according to Louveaux et al. (1973) and free acidity according to AOAC (2000). Moisture content was determined by the refractometric method using a Jena refractometer and a sodium light source (AOAC 2000).

Hydroxymethylfurfural (HMF) content was determined after clarifying samples with Carrez reagents (I and II) and the addition of sodium bisulfate (AOAC 2000); absorbance was determined at 284 and 336 nm in a 1 cm quartz cuvette in a spectrophotometer.

Nine minerals (Ca, Mg, Na, K, Zn, Cu, Fe, P, and S) were quantified by using inductively coupled plasma-atomic emission spectrometry (ICP-AES) in honey solutions 1:5 w/w.

32.2.3 *Statistical Analysis*

The mineral content of honeys from both ecological regions was compared by means of Student's t test.

Principal component analysis (PCA) was used to obtain a reduction of dimensionality of the 14×30 data matrix and determine the relationships between variables (physicochemical characteristics and mineral content) and objects (honey samples) through optimal 2-D graphical display. The search for natural groups in the honey samples was performed by cluster analysis, using complete linkage and Euclidean distance. Cluster analysis also represents a multidimensional space by mapping in two dimensions.

The "extra light amber" upper limit in the Pfund grader scale (50 mm Pfund) was established as the limit between light and dark honeys. Within these two groups, the relationship between color and mineral content was studied by Pearson's correlation and linear regressions.

32.3 Results and Discussion

The analytical results are summarized in Table 32.1. Color values varied from 16 to 109 mm Pfund while electrical conductivity ranged from 0.13 to 0.90 mS cm^{-1}. Comparing both ecological regions, highly significant differences were detected in K, Ca, and Mg levels, but there were no differences in Na content.

The PCA showed a direct relationship between color, conductivity, K, Ca, and Cu. Other physicochemical properties such as moisture, free acidity, and HMF, and some minerals were not related to color (Figure 32.2a). Sample grouping on the basis of similarities (Figure 32.2b) showed that honeys from south of Caldén district were divided into two different groups while those from the Pampean Austral district made up only one.

The palynological analyses of samples coming from the Pampean Austral district (southern Pampas) showed a representative combination of three pollen types: *Diplotaxis tenuifolia*, *Eucalyptus* sp., and *Centaurea solstitialis* (Valle et al. 2007). Nearly all of these samples were light-colored. The color of honeys from southern Caldenal ranged from light to dark. Most of the dark honeys (>50 mm Pfund) contained over 75% of *Condalia microphylla* pollen

Table 32.1 Physicochemical Characteristics and Mineral Content of the Honey Samples from "Pampean Austral" and South of "Caldén" Districts, Argentina

	Pampean Austral District ($n = 15$)				Southern Caldenal ($n = 15$)			
	Mean	SD	Min.	Max.	Mean	SD	Min.	Max.
Color (mm Pfund)	36.47	10.88	16.50	59.20	65.93	28.00	21.50	109.00
Conductivity (mS cm^{-1})	2.02	0.40	1.30	2.80	5.36	1.93	3.20	9.00
Moisture (%)	16.03	1.32	12.50	18.00	15.77	1.33	13.20	18.80
Acidity (meq kg^{-1})	16.10	2.92	12.00	23.00	18.10	3.86	140	26.00
HMF (mg kg^{-1})	2.05	2.30	0.00	6.58	1.78	2.32	0.00	5.97
Minerals (ppm sol.)								
Ca	10.11	3.69	5.20	17.60	23.11	8.32	10.90	38.70
Mg	1.82	0.34	1.37	2.41	2.30	0.36	1.50	2.72
Na	12.02	5.40	5.90	23.40	11.29	6.23	3.70	26.30
K	45.49	14.19	21.10	77.90	199.25	108.39	97.00	422.00
Zn	0.38	0.15	0.19	0.64	0.24	0.09	0.11	0.45
Cu	0.01	0.01	0.01	0.03	0.05	0.04	0.01	0.15
Fe	0.09	0.08	0.01	0.31	0.18	0.10	0.05	0.44
P	6.93	2.35	4.88	13.30	8.45	4.66	2.86	16.10
S	9.30	5.01	3.74	24.00	11.77	5.63	3.88	21.70

SD, standard deviation; Min., minimum; Max., maximum; ppm sol., ppm in honey solution of 1:5.

and showed high conductivity values. Two of them were over the threshold established by the International Honey Commission to differentiate honeydew honeys (>8 mS cm^{-1}) (Figure 32.3a). On the other hand, light samples were multifloral honeys.

Potassium was the most abundant mineral in all samples, in accordance with other authors (Crane 1990, Hernández et al. 2005). Within the same color range (<50 mm Pfund), the southern Caldenal samples contained higher levels of K (Figure 32.3b).

Considering all samples, the correlation between color and K content was the highest ($r = 0.916$), and a high correlation between color and conductivity was also found ($r = 0.895$). However, when light honeys were studied separately, no correlation was found between K content and color. Similar results were obtained between color and conductivity (Figure 32.3a).

Honey color is known to be affected by different compounds. Phenolic compounds, in particular the flavonoids quercetin, kaempferol, and isorhamnetin, were identified in honeys from the "Pampean Austral" district

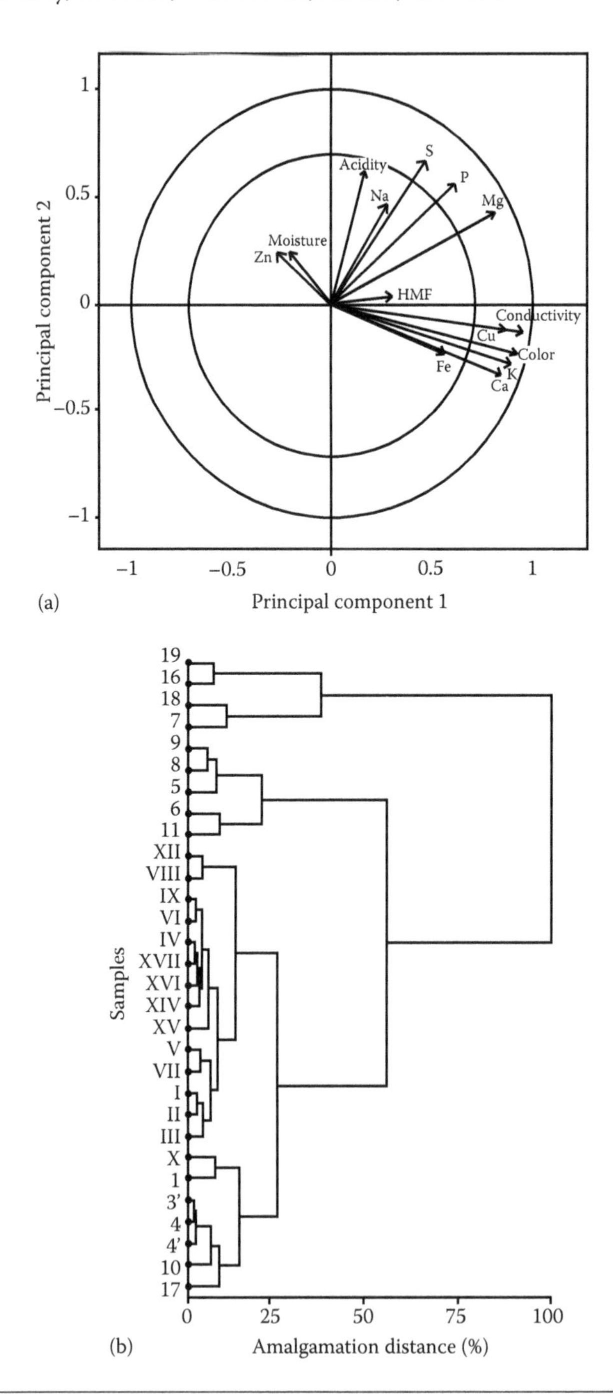

Figure 32.2 Relationship between mineral content and physicochemical honey characteristics. (a) PCA: vectors between concentric circles have more than 70% representation. (b) Dendrogram with complete linkage Euclidean distance: Roman and Arabic numbers belong to samples from Pampean Austral and south of Caldén districts, respectively.

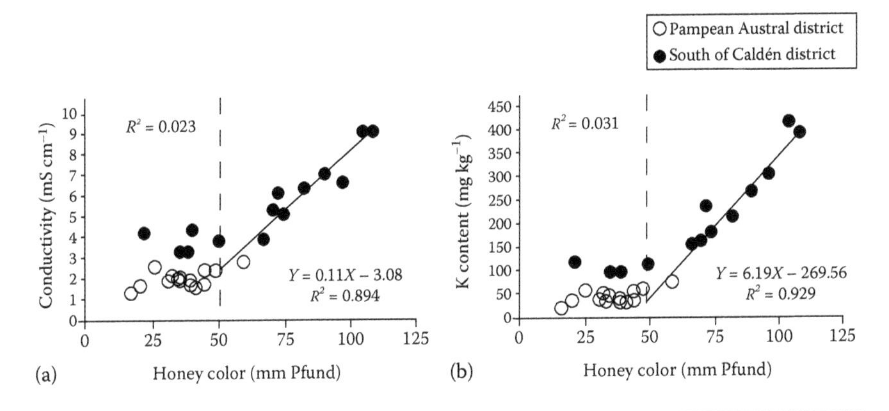

Figure 32.3 Relationship between color and mineral content of the honey samples from the "Pampean Austral" (open symbols) district and the south of the "Caldén" district (solid symbols). (a) Relationship between honey color and electrical conductivity. (b) Relationships between honey color and K content.

(Truchado et al. 2010), and phenolics were found to be positively correlated to color of Slovenian honeys (Bertoncelj et al. 2007). It is remarkable that in this study, no correlation was found between minerals and color in light honeys, while dark ones showed a close relationship, suggesting that some components other than the total mineral content affect the color of light honeys.

Acknowledgments

The authors are grateful to Lic. Miriam Crespo and Dr. Claudio Ferrarello for their contribution with the mineral analysis, and to Dr. Cecilia Pellegrini for her valuable assistance and critical reading of the manuscript.

References

Andrada A. C., J. Coppis, E. Tourn, C. N. Pellegrini, and M. E. Gil 2008. Software para la identificación de pólenes. Parte I. Pólenes en las mieles del sistema serrano Ventania. DOI: http://www.criba.edu.ar/agronomia/laboratorios/labea/polen/polenUNS.htm

Anklam, E. 1998. A review of the analytical methods to determine the geographical and botanical origin of honey. *Food Chemistry* 63 (4): 549–562.

AOAC (Association of Official Analytical Chemists). 2000. *Official Methods of Analysis*, 17th edn. Gaithersburg, MD: AOAC International.

Bertoncelj, J., U. Doberšek, M. Jamnik, and T. Golob. 2007. Evaluation of the phenolic content, antioxidant activity and colour of Slovenian honey. *Food Chemistry* 105 (2): 822–828.

Bianchi, E. M. 1980. La conductibilité électrique des miels de la province de Santiago del Estero en Argentine. *Apidologie* 11 (1): 25–28.

Bogdanov, S., M. Haldimann, W. Luginbühl, and P. Gallmann. 2007. Minerals in honey: Environmental, geographical and botanical aspects. *Journal of Apicultural Research and Bee World* 46 (4): 269–275.

Bogdanov, S., K. Ruoff, and L. Persano Oddo. 2004. Physico-chemical methods for characterization of unifloral honeys: A review. *Apidologie* 35: S4–S17.
Cabrera, A. 1976. Regiones fitogeográficas argentinas. In *Enciclopedia argentina de agricultura y jardinería*, 2nd edn. Buenos Aires, Argentina: ACME.
Crane, E. 1990. *Bees and Beekeeping. Science, Practice and World Resources*. London, U.K.: Heinemann Newnes.
Devillers, J., M. Morlot, M. H. Pham-Delegue, and J. C. Doré. 2003. Classification of monofloral honeys based on their quality control data. *Food Chemistry* 86: 305–312.
Erdtman, G. 1966. *Pollen Morphology and Plant Taxonomy. Angiosperms*. New York: Hafner Publication.
Gallez, L., A. Andrada, E. Galassi, C. Pellegrini, and M. Gil. 2009. Mieles de las sierras de Ventania: hacia su tipificación. In *Ambientes y recursos naturales del sudoeste bonaerense: Producción, contaminación y conservación*, eds. N. J. Cazzaniga and H. M. Arelovich. Bahía Blanca, Argentina: EDIUNS, pp. 289–301.
González Paramás, A. M., J. A. Gómez Bárez, R. J. García-Villanova, T. Rivas Palá, R. Ardanuy Albajar, and J. Sánchez Sánchez. 2000. Geographical discrimination of honeys by using mineral composition and common chemical quality parameters. *Journal of the Science of Food and Agriculture* 80: 157–165.
Hernández, O. M., J. M. G. Fraga, A. I. Jiménez, F. Jiménez, and J. J. Arias. 2005. Characterization of honey from the Canary Islands: Determination of the mineral content by atomic absorption spectrophotometry. *Food Chemistry* 93 (3): 449–458.
Heusser, C. 1971. *Pollen and Spores of Chile*. Tucson, AZ: University of Arizona Press.
Kremp, G. 1968. *Morphologic Encyclopedia of Palynology*. Tucson, AZ: University of Arizona Press.
La Serna Ramos, E. I., B. Mendez Pérez, and C. Gómez Ferreras. 1999. *Aplicación de nuevas tecnologías en mieles canarias para su tipificación y control de calidad*. Islas Canarias, Spain: Confederación de Cajas de Ahorros, Ministerio de Cultura.
Latorre, M. J., R. Peña, C. Pita, A. Botana, S. García, and C. Herrero. 1999. Chemometric classification of honeys according to their type. II. Metal content data. *Food Chemistry* 66 (2): 263–268.
Louveaux, J., A. Maurizio, and G. Vorwohl. 1978. Methods of melissopalynology by International Commission for Bee Botany of IUBS. *Bee World* 59: 139–157.
Louveaux, J., M. Pourtallier, and G. Vorwohl. 1973. Méthodes d'analyses des miels. Conductivité. *Bulletin Technique Apicole* 16: 1–7.
Markgraf, V. and H. D'Anthoni. 1978. *Pollen Flora of Argentina*. Tucson, AZ: University of Arizona Press.
Mateo Castro, R., M. Jiménez Escamilla, and F. Bosch-Reig. 1992. Evaluation of the color of some unifloral honey types as a characterization parameter. *Journal of AOAC International* 75: 537–542.
Nanda, V., B. C. Sarkara, H. K. Sharmaa, and A. S. Bawab. 2003. Physico-chemical properties and estimation of mineral content in honey produced from different plants in Northern India. *Journal of Food Composition and Analysis* 16 (5): 613–619.
Piazza, M., M. Accorti, and L. Persano Oddo. 1991. Electrical conductivity, ash, colour and specific rotatory power in Italian unifloral honeys. *Apicoltura* 7: 51–63.
Pyrzynska, K. and M. Biesaga. 2009. Analysis of phenolic acids and flavonoids in honey. *Trends in Analytical Chemistry* 28 (7): 893–902.

Reille, M. 1992. *Pollen et spores d'Europe et d'Afrique du nord*. Marseille, France: Laboratoire de Botanique Historique Palynologie.

Terrab, A., D. Hernanz, and F. Heredia. 2004. Inductively coupled plasma optical emission spectrometric determination of minerals in thyme honeys and their contribution to geographical discrimination. *Journal of Agriculture and Food Chemistry* 52: 3441–3445.

Truchado P., E. Tourn, L. M. Gallez, D. A. Moreno, F. Ferreres, and F. A. Tomás-Barberán. 2010. Identification of botanical biomarkers in Argentinean diplotaxis honeys: Flavonoids and glucosinolates. *Journal of Agriculture and Food Chemistry* 58: 12678–12685.

Valle, A., A. Andrada, E. Aramayo, M. Gil, and S. Lamberto. 2007. A melissopalynological map of the south and southwest of the Buenos Aires Province, Argentina. *Spanish Journal of Agricultural Research* 5 (2): 172–180.

CHAPTER 33

Importance of Anthocyanic Copigmentation on the Color Expression of Red Wine in Warm Climate

BELÉN GORDILLO, M. LOURDES GONZÁLEZ-MIRET, and FRANCISCO J. HEREDIA

Contents

33.1 Introduction

The expression of the color in red wines depends not only on the anthocyanin concentration and their derived pigments but also on physicochemical phenomena known as copigmentation, which is described to stabilize wine color (Boulton 2001). In red wines, copigmentation reactions consist of noncovalent interactions between original grape anthocyanins among themselves (self-association) or with a wide variety of colorless wine constituents named copigments or cofactors (flavonoids, amino acids, organic acids, polysaccharides, etc.). The global result of copigmentation is based on two positive effects on color (Dangles and Brouillard 1992): (1) changes in the spectral properties of the chromophore group, that is, an increase of absorptivity and frequently a shift of the visible λ_{max} toward higher wavelengths, and (2) the stabilization of the flavylium cation reducing the formation of the other colorless forms of anthocyanins. For these reasons, copigmented anthocyanin solutions show higher chroma values than theoretically could be expected according to the pH value of the media (Goto and Kondo 1991). Additionally, in the next few months of aging, the earlier copigmentation complexes trend to be progressively transformed into polymeric pigments by covalent links (Bakker and Timberlake 1997). This conversion yields to the evolution and stabilization of the wine color, which changes from the initial purple-red hue to red-orange hue, and shows less saturated color. Thus, in enology, the assessment of the copigmentation process is of critical importance in understanding the relationship between grape composition and color in young red wines, and it could also act as a first stage in the formation of more stable pigments that determine the color of aged red wines. Most of the studies about copigmentation in model solutions have shown that the effectiveness and the intensity of the copigmentation effect is dependent upon several factors including the concentrations of pigments and cofactors, their chemical structures, the cofactor/pigment molar ratio, the pH of the medium, or the temperature (Brouillard et al. 1989, González-Manzano et al. 2009). In the same way, an extensive review of studies of copigmentation over the past 30 years in wines from several wineries (Boulton 2001) confirms that the control of these factors during winemaking can exert an important influence on the color chroma and saturation and on the higher or lower stability of color of the red wine. In particular, the production of red wines in winemaking regions that have produced traditionally fortified and young white wines (warm climate) has grown in the last decades in order to diversify the offer in the market. Nowadays, the advance in the knowledge about the repercussion that copigmentation exerts over the final color of these wines has received much attention. In these areas, the production of high-quality red wines, with highly saturated and stable color, is greatly limited due to the stressful climate conditions that do not enable the grapes to reach optimum phenolic maturity at harvest (Mira de Orduña 2010). The most likely reason for this fact is that wines made from grapes low in pigments and cofactors are not able to form much copigmentation at the first steps of the winemaking process (Boulton 2001). Consequently, their color stabilization might not occur correctly, resulting in the so-called "color fall," or color loss. However, it has been recently demonstrated that the magnitude of copigmentation and its effect on the color quality and stability in these red wines are extremely variable and can be affected in a discriminated way by specific factors such as the characteristics of the grape, the winemaking process, or the viticulture practices.

33.2 Variation in Copigmentation regarding Grape Variety

The grape variety used in the elaboration of the red wine affects the concentration and composition of wine anthocyanins and therefore the copigmentation process. In this sense, an exhaustive follow-up of the copigmentation phenomena according to the grape variety has permitted to evaluate the impact of the copigmentation to the color of red wines elaborated in warm climate and its development during the vinification. Wines elaborated in 2008 by conventional vinification in "Condado de Huelva" region (southwest Spain) were submitted to study. The cultivars assayed were Tempranillo (TE), traditionally grown in Spain, and the most commonly introduced varieties in this zone: Syrah (SY) and Cabernet Sauvignon (CS). The measurement of the copigmentation phenomena was carried out using the traditional method developed by Boulton (1996). Analysis of variance was performed to compare the differences in the contribution of each component of the total color: copigmented anthocyanins (%CA), free anthocyanins (%FA), and polymeric pigments (%PP), as well as the anthocyanin composition and color between the wines. Table 33.1 summarizes the results of the anthocyanin concentration (grouped by the substituent) and Boulton's parameters, while Figure 33.1 shows the location of the wines in the (a^*b^*) color diagram.

Considering the whole set of data, it can be seen that the wines were markedly different and that the variety affected significantly both the phenolic composition and the magnitude of copigmentation. SY wines showed significantly higher values for %CA than TE and CS wines, which was in agreement with the highest contents of total monomeric anthocyanin. Moreover, each type of wine had a distinctive anthocyanic pattern. Specifically, significant differences on delphinidin, malvidin, and peonidin acetates and coumarilic derivatives were also found between the wines.

As far as the color was concerned, the wines also showed a clear difference in hue (h_{ab}) and chroma (C^*_{ab}) according to the variety. SY wines resulted in final bluish red hues (0°–350°) and greater color chroma ($C^*_{ab}=24.04$), which is consistent with a higher concentration of anthocyanin and magnitude of copigmentation. On the contrary, lower values of the %CA and pigment concentration in the TE and CS wines produced less saturated, lighter, and less purple colors ($C^*_{ab}=21.63$ and 20.23, respectively, and $h_{ab}=0°–10°$).

Table 33.1 Average Values of the Different Group of Anthocyanins and Percentage of Boulton's Parameters according to the Variety

	TE ($n=3$)	SY ($n=3$)	CS ($n=2$)
ΣGl (mg/L)	84.91 ± 21.48 a	72.47 ± 21.59 a	71.81 ± 17.99 a
ΣAc (mg/L)	9.84 ± 4.30 a	35.68 ± 12.96 b	35.15 ± 8.92 b
ΣCum (mg/L)	11.06 ± 5.40 a	16.54 ± 7.04 a	8.60 ± 1.95 b
AT (mg/L)	105.83 ± 30.24 a	124.70 ± 26.89 a	115.56 ± 28.65 a
%CA	16.32 ± 4.17 a	25.72 ± 9.48 b	15.02 ± 7.22 a
%PP	48.72 ± 5.32 a	33.54 ± 6.31 b	43.42 ± 1.24 c
%FA	36.68 ± 3.35 a	40.77 ± 5.67 a	41.54 ± 2.08 a

Different letters within the same row mean significant differences ($p<0.05$).

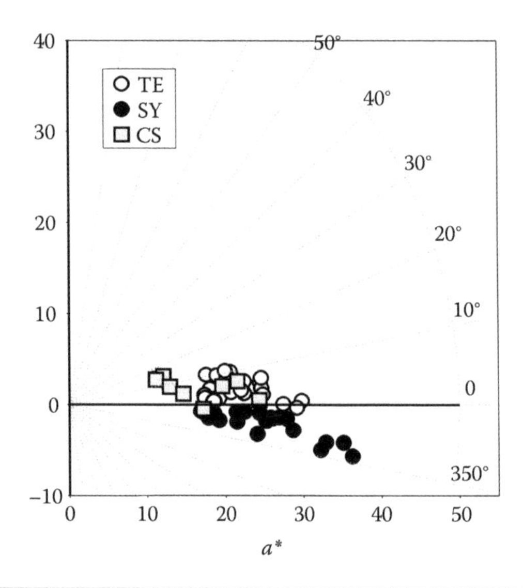

Figure 33.1 Location of wines in the (a^*b^*) color diagram.

The assessment of the color differences (ΔE^*_{ab}) between the beginning and the end of the stabilization period allowed us to evaluate the color stability of the different wines considered. Higher color variation obtained for SY wines ($\Delta E^*_{ab} = 13.34$) compared to TE ($\Delta E^*_{ab} = 10.72$) and CS ($\Delta E^*_{ab} = 4.66$) indicated a lower color stability, being consistent with a lower degree of polymerization at the end of the stabilization (PP = 34% vs. 49% and 43%, respectively). Therefore, this fact reflects a greater difficulty in SY wines to convert the copigmented complexes into more stable pigment.

33.3 Variation in Copigmentation regarding Ripeness Grade

The process of ripening is another factor that influences the quality of the grape because the changes that take place during ripening do not occur simultaneously. The date for harvesting has been traditionally estimated considering only the classical criteria based on "technological maturity" of the grapes, which is the sugar content. However, it is well known that the difference between "technological" and "phenolic" maturity of the grapes in warm climate regions is particularly remarkable due to the sun overexposure during the ripening that inhibits the anthocyanins formation in the grapes (Mori et al. 2005). In this context, one of the most significant advances in warm climate vinifications has been the investments in technology and the application of novel procedures such as low maceration temperatures (5°C–15°C) prior to fermentation, intended to enhance the extraction of the grape components accounting for the color of the wine (Gómez-Míguez et al. 2007, Heredia et al. 2010). In fact, the positive repercussion of the prefermentative cold maceration (CM) on the copigmentation evolution and the color changes of red wines elaborated in warm climate from grapes with different ripeness grade have also been confirmed. In this experience, the wines submitted to study were elaborated in duplicate from *Vitis vinifera* cv. SY grapes in 2008 from two vineyards grown in "Condado de Huelva" region: SyA (12.5°Bé) and SyB (14.0°Bé).

Quantitatively, an important influence of the maturity grade of the grape on the pigment content of the wines was observed. SyA wines, which were elaborated from less matured grapes, showed a significant lower anthocyanic and phenolic content than SyB wines, made from more matured grapes (total anthocyanins [TAs] 149 vs. 204 mg/L; total polyphenols [TPs], 1850 vs. 1900 mg/L) (Figure 33.2). In general, the phenolic content and the

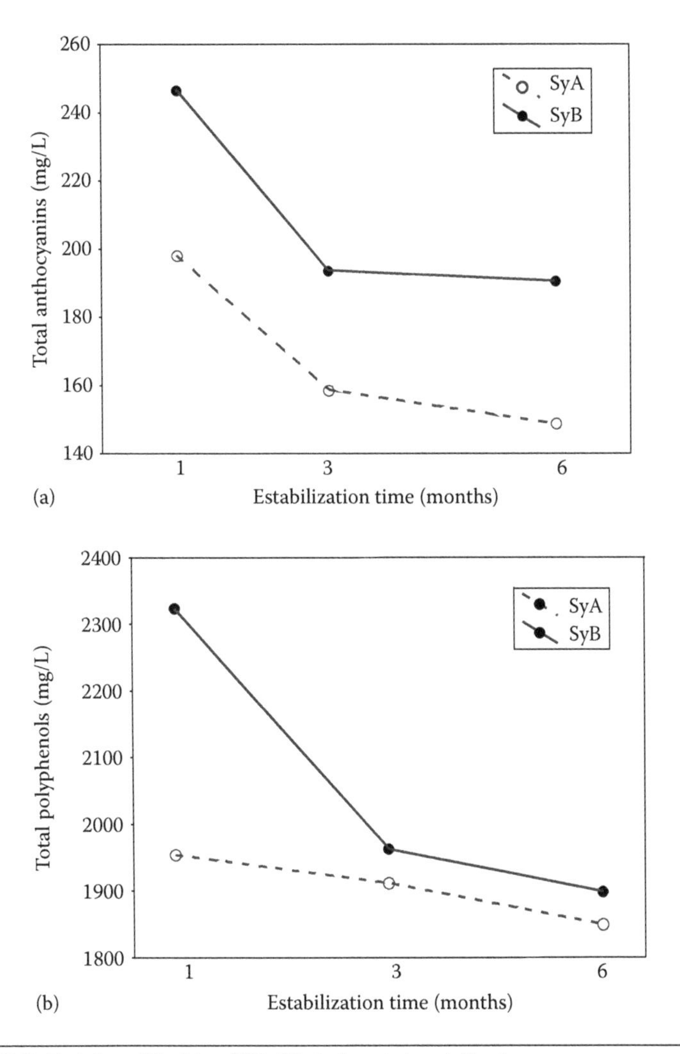

Figure 33.2 Evolution of TAs (a) and TPs (b) of wines during vinification.

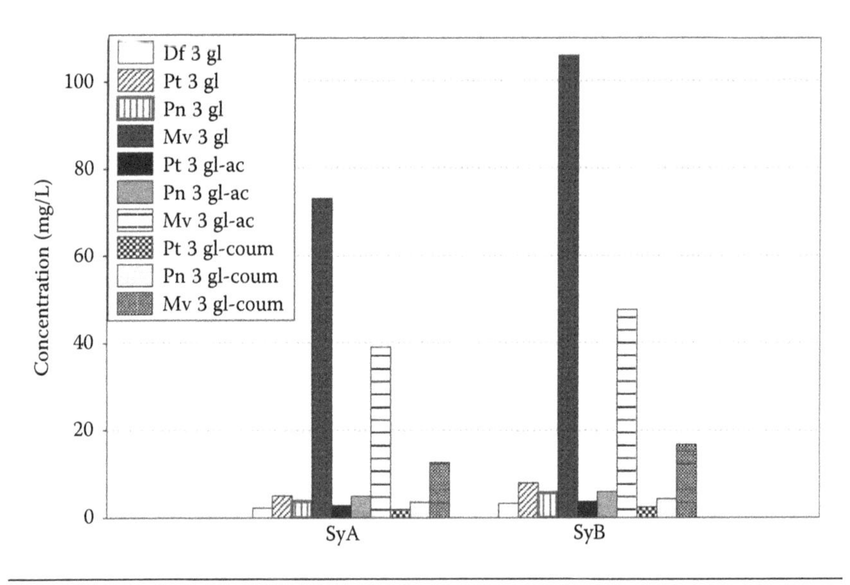

Figure 33.3 Anthocyanic profile of SyA and SyB wines, corresponding to 6 months of stabilization.

extractability index of grape increase throughout ripening process, which explain the results obtained. However, the intensive maceration conditions applied during prefermentative phase favored that SyA wines presented adequate phenolic potential and pigment stability.

It was confirmed that the ripeness grade of the grape also influenced qualitatively the anthocyanic profile (Figure 33.3), since significant differences ($p < 0.05$) were found for the most individual anthocyanins between the two SY wines elaborated (except for peonidin acetate and petunidin coumarate). An accurate differentiation and classification of the wines into each respective group was obtained by applying a discriminant analysis, being petunidin, petunidin acetate, and TPs, the variables which make possible the differentiation according to the ripeness grade of grapes. Considering only these variables, a 100% correct classification was achieved.

Furthermore, the relative contribution of each group of pigments to the total color of wines was also different (Figure 33.4), which determined particular colorimetric changes of the wines studied. During the storage period (6 months), the %PP increased in all wines; however, the %CA was quite variable. Despite the pigment degradation, SyB wines reached the highest polymerization grade values (49% vs. 37% in SyA wines), indicating a higher proportion of more stable pigment in those wines. The increase of the %CA experimented in SyB wines (approx. 20%) favored a more intense development of the polymerization process, which evidences a better cofactor/pigment ratio than SyA wines.

As a consequence of a better phenolic potential, SyB wines experimented optimum development of copigmentation simultaneously to

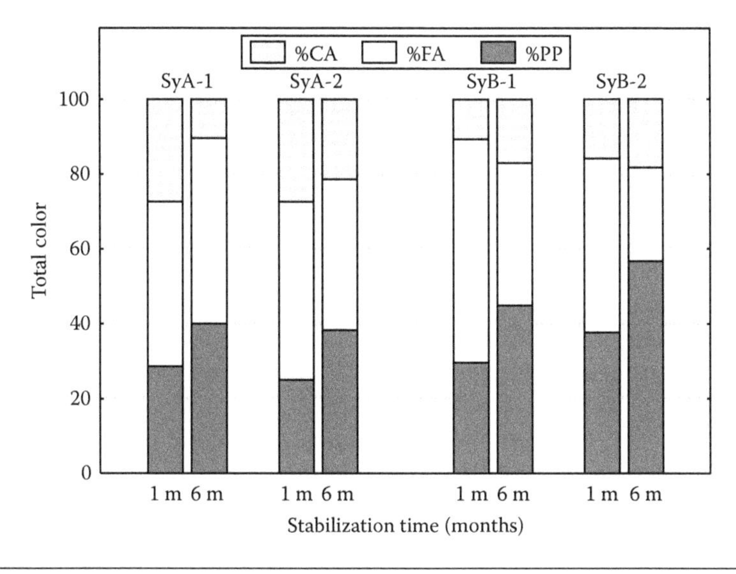

Figure 33.4 Evolution of the %CA, %FA, and %PP during the vinification.

Table 33.2 Colorimetric Parameters of SyA and SyB Wines during Vinification

	SyA-1		SyA-2		SyB-1		SyB-2	
	1 Month	**6 Months**	**1 Month**	**6 Months**	**1 Month**	**6 Months**	**1 Month**	**6 Months**
L^*	74.1	78.4	76.9	81.3	68.8	72.5	65.3	70.8
C^*_{ab}	32.5	24.2	29.1	20.9	34.9	31.2	37.2	33.5
h_{ab}	−6.4	−3.8	−6.6	−3.3	−5.9	−2.5	−2.4	2.4
s^*_{uv}	0.6	0.5	0.5	0.4	0.8	0.6	0.8	0.7

intensive polymerization process; hence, finally, red wines elaborated with more matured grapes exhibited darker, highly saturated color, with red hue (Table 33.2). However, the application of low maceration temperatures prior to fermentation confirmed to be an interesting strategy to improve and protect the phenolic potential as well as the color of red wines made from less matured grapes.

33.4 Variation in Copigmentation regarding Vinification Technique in Organic Wines

Among the agronomical practices, the organic viticulture constitutes an interesting strategy to improve the phenolic potential of red grapes in warm climate regions since organic vineyards usually have higher natural resistance to the weather inclemency. Since no fungicides are used, microbes are more abundant, which lead to an increase in the synthesis of phenolic compounds acting as antioxidants (Zafrilla et al. 2003). As it was previously commented, although the application of prefermentative CM implies an important investment in technology, one of the main advantages of this vinification technique compared to traditional vinification is the rapid cooling down of the must. This fact inhibits the activity of some enzymes such as polyphenol oxidase and microorganism development, which protect pigments and other aromatic compounds from degradation in a nontoxic way (Gil-Muñoz et al. 2009). Additionally, the excessive traditional chemical treatments which may produce losses of pigments as well as potential health problems are avoided. However, there are no previous studies about the application of this vinification technique to organic wines as a useful practice to increase the extraction of phenolic compounds as well as the safety of the product, environment, and consumer. Thus, the phenolic composition, magnitude of copigmentation, and colorimetric characteristics of organic wines obtained by CM were compared to control wines produced by traditional maceration (TM) to assess the impact of the winemaking technique (Gordillo et al. 2010). In this experience, the TE cv. grapes were grown in a vineyard located in southeastern Spain. The samples of the wines were specifically selected during the first few months of storage since the short shelf life of red wines produced in warm climates requires careful control of their color characteristics when the main mechanism of the stabilization of color occurs.

Evolution of TAs and TPs in the two vinification protocols, as well as for the different anthocyanin fractions, is shown in Figure 33.5. It might be stated that CM wines were richer for both kinds of compounds than TM wines during the whole vinification process. As expected, the concentration of monomeric anthocyanins decreased from skin removal to the moment of bottling in both kinds of wines studied, being especially remarkable for TM wines (77% vs. 38%, in CM wines). These observations confirm that red wines produced in warm climate regions easily suffer a considerable loss of pigment during the first stage of vinification, especially by traditional vinification. In this sense, the application of prefermentative CM represented an effective enological alternative to prevent an excessive pigment loss, improving the global quality of these wines. At the moment of bottling, both types of wines could be statistically differentiated regarding their chemical composition, with higher level of significance ($p < 0.01$) for malvidin (384.68 vs. 311.44 mg/L), delphinidin (61.16 vs. 40.28 mg/L), and petunidin (81.00 vs. 56.64 mg/L). An interesting observation is that the CM technique seemed to protect to a larger extent the presence of methylated anthocyanins (malvidin, petunidin, and peonidin) and acylated anthocyanins which chemical characteristics have a great effect on the copigmentation phenomenon. Therefore, cold prefermentative maceration increased not only the extraction of anthocyanins but also their initial stability in the organic wines.

At the end of the maceration period (day 6), the high polyphenol content reached by low-temperature treatment caused CM wines to have a higher grade of copigmentation and polymerization than TM wines (31% vs. 27% and 30% vs. 21%, respectively) (Figure 33.6). Despite the pigment degradation, at the end of malolactic fermentation, CM wines showed the highest percent of copigmentation values (18% vs. 10.5% in TM wines). The highest copigmentation degree observed in CM wines evidences that they presented a better cofactor/pigment ratio than TM probably due to the lower loss of pigment experienced. During the last stage of the maturation phase (15–45 days), an important decrease of %CA was produced, which was near zero at the moment of bottling (2%–4%). The gradual formation of new and more stable pigments during this period in the two kinds of wines was confirmed by a notable increase of the %PP, which was slightly higher in TM wines (46% vs. 42%, in CM wines), although this difference was not significant.

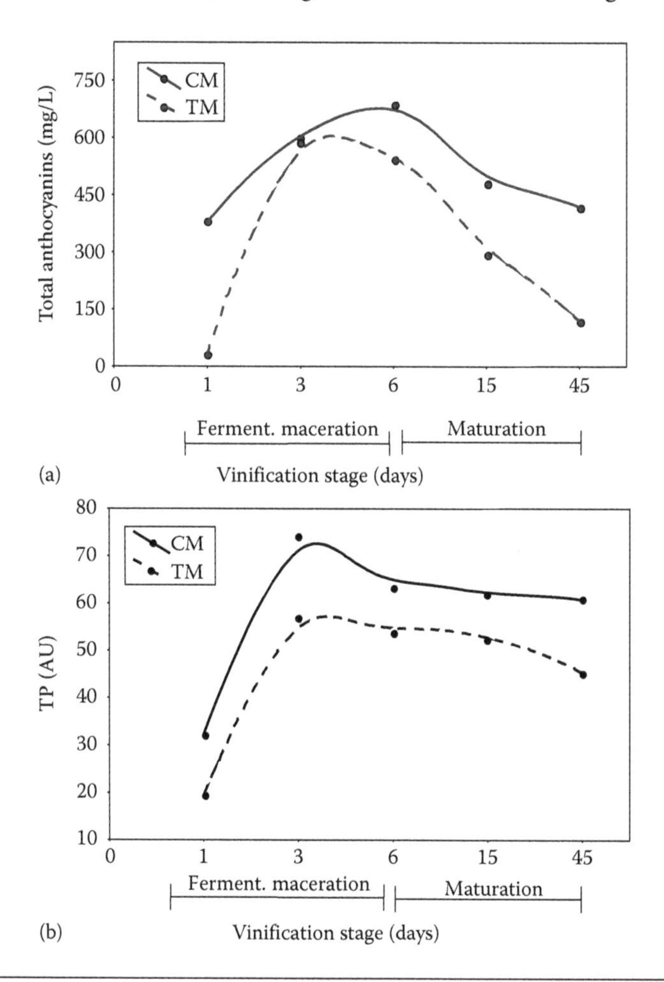

Figure 33.5 Evolution of TAs (a) and TPs (b) during vinification in CM wines and TM wines. Evolution of the anthocyanin fractions (mg/L) during vinification for both TM (c) and CM (d) wines.

(continued)

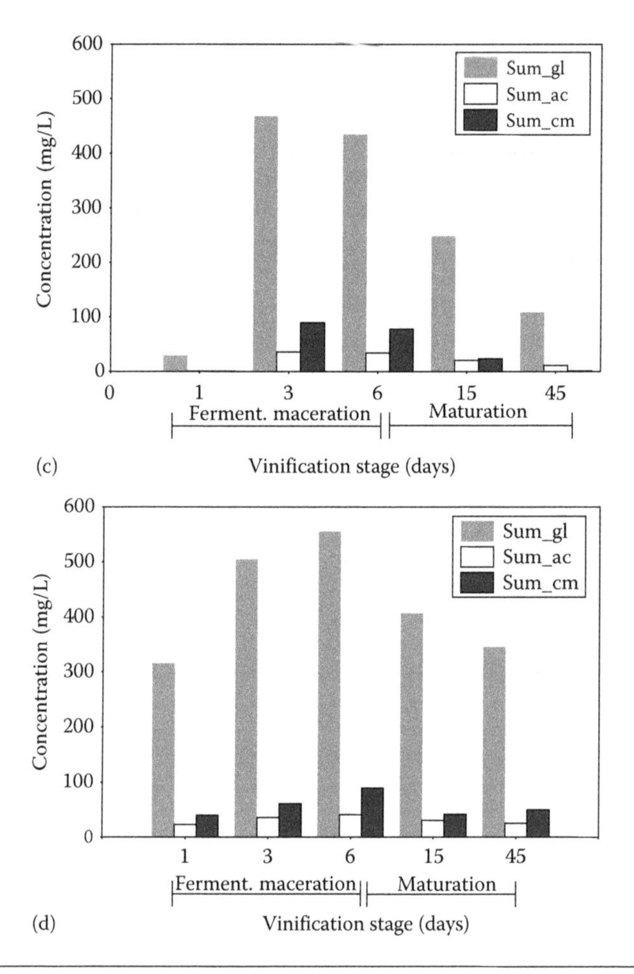

Figure 33.5 (continued)

Considering the evolution of CIELAB (L^*, C^*_{ab}, and h_{ab}) and CIELUV (s^*_{uv}) psychometric color parameters in the course of the two vinification protocols (Figure 33.7), it is observed that CM wines tended to show higher colorimetric stability than TM wines. It was confirmed that the application of a more intensive maceration, achieved by CM technique, induced higher color extraction in CM wines at the end of the alcoholic fermentation. At the moment of bottling, although higher grade of polymerization was obtained in TM wines, lighter and less saturated wines were finally obtained by this technique ($L^* = 83.52$ and $s^*_{uv} = 0.29$) vs. ($L^* = 76.36$ and $s^*_{uv} = 0.44$) CIELUV units, respectively, being these differences statistically significant ($p < 0.05$). In relation to the qualitative attribute of color (h_{ab}), due to a less intense evolution, CM wines kept their bluer tonalities for a longer time than the TM wines ($h_{ab} = -5.9°$ vs. $-1.5°$, respectively). The higher amount in bluish forms

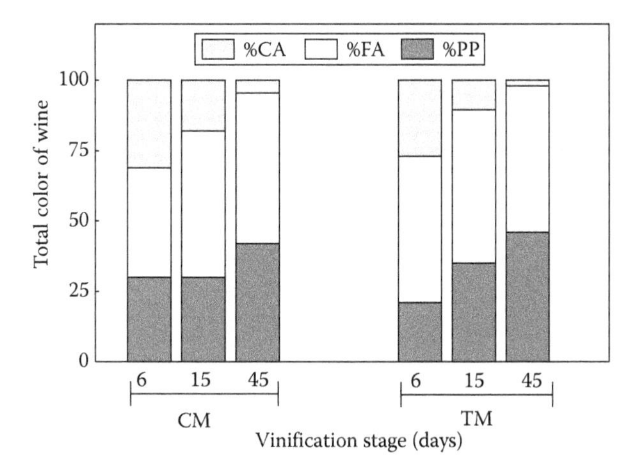

Figure 33.6 Evolution of the %CA, %FA, and %PP during the two vinification techniques.

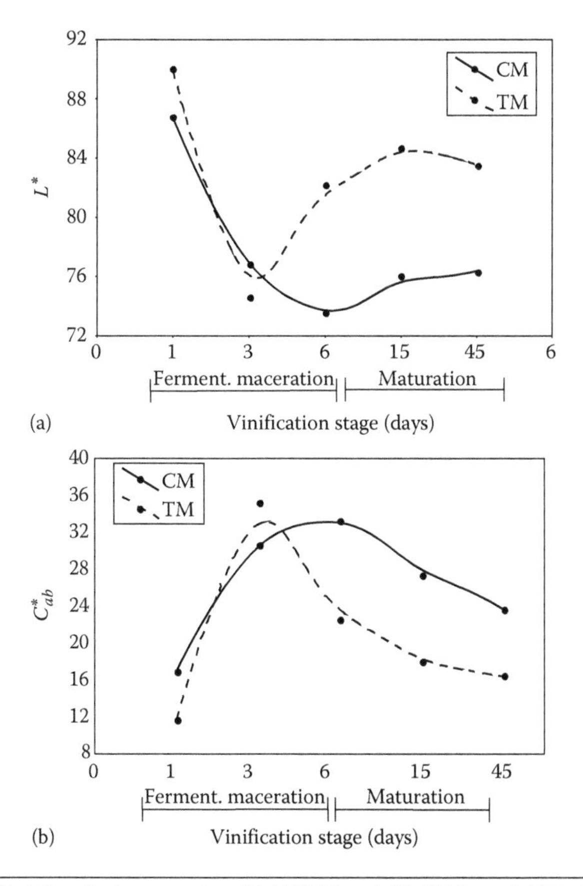

Figure 33.7 Evolution of color parameters: (a) L^* (lightness), (b) C^*_{ab} (chroma), (c) h_{ab} (hue angle), and (d) s^*_{uv} (saturation) during vinification. CM vs. TM wines.

(*continued*)

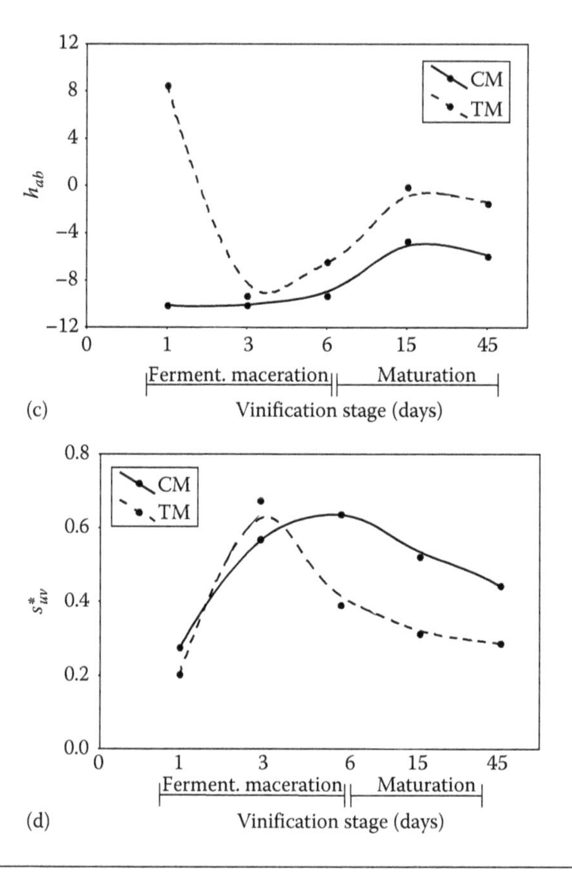

Figure 33.7 (continued)

of anthocyanins (malvidin, petunidin, and delphinidin) and the higher %CA could explain this finding (Heredia et al. 1998). The mean value of the color differences (ΔE^*_{ab}) calculated between the final wines was 9.6 CIELAB units, that is, perceptible by the human eye (Martínez et al. 2001). The differences of lightness (ΔL^*), chroma (ΔC^*_{ab}), and hue (Δh_{ab}) calculated among them showed that CM yields to darker, more saturated, and with more bluish color wines than those submitted to TM.

33.5 Final Remarks

The results obtained from the full set of observations confirmed the importance of both the quantitative and qualitative phenolic composition of grapes to achieve an effective magnitude and development of the copigmentation, which is required to produce high-quality red wines in warm climate, especially in terms of color. In general, it has been shown that SY wines reached better chemical and sensory characteristics than the other two grape varieties studied (TE and CS), but they show a greater difficulty to convert the copigmented complexes into more stable pigment. Concerning agronomical practices and vinification techniques, the copigmentation and the quality and stability of color were improved in SY wines made from more matured grapes and TE wines made from organic grapes, both of them elaborated by CM. These studies have practical interest for wineries, but an interesting observation is that the copigmentation measurement as an evaluation of changes at a unique wavelength does not show the integral color behavior of this phenomena.

Acknowledgments

We are indebted to Consejería de Innovación, Ciencia y Empresa (Junta de Andalucía, Spain) for the financial support (P07-AGR-02893 Research Project).

References

Bakker, J. and C. F. Timberlake. 1997. Isolation, identification, and characterization of new color-stable anthocyanins occurring in some red wines. *Journal of Agricultural and Food Chemistry* 45: 35–43.

Boulton, R. B. 1996. A method for the assessment of copigmentation in red wines. In *The 47th Annual Meeting of the American Society for Enology and Viticulture*, Reno, NV, June 1996.

Boulton, R. B. 2001. The copigmentation of anthocyanins and its role in the color of red wine: A critical review. *American Journal of Enology and Viticulture* 52: 67–87.

Brouillard, R., G. Mazza, Z. Saad, A. M. Albrecht-Gary, and A. Cheminatt. 1989. The copigmentation reaction of anthocyanins: A microprobe for the structural study of aqueous solutions. *Journal of the American Chemical Society* 111: 2604–2610.

Dangles, O. and R. Brouillard. 1992. Polyphenol interactions. The copigmentation case: Thermodynamic data from temperature variation and relaxation kinetics. Medium effect. *Canadian Journal of Chemistry* 70 (8): 2174–2189.

Gil-Muñoz, R., A. Moreno-Pérez, J. I. Fernández-Fernández, A. Martínez-Cutillas, and E. Gómez-Plaza. 2009. Influence of low temperature prefermentative techniques on chromatic and phenolic characteristics of Syrah and Cabernet Sauvignon wines. *European Food Research and Technology* 228: 777–788.

Gómez-Míguez, M., M. L. Gónzalez-Miret, and F. J. Heredia. 2007. Evolution of colour and anthocyanin composition of Syrah wines elaborated with pre-fermentative cold maceration. *Journal of Food Engineering* 79: 271–278.

González-Manzano, S., M. Dueñas, J. Rivas-Gonzalo, M. T. Escribano-Bailón, and C. Santos-Buelga. 2009. Studies on the copigmentation between anthocyanins and flavan-3-ols and their influence in the colour expression of red wine. *Food Chemistry* 114 (2): 649–656.

Gordillo, B., M. I. López-Infante, P. Ramírez-Pérez, M. L. González-Miret, and F. J. Heredia. 2010. Influence of prefermentative cold maceration on the color and anthocyanic copigmentation of organic tempranillo wines elaborated in a warm climate. *Journal of Agricultural and Food Chemistry* 58 (11): 6797–6803.

Goto, T. and T. Kondo. 1991. Structure and molecular stacking of anthocyanins—Flower color variation. *Angewandte Chemie International English Edition* 30: 17–33.

Heredia, F. J., M. L. Escudero-Gilete, D. Hernanz, B. Gordillo, A. J. Meléndez-Martínez, I. M. Vicario, and M. L. González-Miret. 2010. Influence of the refrigeration technique on the colour and phenolic composition of syrah red wines obtained by pre-fermentative cold maceration. *Food Chemistry* 118: 377–383.

Heredia, F. J., E. M. Francia-Aricha, J. C. Rivas-Gonzalo, I. M. Vicario, and C. Santos-Buelga. 1998. Chromatic characterization of anthocyanins from red grapes—I. pH effect. *Food Chemistry* 63: 491–498.

Martínez, J. A., M. Melgosa, M. Pérez, E. Hita, and A. I. Negueruela. 2001. Visual and instrumental color evaluation in red wines. *Food Science and Technology International* 7: 439–444.

Mira de Orduña, R. 2010. Climate change associated effects on grape and wine quality and production. *Food Research International* 43: 1844–1855.

Mori, K., S. Suyaga, and H. Gemma. 2005. Decrease anthocyanin biosynthesis in grape berries grown under elevated night temperature condition. *Scientia Horticulturae* 5: 319–330.

Zafrilla, P., J. Morillas, J. Mulero, J. M. Cayuela, A. Martinez-Cachá, F. Pardo, and J. M. López Nicolás. 2003. Changes during storage in conventional and ecological wine: Phenolic content and antioxidant activity. *Journal of Agricultural and Food Chemistry* 51: 4694–4700.

CHAPTER 34

Color–Composition Relationships of Seeds from Red Grape Varieties

M. JOSÉ JARA-PALACIOS, JUAN MANUEL ZALDÍVAR-CRUZ, FRANCISCO JOSÉ RODRÍGUEZ-PULIDO, DOLORES HERNANZ, FRANCISCO J. HEREDIA, and M. LUISA ESCUDERO-GILETE

Contents

34.1 Introduction

Grape (*Vitis vinifera*) is one of the world's largest fruit crops with a global production of around 55 million tones and 26 million tones of wine in 2008. This generates a large amount of pomace, including grape stems, skins, and seeds (OIV 2009). Recent investigations have stressed the importance of vinification by-products as plant materials, particularly rich in a wide range of polyphenols. Phenolics are chemicals that have different structures and biological activity and can be divided into two groups: nonflavonoids (hydroxybenzoic and hydroxycinnamic acids and stilbenes) and flavonoids (anthocyanins, flavonols, and flavanols) (Rodríguez Montealegre et al. 2006). Red grape seeds are excellent sources of monomeric phenolic compounds such as catechin, epicatechin, epicatechin gallate, and dimeric, trimeric, and tetrameric procyanidins as well as highly polymerized proanthocyanidins. Phenolic compounds of wine have also attracted much interest due to their antioxidant properties and their potentially beneficial effects for human health (Scalbert et al. 2005), and they are important determinants in sensory and nutritional quality of fruits and vegetables. The study of polyphenols and its relationship with the antioxidant power is very interesting for the wine sector (Borbalán et al. 2003) because by-products are cheap sources of natural antioxidants.

The aim of this study is to relate the chemical composition of the seeds of *V. vinifera* with the color of the extracts obtained. Recently, extraction of polyphenols from grape seeds has emerged as an opportune and vital business for the wine industry. Color is a very important factor in the quality of wine and is related to phenolic compounds, and CIE (*Commission Internationale de l'Éclairage*) colorimetry is widely used for evaluating the quality and composition of foods. The color of the extracts is related to the total phenolic content (TPC); therefore, the CIE colorimetry appears as an adequate technique to control the extraction process of the phenolic content in grape seeds. Extraction conditions influence the color of the extracts, and therefore, the color–composition relationships allow optimizing the extraction process.

34.2 Phenolic Compounds in Red Grape Seeds

Grape pomace (rich source of polyphenols) consists of the skin, stems, and seeds of grapes that remain after processing in the wine and juice industry. The total extractable phenolics in grape are present over 10% in the pulp, 60%–70% in the seeds, and 28%–35% in the skin. Studies regarding vinification by-products are mainly focused on the polyphenolic composition of seeds. Catechin and epicatechin are compounds very important in grape seeds, but the frequent presence of procyanidin oligomers and polymers (procyanidin B1, B2, B3, and B4) and of hydroxybenzoic acids such as gallic acid can also be seen. According to Iacopini et al. (2008), Merlot and Cabernet Sauvignon varieties showed high level of catechin (138.8 and 141.8 mg/100 g seeds) and epicatechin (131.8 and 127.6 mg/100 g seeds) in seed extracts. These results were confirmed by other authors in other varieties (Santos-Buelga et al. 1995; Guendez et al. 2005; Rodríguez Montealegre et al. 2006). Whereas Monagas et al. (2003) analyzed two Spanish original *V. vinifera* varieties (Tempranillo and Graciano) in which epicatechin was the main compound (62 and 312 mg/100 g seeds, respectively). Bucic-Kojic et al. (2009) observed that catechin was the most abundant individual compound in all tested grape seed extracts (average 45.11%) followed by epicatechin (34.45%), procyanidin B2 (12.90%), and gallic acid (5.34%).

In this study, seeds from two widely cultivated wine grape varieties were chosen: Tempranillo and Syrah. All samples were taken in the period of grape ripening and obtained from vineyards located in "Condado de Huelva" (designation of origin), in southwestern Spain, with the typical climatological conditions of warm climate regions.

Conditions in the extraction process are very important for the recovery of polyphenols from the sample. Plant polyphenols have very different structures, so it is practically impossible to find an only method for the extraction of all polyphenols. Besides, extraction of phenolic compounds from plant materials is influenced by several factors: solvent, temperature, solid–liquid ratio, time of extraction, etc. (Robards 2003). In our study, ethanol showed to be an efficient solvent to extract polyphenols from grape seeds (Nawaz et al. 2006). The individual phenolic compounds were analyzed by HPLC system (Agilent 1200 series) equipped with diode-array detector (280, 320, and 360 nm) following the method of Hernanz et al. (2007) with some modifications. In the analysis of individual phenolic compounds were identified 10 polyphenols before the 30 min of retention time. Catechin was found at the highest levels; its concentration decreased with the degree of ripeness from 1.10 to 0.39 mg/g and from 4.79 to 0.65 mg/g in Tempranillo and Syrah seeds, respectively. To assess the existence or absence of statistically significant differences among the varieties studied (Tempranillo and Syrah) regarding their phenolic compounds, ANOVA was applied. For this purpose, the phenolics are grouped in benzoic acids and flavonoids (considering the sum of all the individual phenolic compounds). Significant differences were found between the flavonoid content of each variety. When two-way ANOVA was applied, significant interactions ($p < 0.05$) between varieties and degree of ripeness were observed for flavonoid content. Phenolic content, mainly flavonoid content, decreases during ripening period, although this decrease is not statistically significant (Figure 34.1).

The amount of polyphenolic compounds in grapes depends on the variety of grapevine and is highly influenced by viticultural and environmental factors such as light, temperature, altitude, soil type, water, nutritional status,

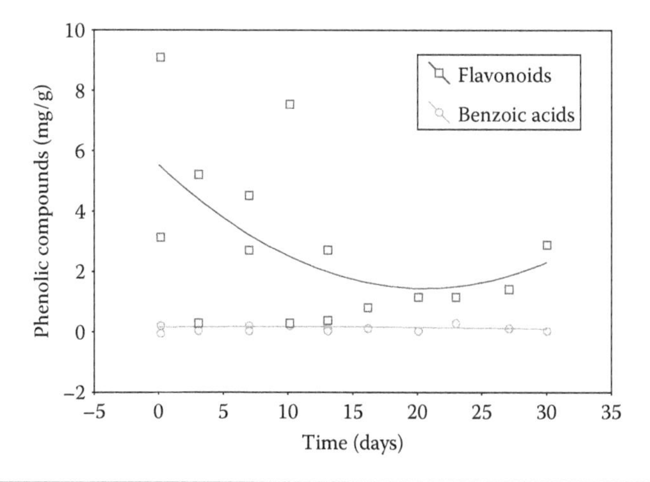

Figure 34.1 Evolution of flavonoids and benzoic acids.

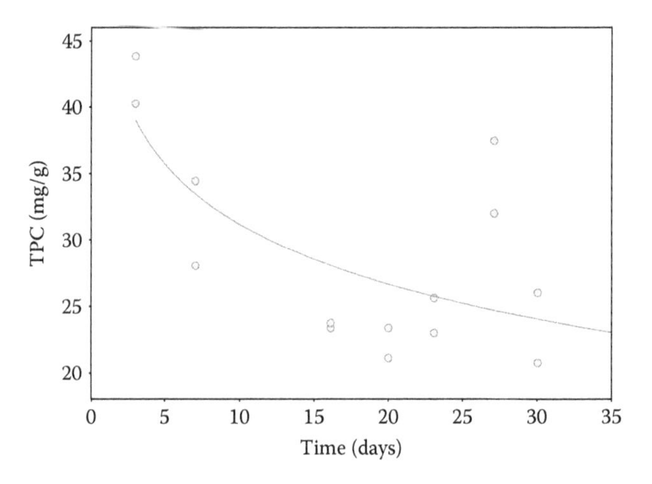

Figure 34.2 Evolution of TPC.

pathogenesis, and various developmental processes (Downey et al. 2006). Bozan et al. (2008) analyzed the TPC in the seeds of red grape international varieties (Merlot, Cabernet Sauvignon, Cinsault, and Alphonso Lavallee) widely cultivated in Turkey. Seeds from these varieties had different TPCs (88.11–105.7 mg/g seeds). In other studies (Guendez et al. 2005), TPC ranged from 1.43 to 22.28 mg/g seeds. In addition, polyphenolic composition is not stable during the ripening of the grapes but is modified qualitatively and quantitatively along the same. Phenolic compounds are synthesized from the early stage of berry development and decline toward ripening (Conde et al. 2007).

In this study, TPC in the extracts was determined by the Folin–Ciocalteu spectrophotometric method and expressed as gallic acid equivalents (mg GAE/g dry seeds). With regard to the evolution of the TPC, a decrease is detected during the degree of ripeness (Figure 34.2). TPC varied from 43.9 to 20.7 mg/g of seeds, although this decrease is not significant ($p<0.05$). The extraction procedures can also affect the determination of the phenolic content of the samples (Downey et al. 2007) and be the cause of the variability in the amount of total phenols.

The differences with respect to the results reported by other authors in relation to TPC were attributable to climatic differences, in our case very hot summers with very high environmental temperatures, typical of "Condado de Huelva."

34.3 Color–Composition Relationships of Grape Seed Extracts

Color is a very important factor in the quality of wine and is related to phenolic compounds. The interest of winemakers in the polyphenol content of grapes is increasing, as it offers ways of influencing the color, bitterness, astringency, mouth-feel, and ageability of wines. Sensorial qualities are related to the content in phenolic compounds because they are important determinants in sensory and nutritional quality of fruits and vegetables and play an important role in the quality of grapes and wines. Anthocyanins are directly responsible for color; flavonols and hydroxycinnamic acid derivatives are involved in the stabilization of anthocyanins in young red wines through copigmentation (Darías-Marín et al. 2001); moreover, flavonols seem to contribute to bitterness and are mainly responsible for astringency, bitterness, and "structure" of wines (Kennedy et al. 2006).

The application of colorimetric systems, based on uniform color spaces (CIELUV and CIELAB) and nonuniform color spaces (CIE *XYZ*), is of great value in the quantification and characterization of the color properties of pigments and foods. In this study, the relationship of phenolic compounds with color of the extracts was estimated. The phenolic compounds quantified by spectrophotometric methods were related to chroma (C^*_{ab}) and lightness (L^*). The spectral transmittances of the extracts were measured. The CIELAB parameters (L^*, a^*, b^*, C^*_{ab}, and h_{ab}) were determined by using the original software CromaLab® (Heredia et al. 2004), following the CIE's recommendations (CIE 2004): the 10° standard observer and the standard illuminant D65.

Figure 34.3 shows the distribution of the seeds grouped according to the grape varieties (Tempranillo and Syrah) in the (a^*b^*) plane, in which the color points are represented regarding the axes green-red ($-a^* + a^*$) and blue-yellow ($-b^* + b^*$). It can be observed that most of the samples are located inside a defined area between 80° and 110° of hue angle (h_{ab}). The seed extracts from Tempranillo tend to be located in the first quadrant of (a^*b^*)

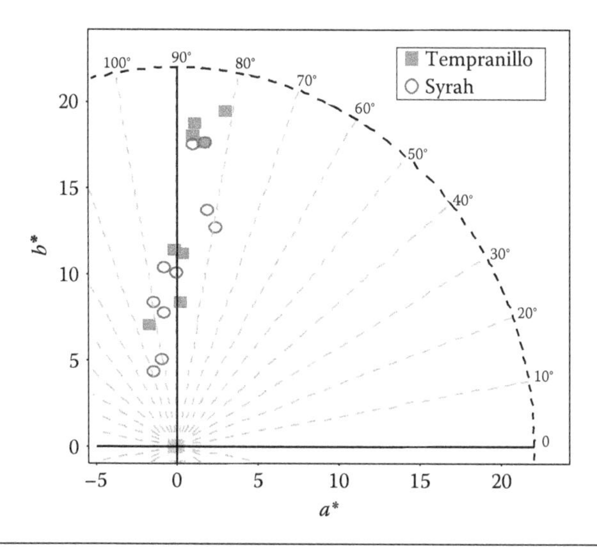

Figure 34.3 (a^*b^*) diagram of seed extracts from two red grape varieties.

diagram (80°–90°). On the other hand, the samples from Syrah tend to be located between the first and the second quadrants (80°–100°). Certain differences were observed for the color of the extracts from different varieties, although these differences were not significant ($p < 0.05$). However, there was a significant difference ($p > 0.05$) for L^* when the ripeness evolution time was considered.

The relationships between the color parameters and the phenolic compounds were explored by means of simple correlations. Significant correlations ($p < 0.05$) were found between b^* and the gallic acid levels and between C^*_{ab} and gallic acid.

In order to observe the contribution of color parameters to differentiate between the varieties, the results are subjected to a discriminant analysis. Forward stepwise method selected b^* and L^* as the variables of highest discriminant capacity (according to Fisher's test). The calculated discriminant function allows discrimination between Tempranillo and Syrah, as observed in Figure 34.4.

So far, there have been some studies that relate the color and phenolic content of a sample. Montes et al. (2005) studied correlations between the yield of anthocyanins and the color parameters. A high and significant ($p < 0.05$; $r = 0.84$) correlation was found between the yield anthocyanins and chroma. In a study realized by Bucic-Kojic et al. (2009), the relationship of determined compounds from grape seeds with the color of the extracts was also estimated. Results indicated a high positive correlation between total polyphenols and color parameters ($r^* > 0.93$).

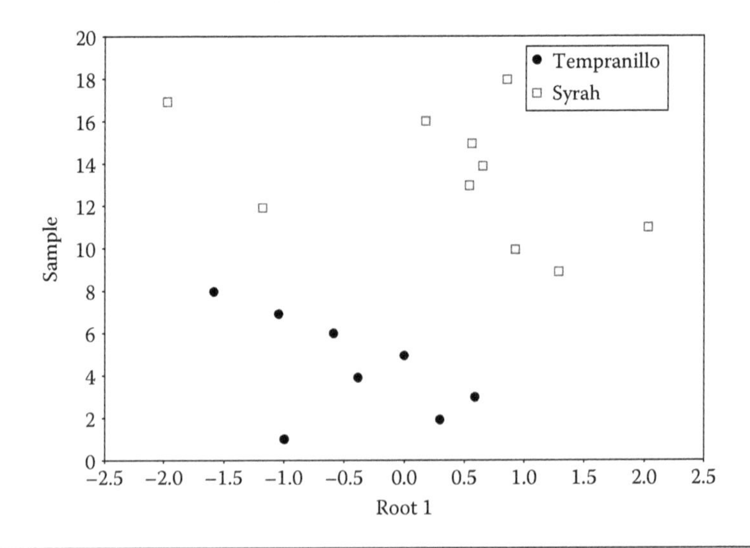

Figure 34.4 LDA scatterplot of the canonical functions.

34.4 Conclusions

The main polyphenolic components identified in seeds grapes of Tempranillo and Syrah are gallic acid, catechin, and epicatechin. The phenolic composition of grape seeds shows a direct relation with the color parameters of the extracts, so colorimetry is a suitable method for the extraction process control and provides an approximate measure of the phenolic content in grape seeds during the period of ripening.

Acknowledgment

M. José Jara-Palacios holds a predoctoral research grant from the Ministerio de Educación, Spain.

References

Borbalán A., L. Zorro, D. Guillén, and C. Barroso. 2003. Study of the polyphenol content of red and white grape varieties by liquid chromatography–mass spectrometry and its relationship to q antioxidant power. *Journal of Chromatography A* 1012: 31–38.

Bozan, B., G. Tosun, and D. Ozcan. 2008. Study of polyphenol content in the seeds of red grape (Vitis vinifera L.) varieties cultivated in Turkey and their antiradical activity. *Food Chemistry* 109: 426–430.

Bucic-Kojic A., M. Planinic, S. Tomas, L. Jakobek, and M. Seruga. 2009. Influence of solvent and temperature on extraction of phenolic compounds from grape seed, antioxidant activity and colour of extract. *International Journal of Food Science and Technology* 44: 2394–2401.

CIE (Commission Internationale de l'Éclairage). 2004. *Colorimetry*, 3rd edn. Vienna, Austria: CIE Central Bureau, Technical Report 15.2.

Conde, C., P. Silva, N. Fontes, A. C. P. Dias, R. M. Tavares, A. Agasse, S. Delrot, and H. Gerós. 2007. Biochemical changes throughout grape berry development and fruit and wine quality. *Food* 1: 1–22.

Darias-Martín, J., M. Carrillo, E. Díaz, and R. Boulton. 2001. Enhancement of red wine colour by pre-fermentation addition of copigments. *Food Chemistry* 73: 217–220.

Downey, M. O., N. K. Dokoozlian, and M. P. Krstic. 2006. Cultural practice and environmental impacts on the flavonoid composition of grapes and wine: A review of recent research. *American Journal of Enology and Viticulture* 57 (3): 257–268.

Downey, M. O., M. Mazza, and M. P. Krstic. 2007. Development of a stable extract for anthocyanins and flavonols from grape skin. *American Journal of Enology and Viticulture* 58: 358–364.

Guendez, R., S. Kallithraka, D. P. Makris, and P. Kefalas. 2005. Determination of low molecular weight polyphenolic constituents in grape (Vitis vinifera sp.) seed extracts: Correlation with antiradical activity. *Food Chemistry* 89: 1–9.

Heredia, F. J., C. Álvarez, M. L. González-Miret, and A. Ramírez. 2004. CromaLab®, análisis de color. Registro General de la Propiedad Intelectual: SE-1052-04 Sevilla, Spain.

Hernanz D., Á. F. Recamales, A. J. Meléndez-Martínez, M. L. González-Miret, and F. J. Heredia. 2007. Assessment of the differences in the phenolic composition of five strawberry cultivars (Fragaria x ananassa Duch.) grown in two different soilless systems. *Journal of Agricultural and Food Chemistry* 55: 1846–1852.

Iacopini P., M. Balde, P. Storchi, and L. Sebastian. 2008. Catechin, epicatechin, quercetin, utin and resveratrol in red grape: Content, in vitro antioxidant activity and interactions. *Journal of Food Composition and Analysis* 21: 589–598.

Kennedy, J. A., C. Saucier, and Y. Glories. 2006. Grape and wine phenolics: History and perspective. *American Journal of Enology and Viticulture* 57 (39): 239–248.

Monagas, M., C. Gómez-Cordovés, B. Bartolomé, O. Laureano, and J. M. Ricardo da Silva. 2003. Monomeric, oligomeric, and polymeric flavan-3-ol composition of wines and grapes from Vitis vinifera L. Cv. Graciano, Tempranillo, and Cabernet Sauvignon. *Journal of Agricultural and Food Chemistry* 51: 6475–6481.

Montes C., I. M. Vicario, M. Raymundo, R. Fett, and F. J. Heredia. 2005. Application of tristimulus colorimetry to optimize the extraction of anthocyanins from Jaboticaba (Myricia Jaboticaba Berg). *Food Research International* 38: 983–988.

Nawaz, H., J. Shi, G. Mittal, and Y. Kakuda. 2006. Extraction of polyphenols from grape seeds and concentration by ultrafiltration. *Separation and Purification Technology* 48: 176–181.

OIV, 2009. Situation and statistics of the world vitiviniculture sector in 2007. Organisation Internationale de la Vigne et du Vin 2009. http://www.oiv.int.

Robards, K. 2003. Strategies for the determination of bioactive phenols in plants, fruit and vegetables. *Journal of Chromatography A* 1000: 657–691.

Rodríguez Montealegre, R., R. Romero, J. L. Chacón, J. Martínez, and E. García. 2006. Phenolic compounds in skins and seeds of ten grape *Vitis vinifera* varieties grown in a warm climate. *Journal of Food Composition and Analysis* 19: 687–693.

Santos-Buelga, C., E. M. Francia-Aricha, and M. T. Escribano-Bailon. 1995. Comparative flavan-3-ol composition of seeds from different grape varieties. *Food Chemistry* 53: 197–201.

Scalbert, A., C. Manach, C. Morand, C. Rémésy, and L. Jiménez. 2005. Dietary polyphenols and the prevention of diseases. *Critical Reviews in Food Science and Nutrition* 45: 287–306.

CHAPTER 35

Color as an Indicator for the Maillard Reaction at Mild Temperatures

The Effect of Reducing Sugars

GRACIELA LEIVA, LETICIA GUIDA, GABRIELA NARANJO, and LAURA MALEC

Contents

35.1 Introduction

Maillard browning is one of the main chemical reactions occurring during the processing and storage of foods containing reducing sugars and proteins. This reaction affects protein quality and gives rise to compounds responsible for color and flavor changes. It is responsible for undesirable attributes in foods such as off-flavors, loss of nutritional value, and generation of toxic compounds, which have to be limited, but also for pleasant sensory characteristics and formation of biologically active molecules which have, on the opposite, to be promoted.

It is useful to consider three stages in the Maillard reaction: initial, advanced, and final stages. In the early stage, the reducing sugar condenses with free amino group of amino acids or proteins to give a condensation product, N-substituted glycosylamine, via the formation of a Schiff's base and the Amadori rearrangement. When heating conditions are not too drastic, the Amadori product formed in this stage is rather stable (Mauron 1981). At this point, neither color nor flavor changes arise, while amino acid availability is reduced, lysine being the most affected amino acid due to the blockage of its free ε-amino group (Hurrell 1990). Therefore, the estimation of available lysine loss can be used as an indicator of the extent of the reaction.

Milk proteins are one of the best sources of lysine for human nutrition. However, the presence of sugars in dairy foods makes them very vulnerable to protein damage by Maillard reaction, lessening the availability of this amino acid (van Boekel 1998). This reaction is mainly induced by heat treatment. Nevertheless, some authors have demonstrated that lactosylation also occurs during storage of milk powder (De Block et al. 1998; Guyomarc'h et al. 2000).

The advanced Maillard reaction consists of dehydration and fission of the Amadori product into colorless reductones as well as furfurals (Hodge 1953) and fluorescent compounds (Morales and van Boekel 1997). In the final stage, most of the color is produced due to the formation of brown polymers and copolymers called melanoidins. These compounds have significant effect on the quality of food, since color is an important food attribute and a key factor in consumer acceptance (Mauron 1990). Quantitative measurement of browning rate (brown compounds or color) is considered an indicator of severity of heat treatment (Andrews and Morant 1987; Buera et al. 1985; Horak and Kessler 1981; Rhim et al. 1988).

The rate of the reaction is strongly dependent on concentration, ratio, and chemical nature of reactants, temperature, time of heating, water activity and pH (Labuza and Baisier 1992). The regulation of these factors is one of the means to control the Maillard reaction progress. Kinetic studies may be useful for their predictive value on the characteristics of the products generated in the Maillard reaction.

The purpose of this work was to study the effect of heat treatment at mild temperatures in milk protein–sugar models on the development of color and the loss of lysine by nonenzymatic browning. The kinetic behavior of the Maillard reaction of casein (C) was compared to that of whey proteins (WP), and the influence of the nature of reducing sugar on the extent of the reaction was compared at the initial and final stages.

35.2 Materials and Methods

35.2.1 Materials

WP isolate (Davisco Foods International, Le Sueur, MN) contained 93.1% protein, 2.3% ash, 4.3% moisture, and 0.1% lactose. All chemicals were of analytical grade.

35.2.2 Preparation of Samples

Four model systems containing C or WP and glucose or lactose as reducing sugars were prepared. The initial sugar:available lysine molar ratio was the same as that of milk (9:1). Samples were prepared by freeze-drying dispersion of the dry ingredients (20% w/w) in phosphate buffer solution, 0.1 M, pH 6.5, containing 0.06% (w/w) potassium sorbate as an antimicrobial agent.

A freeze-dryer model ALPHA 1-4 LD2 (Martin Christ Gefriertrocknungsanlagen, GMBH, Germany) was used, which operated at a −55°C condenser plate temperature and a chamber pressure of less than 100 μm Hg during 48 h. Each system was equilibrated at 25°C in vacuum desiccators over saturated solution of $Mg(NO_3)_2$ at water activity (a_w) of 0.52, until weight was constant.

35.2.3 Kinetic Study

Several 1.5 g samples were sealed in glass flasks and stored at 37°C and 50°C. A pair of duplicate flasks was periodically removed and held at −18°C until available lysine content and color were analyzed. Control samples of the systems without thermal treatment were analyzed in triplicate and designated as time 0.

Kinetic rate constants for lysine loss (k) and their standard error were calculate by linear regression analysis of

$$L_t = L_0 \exp(-kt) \quad (35.1)$$

where

L_t is the lysine concentration at time t
L_0 is the lysine concentration at $t = 0$

The confidence intervals were estimated for a significance level of 95% by means of the Student t-test. Analysis of variance (ANOVA) was performed to check if the rate constants were significantly different from 0. The validity of the linear equations was tested by F-test for lack of fit.

35.2.4 Analytical Methods

The water activity of the milk systems was measured using an AquaLab Water Activity Meter Series 3TE with internal temperature control (Decagon Devices, Inc., Pullman, WA).

The kinetic of the initial stage in the nonenzymatic browning reaction was studied by measuring the extent of lysine loss over time using the *o*-phthalaldehyde/*N*-acetyl-L-cysteine spectrophotometric method reported by Medina Hernández and García Alvarez-Coque (1992). For analysis, samples were dissolved in 5% w/v sodium dodecyl sulfate solution. To 10 mL of the OPA-NAC reagent (25 mL of a 0.05 M ethanolic OPA solution, 25 mL of a 0.05 M aqueous NAC solution, and 200 mL of pH 9.5 boric acid-borate buffer solution [0.02 M] in 1 L of water), 2.0 mL of the sample solution was added and diluted to 25 mL with water. Absorbance was measured at 335 nm with a Hewlett Packard spectrophotometer HP 8453 (Hewlett Packard, Palo Alto, CA). Three replicates of each sample were analyzed. The coefficient of variation for this assay was <3%. The available lysine content was obtained from a standard curve plotted using C or WP as was previously reported by Malec et al. (2002).

Total nitrogen content was determined in duplicate by the Kjeldahl method with a digester Bloc Digest 6 P-Selecta (J. P. Selecta SA, Spain) and a nitrogen distillation unit Pro-Nitro M P-Selecta (J. P. Selecta SA, Spain).

The degree of browning (final stage) was evaluated using a handheld tristimulus reflectance spectrocolorimeter with an integrating sphere (Minolta CM-508-d, Minolta Corp., Ramsey, NJ). The colorimetric parameters L^*, a^*, and b^* were referred to illuminant D65 at 2° standard observer and in the CIELAB uniform color space. The color function $L_0^* - L^*$ was found to be an adequate parameter to evaluate the nonenzymatic browning reactions, being L_0^* and L^*, the sample color attribute before and after heat treatment, respectively. Two replicates were analyzed for each storage time.

35.3 Results and Discussion

The extent of the Maillard reaction at the initial stage was analyzed by the loss of available lysine and followed pseudo first-order reaction kinetics in all systems analyzed. This reaction order was in agreement with previous reports (Baisier and Labuza 1992; Naranjo et al. 1998; Pereyra Gonzáles et al. 2010; Warren and Labuza 1977). The rate constants for available lysine loss (k) in the model systems with C and WP stored at 37°C and 50°C are summarized in Table 35.1. At 37°C, the loss of lysine was approximately 50% higher in the WP-lactose system than in the C-lactose system. It should be noted that the difference between the reaction rates in the systems with WP and C decreases with rise of temperature, since at 50°C, there were no significant differences between the rate constants of both systems ($p > 0.05$). In contrast, the loss of lysine in the systems containing glucose was significantly higher ($p < 0.05$)

Table 35.1 First-Order Rate Constants (k) with 95% Confidence Limits for Available Lysine Loss from Model Systems with C and WP Stored at 37°C and 50°C

	$k \times 10^3$ (h^{-1})	
Model Systems	**37°C**	**50°C**
C-lactose	1.47 ± 0.08	10.0 ± 1.0
WP-lactose	2.14 ± 0.08	11.2 ± 0.4
C-glucose	9.5 ± 0.5	55 ± 2
WP-glucose	21 ± 1	146 ± 4

with WP than with C at both temperatures analyzed. In concordance with previous reports (van Boekel 1996; Chevalier et al. 2001; Naranjo et al. 1998), the reaction occurred at a considerably faster rate in the systems with glucose than in those containing lactose, and the differences between the rate constants of both sugars were higher in the systems containing WP than in those with C.

The final stage of the reaction was analyzed by the development of color using the color attribute lightness (L^*). This parameter was also employed to follow the rate of browning in dairy systems by other authors, and its decrease was attributed to the augmentation of pigment concentration (Dattatreya and Rankin 2006; Morales and van Boekel 1998; Sithole et al. 2005). At 37°C, in the systems containing lactose, no color was detected along the experiment, and in those with glucose, a slight decrease in lightness was noticeable only at prolonged times of storage, when the loss of lysine was about 50%. Hence, at storage temperatures, the final stage of Maillard reaction is not significant in systems containing milk proteins.

At a higher temperature, the effect was more pronounced. Figure 35.1 shows lysine loss and the progress of color in systems stored at 50°C. In the

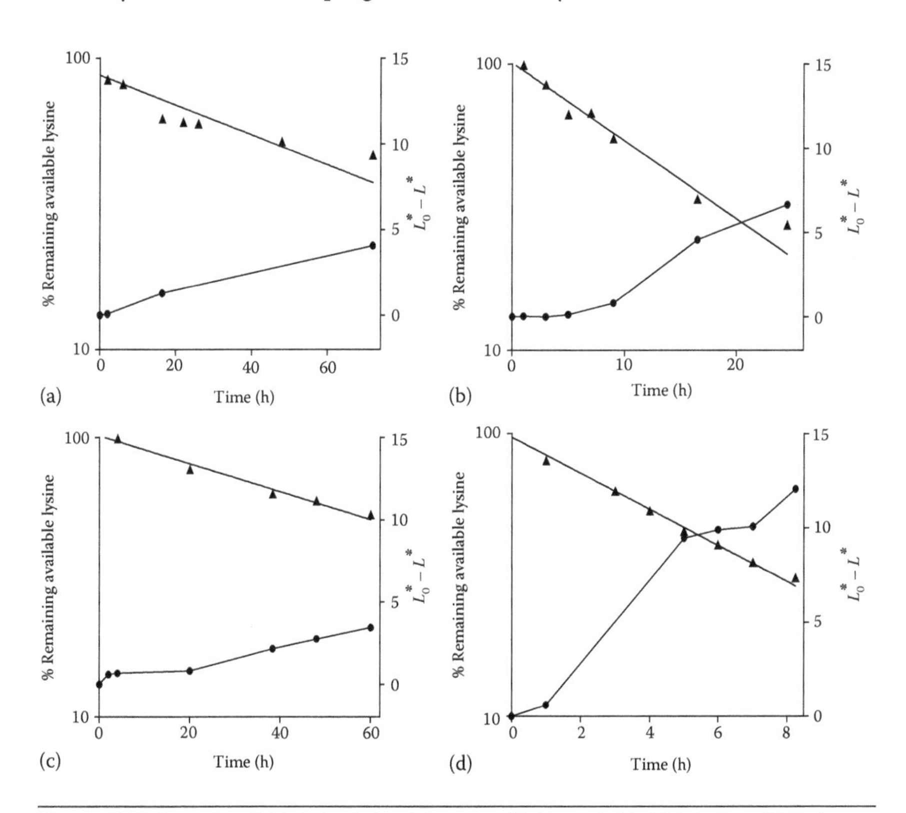

Figure 35.1 Loss of available lysine (▲) and decrease of lightness (●) in (a) C-lactose, (b) C-glucose, (c) WP-lactose, and (d) WP-glucose systems stored at 50°C.

systems with lactose, the decrease of lightness was similar for both proteins, whereas in the systems with glucose, browning developed fivefold faster with WP than with C. In agreement with Morales and van Boekel (1998), lightness decreased more rapidly in these systems than in those containing lactose. However, when the progresses of the initial and final stages of the reaction were compared, it could be seen that in most analyzed systems, color was detected only at prolonged storage times. The WP-glucose system was the only system where color was noticed when losses of lysine were lower than 40%.

Several formulas have been described by applying the parameters of the CIELAB color space. In the present study, besides lightness, the evolution of color was also assessed by the total color difference (ΔE^*) as was reported by many authors in dairy systems (Baechler et al. 2005; Ferrer et al. 2005; Guerra-Hernández et al. 2002; Morales and van Boekel 1998):

$$\Delta E^* = [(\Delta L^*)^2 + (\Delta a^*)^2 + (\Delta b^*)^2]^{1/2}. \tag{35.2}$$

Figure 35.2 shows the evolution of the total color difference as a function of time for each system stored at 50°C. Results were comparable to those obtained with $L_0^* - L^*$. In the systems with lactose, the increase of color was similar for both proteins and in those with glucose, WP were more reactive than C. Also, the increase of color was much higher in the systems with glucose than in those with lactose. When the evolutions of ΔE^* and $L_0^* - L^*$ against time were compared, it could be seen that the former reached higher values. Hence, the difference of color proved to be more sensitive, as it includes the variation of two other parameters. In addition, the differences between the reaction rates of the different systems observed for the progress of the total color (ΔE^*) were the same than those calculated for the loss of lysine.

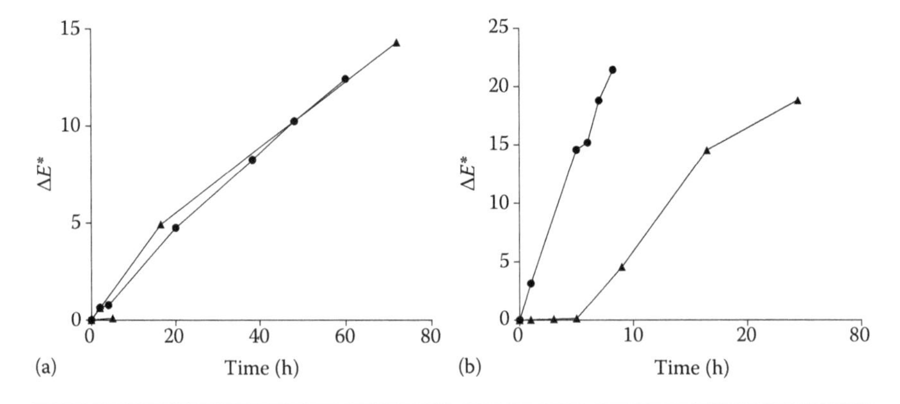

Figure 35.2 Change of color (ΔE^*) in systems containing C (▲) or WP (●) and (a) lactose or (b) glucose stored at 50°C.

It can be concluded that the influence of temperature and of the nature of the reactants was stronger on the first stage of the reaction than when color was analyzed. Therefore, at mild temperatures, parameters related to the early stage of the reaction, such as lysine loss, are better indicators of heat treatment than color. The latter would be suitable for the analysis of sensory changes caused by Maillard reaction but not to evaluate nutritional deterioration.

Acknowledgments

This work was supported by grants UBACyT X062 and X126 from Secretaría de Ciencia y Técnica, Universidad de Buenos Aires.

References

Andrews, G. R. and S. V. Morant. 1987. Lactulose content, colour and organoleptic assessment of UHT and sterilized milk. *Journal Dairy Research* 54: 493–507.

Baechler, R., M.-F. Clerc, S. Ulrich, and S. Benet. 2005. Physical changes in heat-treated whole milk powder. *Lait* 85: 305–314.

Baisier, W. M. and T. P. Labuza. 1992. Maillard browning kinetics in a liquid model system. *Journal of Agricultural and Food Chemistry* 40: 707–713.

van Boekel, M. A. J. S. 1996. Kinetic modelling of sugar reactions in heated milk-like systems. *Netherlands Milk and Dairy Journal* 50: 245–266.

van Boekel, M. A. J. S. 1998. Effect of heating on Maillard reactions in milk. *Food Chemistry* 62: 403–414.

Buera, M. P., R. D. Lozano, and C. Petriella. 1985. Definition color in the non-enzymatic browning process. *Die Farbe* 32: 316–326.

Chevalier, F., J. Chobert, Y. Popineau, M. G. Nicolas, and T. Haertlé. 2001. Improvement of functional properties of β-lactoglobulin glycated through the Maillard reaction is related to the nature of the sugar. *International Dairy Journal* 11: 145–152.

Dattatreya, A. and S. A. Rankin. 2006. Moderately acidic pH potentiates browning of sweet whey powder. *International Dairy Journal* 16: 822–828.

De Block, J., M. Merchiers, and R. Van Renterghem. 1998. Capillary electrophoresis of the whey protein fraction of milk powders. A possible method for monitoring storage conditions. *International Dairy Journal* 8: 787–792.

Ferrer, E., A. Alegria, R. Farré, G. Clemente, and C. Calvo. 2005. Fluorescence, browning index, and color in infant formulas during storage. *Journal of Agricultural and Food Chemistry* 53: 4911–4917.

Guerra-Hernández, E., C. Leon, N. Corzo, B. García-Villanova, and J. M. Romera. 2002. Chemical changes in powdered infant formulas during storage. *International Journal of Dairy Technology* 55: 171–176.

Guyomarc'h, F., F. Warin, D. D. Muir, and J. Leaver. 2000. Lactosylation of milk proteins during the manufacture and storage of skim milk powders. *International Dairy Journal* 10: 863–872.

Hodge, J. E. 1953. Dehydrated foods: Chemistry of browning reaction in model systems. *Journal of Agricultural and Food Chemistry* 1: 928–943.

Horak, F. P. and H. G. Kessler. 1981. Colour measurement as an indicator of heat treatment of foods, as exemplified by milk products. *Zeitschrift für Lebensmittel—Technologie und Verfahrenstechnik* 32: 180–184.

Hurrell, R. F. 1990. *The Maillard Reaction in Food Processing, Human Nutrition and Physiology*. Basel, Switzerland: Birkhäuser.
Labuza, T. P. and W. M. Baisier. 1992. *Physical Chemistry of Foods*. New York: Marcel Dekker.
Malec, L. S., A. S. Pereyra Gonzáles, G. B. Naranjo, and M. S. Vigo. 2002. Influence of water activity and storage temperature on lysine availability of a milk like system. *Food Research International* 35: 849–853.
Mauron, J. 1981. The Maillard reaction in food: A critical review from the nutritional standpoint. *Progress in Food and Nutrition Science* 5: 5–35.
Mauron, J. 1990. Influence of processing on protein quality. *Journal of Nutritional Science and Vitaminolog* 36 (1): 57–69.
Medina Hernández, M. J. and M. C. García Alvarez-Coque. 1992. Available lysine in protein, assay using o-phthalaldehyde/N-acetyl-L-cysteine spectrophotometric method. *Journal of Food Science* 57: 503–505.
Morales, F. J. and M. A. J. S. van Boekel. 1997. A study on advanced Maillard reaction in heated casein/sugar solutions: Fluorescence accumulation. *International Dairy Journal* 7: 675–683.
Morales, F. J. and M. A. J. S. van Boekel. 1998. A study on advanced Maillard reaction in heated casein/sugar solutions: Colour formation. *International Dairy Journal* 8: 907–915.
Naranjo, G. B., L. S. Malec, and M. S. Vigo. 1998. Reducing sugars effect on available lysine loss of casein by moderate heat treatment. *Food Chemistry* 62: 309–313.
Pereyra Gonzáles, A. S., G. B. Naranjo, G. E. Leiva, and L. S. Malec. 2010. Maillard reaction kinetics in milk: Effect of water activity at mild temperatures. *International Dairy Journal* 20: 40–45.
Rhim, J. W., V. A. Jones, and K. R. Swartzel. 1988. Kinetics studies in the colour changes of skim milk. *Lebensmittelwissenschaft und—Technologie* 21: 334–338.
Sithole, R., M. R. McDaniel, and L. Meunier Goddik. 2005. Rate of Maillard browning in sweet whey powder. *Journal of Dairy Science* 88: 1636–1645.
Warren, R. M. and T. P. Labuza. 1977. Comparison of chemically measured available lysine with relative nutritive value measured by a *Tetrahymena* bioassay during early stages of nonenzymatic browning. *Journal of Food Science* 42: 429–431.

CHAPTER 36

Statistical Relationships between Soil Color and Some Factors of Soil Formation

ÁNGEL MARQUÉS-MATEU, SARA IBÁÑEZ, HÉCTOR MORENO, JUAN M. GISBERT, SEBASTIÀ BALASCH, and MARIANO AGUILAR

Contents

36.1 Introduction

Soil is a key element in crop development and has always been a topic of great interest to humankind and its development. This interest is the foundation of soil science, which is the study of the formation, properties, ecology, and classification of soils.

The classical theory of soil science establishes the following equation of soil formation:

$$s = f(cl, o, r, p, t)$$

The formation factors are climate (cl), organisms (o), topography (r), parent material (p), and time (t). The outcome of the equation is what we know as soil. It seems clear that the combined action of the previous factors results in very complex processes.

There is not a generally accepted definition for the term soil. Jenny (1941) provides a number of definitions given by different authors. In the context of agriculture and soil science, soil is the body where plants find foothold, nourishment, and other conditions of growth.

One of the most important features of soil is anisotropy which makes it different from other geological formations. The direction of maximum anisotropy is the vertical or z-axis, which allows for visual identification of different layers from texture, structure, color, and other characteristics. The detailed description of the different soil layers is known as a soil profile.

Color provides valuable information on the process of soil formation as well as on constituent elements and other properties. Soil color is the variable of interest in this study. We try to obtain significant statistical relationships between soil color and other agricultural variables which are related to formation factors in some way.

36.1.1 Study Area

The study area is located in the province of Alicante (southeastern Spain). The statistical analyses were performed on data from several agricultural fields belonging to the municipality of Sax (Figure 36.1). The total area of Sax is 6350 ha, and the geographical coordinates are 38°34′ N latitude and 0°48′ W

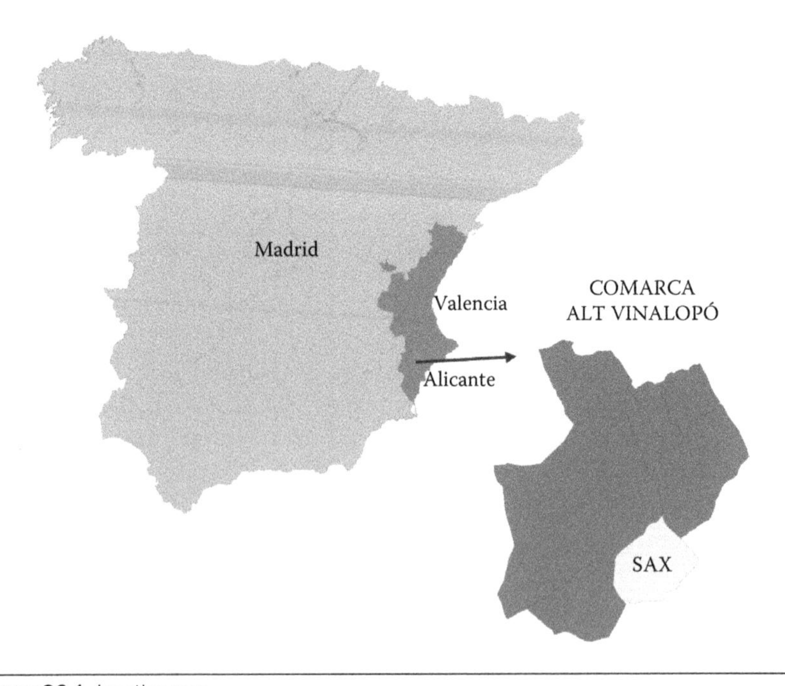

Figure 36.1 Location map.

longitude. The study area is relatively small in size; therefore, factors such as climate, parent material, or topography are well defined.

The climate is Mediterranean with clear continental features. Summers are dry and winters are cold. Rain is abundant in the equinoxes. The average annual rainfall is 305.5 mm, and the average annual temperature is 15.2°C.

The most prominent geomorphological features are the mountain ranges of Cabreras, Peña Rubia, La Argueña, and Cámara which define two north–south oriented valleys (Figure 36.2). The elevation of the terrain ranges from 470 to 800 m above mean sea level.

The agricultural lands are located in the hillsides of the valleys, mainly in terraces. The principal crops are vineyards, olive trees, and almond trees, either under irrigation or not. Some fields are cultivated using conservation agriculture techniques, and the stoniness varies greatly. Natural vegetation is mainly composed of Aleppo pines (*Pinus halepensis Mill.*) and various species of Mediterranean bush.

Parent materials are Tertiary limestone and marls, and Triassic clay materials. Limestone above 600 m of elevation is subject to severe water erosion which greatly influences the processes of soil formation.

The variability of soil color is basically determined by the parent materials. There are red and brown soils, modified by different contents of calcium carbonate and organic matter, which produce lighter and darker colors, respectively.

The most common soil groups are Aridisols and Entisols (Soil Survey Staff 2010). There are also less evolved soils, called Orthents, in steep slopes above 600 m. These cannot be used for agriculture. Calcids, Argids, Cambids, and Salids, with calcic, argillic, cambic, and gypsic horizons, are found in areas of

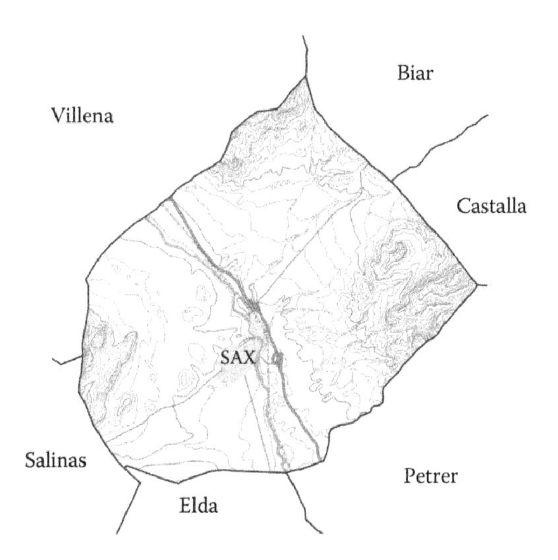

Figure 36.2 Topographic map.

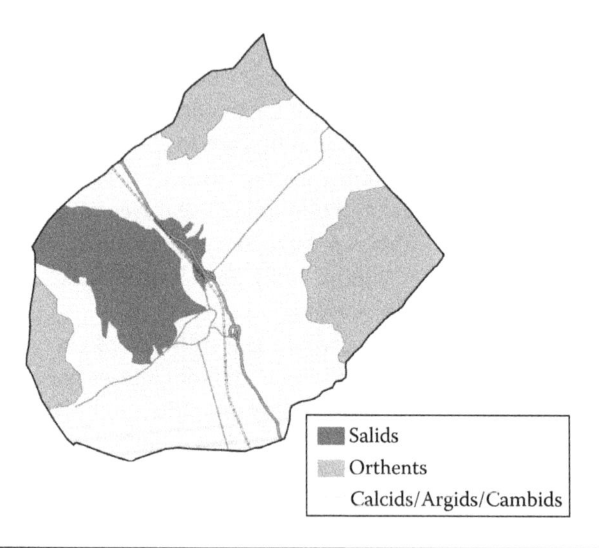

Figure 36.3 Soil map.

gentle slope with elevations between 450 and 600 m. Calcids are located in the foothill areas and support most of the unirrigated crops. Sometimes Calcids have crusts and calcium carbonate cementations. Salids form next to an outcrop of Keuper salts. Alluvial and colluvial soils are located near the banks of river Vinalopó. These young soils have excellent physical properties which make them suitable for agricultural purposes (Figure 36.3).

36.2 Preparation and Processing of Data

The data were gathered using a geographical information system (GIS). GIS can be defined as a system of hardware, software, and procedures designed to support the capture, management, manipulation, analysis, modeling, and display of spatially referenced data for solving complex planning and management problems (NCGIA 1990).

In a GIS context, there are two basic approaches to represent geographical data: the field and object models (Worboys and Duckham 2004). The field-based model treats geographical information as a collection of spatially distributed values. A typical example of field-based information is a pattern of terrain elevations. The object-based model represents the space by means of discrete elements with geographical reference. Basic objects are points, lines, and polygons.

The object model is usually associated with vector representation, whereas the field model is associated with raster representation (Burrough and McDonnell 1998). Both models were used in this study.

The data were obtained in two stages. The first stage consisted of field work carried out to locate and collect soil samples. It is very important to record sample coordinates as accurate as possible in order to properly relate soil samples to other variables. It is also necessary to assign a unique identifier to each sample. The sample coordinates, together with their identifiers, form the sample point map.

The geographical locations of samples were collected with handheld GPS receivers. Other variables were obtained from existing maps. The two data sets, i.e., the sample points and the maps, have different coordinate systems and must be transformed into a common reference system. The goal of this transformation is to line up the different data sets.

Once the spatial data sets are stored in the same coordinate system, samples can be overlaid on existing maps or images to capture spatially distributed variables directly from digital maps (Figure 36.4). Spatially distributed variables include elevation, slope, aspect, and parent material, among others.

The second stage of data collection consisted of laboratory measurements. A number of variables, including soil color, were measured in the laboratory from soil samples. The numerical values of these variables were arranged in tables, which were subsequently linked to the point map using the relational join operator.

The resulting data set can be divided into four groups of variables:

1. Geomorphological variables. The values of elevation above sea level, slope, and aspect were collected from the digital terrain model of the study area.
2. Edaphic variables. This group includes carbonate content, pH, electrical conductivity, texture, and stoniness. Their values were obtained by using specific laboratory procedures except in the case of stoniness, whose values were obtained by direct photographic interpretation.
3. Management variables. These variables include crop and irrigation system and were collected during field visits.
4. Color variables. Chromaticity coordinates were collected using a colorimeter under standard conditions. We used illuminant D65 as the light source and 45/0 geometry. The measurements were performed with a field of view of 2°.

This data set is suitable for statistical processing. There are different statistical techniques available to process multivariate data (McGarigal et al. 2000). In this study, we used the method of stepwise regression. This method is suitable in situations with collinearity between independent variables, which is common in data sets that do not come from experimental design.

36.3 Elements of Colorimetry

In the context of soil science, color is one of the basic properties reported in soil profile descriptions (Bigham and Ciolkosz 1993). The common approach

Figure 36.4 Point map overlaid onto an aerial photograph.

to communicating soil color is based on the use of specific Munsell charts under natural illumination (Munsell Color Co. 1980, Soil Survey Division Staff 1993). According to the Munsell soil color chart, color is reported in terms of the so-called hue, chroma, and value. However, we used a different laboratory procedure based on CIE standards and colorimeter measurements.

Colorimeter measurements provided chromaticity coordinates x and y and the tristimulus value Y of both soil samples and the reference white target. Chromaticity coordinates of all samples were then transformed into tristimulus values XYZ using well-known formulas (CIE 2004):

$$X = x \cdot \left(\frac{Y}{y}\right)$$

$$Z = (1 - x - y) \cdot \left(\frac{Y}{y}\right)$$

Then, tristimulus values were converted into CIELAB (*Commission Internationale de l'Éclairage*, L^*, a^*, and b^*) coordinates:

$$L^* = 116 \times \left(\frac{Y}{Y_n}\right)^{1/3} - 16$$

$$a^* = 500 \times \left[\left(\frac{X}{X_n}\right)^{1/3} - \left(\frac{Y}{Y_n}\right)^{1/3}\right]$$

$$b^* = 200 \times \left[\left(\frac{Y}{Y_n}\right)^{1/3} - \left(\frac{Z}{Z_n}\right)^{1/3}\right]$$

where

X, Y, and Z are the tristimulus values of the soil sample

X_n, Y_n, and Z_n are the tristimulus values of the reference white

There are alternative formulas when the quantity X/X_n, Y/Y_n, or Z/Z_n is less than 0.008856 (CIE 2004). However, this rarely happens in soil color measurements.

Finally, CIELAB coordinates were transformed into the quantities L^*, C^*_{ab}, and h^*_{ab}. These quantities are expressed in polar coordinates in what is known as the CIELCH system. The new quantities are the chroma

$$C^*_{ab} = \left[a^{*2} + b^{*2}\right]^{1/2}$$

and the angle of hue

$$h_{ab}^* = \arctan\left(\frac{b^*}{a^*}\right)$$

CIELAB hue angle is 0 for any stimulus along the positive direction of the a^* axis, also known as the red-green axis.

We prefer CIELCH representation over CIELAB coordinates since C_{ab}^* and h_{ab}^* are physical correlates of the perceptual attributes of color. The CIELCH system is also easier to use than classical Munsell notations when performing numerical computations, as is the case in this study.

36.4 Results and Discussion

The procedures described so far result from a number of relationships which are summarized in this section. It is necessary to note that the results presented here are not based on experimental design but on the sample values themselves. Therefore, we cannot establish true cause-and-effect relationships, and the results should be considered as a first approach to the problem. Nevertheless, the use of GIS together with multivariate statistical methods provides an interesting starting point in the study of soil color from a geospatial perspective.

36.4.1 Lightness (L^*)

There are seven significant variables related to L^*; five of them are positively correlated and two are negatively correlated (Table 36.1). In this context, positively correlated means that L^* has greater values, i.e., soil color is lighter, and negatively correlated means that soil color is darker. The value of the determination coefficient (R^2) was 39%.

Table 36.1 Regression Analysis of Lightness (L^*)

Parameter	Estimate	*p*-Value
Constant	19.543	0.2865
Aspect	0.033	0.0194
Geo28: marl	5.093	0.0237
Sand	−0.124	0.0178
Clay	−0.137	0.0637
Carbonate	0.092	0.0586
pH	4.207	0.0456
EC	0.002	0.0185

There are some relationships which were expected. For instance, high carbonate content and parent material no. 28 (white marl) should produce lighter soils. Results referring to aspect, which determines the amount of solar radiation that falls on the terrain surface, are not so obvious. The most surprising results are those of pH and EC (electrical conductivity). They are very difficult to explain because their values can be greatly affected by tillage practices, or water quality, which were not taken into account in this study.

36.4.2 Hue (h^*_{ab})

There are three significant variables related to hue (Table 36.2). The value of R^2 is 34%.

Slope is positively correlated to hue. This suggests that soils located on higher slopes tend to be less red. Sand and clay are negatively correlated, which suggests that high contents of sand and clay produce lower values of hue. In this context, the soil color approaches red.

36.4.3 Chroma (C^*_{ab})

There are four significant variables related to chroma. Parent material no. 28 and stoniness are negatively correlated, whereas sand and pH are positively correlated (Table 36.3). The value of R^2 is 39%.

It seems that the color of samples located on white marls is less saturated than other soils. This is in agreement with the expected result. The influence of stoniness (i.e., the percentage of soil surface covered by stones or rocks), sand, and pH on chroma is harder to explain.

Table 36.2 Regression Analysis of Hue (h^*_{ab})

Parameter	Estimate	*p*-Value
Constant	82.384	0.0000
Slope	0.324	0.0621
Sand	−0.199	0.0000
Clay	−0.193	0.0010

Table 36.3 Regression Analysis of Chroma (C^*_{ab})

Parameter	Estimate	*p*-Value
Constant	−7.776	0.3496
Geo28: marl	−2.352	0.0279
Sand	0.102	0.0000
Stoniness	−0.023	0.0282
pH	4.428	0.0217

Acknowledgment

This work was partially supported by the Generalitat Valenciana (Grant GVPRE/2008/186), Spain.

References

Bigham, J. M. and E. J. Ciolkosz. 1993. *Soil color*. Madison, WI: SSSA Special Publication Number 31.

Burrough, P. A. and R. A. McDonnell. 1998. *Principles of Geographical Information Systems*. New York: Oxford University Press.

CIE (Commission Internationale de l'Éclairage). 2004. *Colorimetry*, 3rd edn. Vienna, Austria: CIE Central Bureau, Publ. 15:2004.

Jenny, H. 1941. *Factors of Soil Formation. A System of Quantitative Pedology*. New York: Dover Publications.

McGarigal, K., S. Cushman, and S. Stafford. 2000. *Multivariate Statistics for Wildlife and Ecology Research*. New York: Springer.

Munsell Color Co. 1980. *Munsell Soil Color Charts*. New York: Munsell Color Company Inc.

NCGIA (National Center for Geographic Information and Analysis). 1990. Core curriculum in geographic information science. University of California. http://www.geog.ubc.ca/courses/klink/gis.notes/ncgia/u01.html. Accessed March 20, 2010.

Soil Survey Division Staff. 1993. *Soil Survey Manual*. Washington, DC: USDA-SCS.

Soil Survey Staff. 2010. *Keys to Soil Taxonomy*, 11th edn. Washington, DC: USDA-NRCS.

Worboys, M. and M. Duckham. 2004. *GIS: A Computing Perspective*. Boca Raton, FL: CRC Press.

CHAPTER 37

Color Characteristics of Raw Milk from Silage and Alfalfa-Fed Cows

LEANDRO LANGMAN, LUCIANA ROSSETTI, ANA MARÍA SANCHO, EDUARDO COMERÓN, ADRIANA DESCALZO, and GABRIELA MARÍA GRIGIONI

Contents

37.1 Introduction

Food color is the result of natural products present in the raw material from which it is processed or of colored compounds generated as a result of processing (Morales and van Boekel 1998). It is influenced by how the food matrix interacts with light, regarding as its reflecting, absorbing, or transmitting characteristics, which in turn is related to its physical structure and chemical nature (Kaya 2002).

In some foods, as milk and dairy products, color is the first criterion to be perceived by the consumer. As stated by Burrows (2009), the repeated recognition of a particular brand of a food commodity largely depends on its typical color.

The white appearance of raw milk is the result of its physical structure. The casein micelles and fat globules disperse the incident light, and consequently, milk exhibits a high value of L^* parameter (lightness). Technological treatments that influence the physical structure of milk also have an effect on L^*. The other color components (parameters a^* and b^*) are influenced by factors related to natural pigment concentration of milk.

As stated by Noziére et al. (2006), carotenoids are pigments involved in sensory characteristics of dairy products, either indirectly through their antioxidant properties or directly through their yellowing properties. As a result, the color of dairy products highly depends on their carotenoid concentration.

Several dietary factors have been identified as being responsible for the characteristics of the raw milk obtained. In grazing systems, a change in carotenoids in milk in the course of time may depend on both the amount of carotenoid intake and milk yield (Calderón et al. 2006). The carotenoid content and color of milk and dairy products can be controlled by feeding management. In this context, fresh forages are the richest source followed by the silage diets (Kalač 2011). Among about 800 known carotenoids, up to 10 of them have been determined in forages, namely, oxygen-containing xanthophylls and carotenes (mainly ß-carotene). The ß-carotene concentration in cow's milk is variable, depending mainly on its dietary supply (Kalač 2011).

Procedures used to describe color are based on the specification of the three stimuli. Due to the phenomenon of trichromacy, any color stimulus can be matched by a mixture of three primary stimuli in adequate amounts (Guirao 1980). This involves a process of integration. Clarity, tone, and saturation can be discriminated by an observer when seeing the samples' color. But in contrast, the observer cannot detail the spectral composition of the stimulus.

A colorimeter is used to evaluate in physical terms psychological feelings. When a color is described, the observer usually refers to attributes of chromatic sensation as hue, lightness, and saturation. In colorimetry, these three aspects are considered psychological correlates of the physical dimensions of the stimulus.

37.2 Materials and Methods

37.2.1 Assay

The experiment was conducted during spring (October–December) at the National Institute of Agricultural Technology in Rafaela (province of Santa Fe, Argentina) throughout 60 days.

During a first 4 week preexperimental period, 10 Holstein cows were fed on a silage-only diet with at least 50% of forage (SS diet). This diet also contained soy expeller and sunflower pellets (3.5 and 1.1 kg/day per cow, respectively) and hay (1.5 kg/day per cow). Thereafter, five cows were randomly assigned to an alfalfa diet (ALF diet) with at least 60% alfalfa of dry matter on dietary basis, while the other group remained as control in SS diet.

Diets were isoenergetic and evidenced differences concerning the presence of lucerne pasture. They showed distinctive fat-soluble vitamin profiles: ALF was higher in α-tocopherol, β-carotene, and retinol than SS diet. In contrast, silage diet was higher in the gamma and delta isomers of tocopherol due to the soy expeller contribution in these isomers (Rossetti et al. 2010). ALF diet resembles productive conditions applied within the central region of Argentina in the spring–summer period, whereas the SS diet responds to a typical winter productive system, without access to fresh lucerne.

37.2.2 Color Assessment

Color measurements were carried out using a reflectance spectrophotometer (BYK Gardner Color View model 9000) according to CIELAB scale, with a 5 cm port area and D65 illuminate. The methodology used was based on that described by Celestino et al. (1997).

At any experimental point (0, 20, 40, and 60 days, after the initial switch to ALF or SS diet) and for each cow, an aliquot of 30 mL of raw milk was collected and stored for 24 h at 4°C ± 1°C in the dark until analysis. Each sample was measured using an optical cylindrical cell (CC-6136, BYK Gardner, Silver Spring, MD) of 2.4 cm of internal diameter, obtaining a 3 cm milk thickness. Under this experimental condition, no differences associated to sample thickness were observed (Langman 2009). Samples were measured in triplicate.

In order to avoid possible interference of ambient light, milk sample was covered with an opaque white cube during color measurement.

37.2.3 β-Carotene Determination

β-carotene was extracted as described by Buttriss and Diplock (1984). Sample aliquots (3.0 mL) were mixed with 1% pyrogallol (Sigma-Aldrich, United States) in ethanol. Saponification was performed for 30 min at 70°C with 10 N KOH (Merk Química, Argentina). Samples were then extracted twice with *n*-hexane high-performance liquid chromatography (HPLC) grade (J. T. Baker, United States), evaporated under nitrogen flow, dissolved in absolute ethanol HPLC grade (J. T. Baker, Mexico), and filtered through a 0.45 μm pore nylon membrane before injection. All samples were analyzed by reversed-phase HPLC.

37.2.4 Statistic Analysis

Data color and β-carotene values were examined applying a factorial design with two factors: treatment and time, with two (SS and ALF) and four (0, 20, 40, and 60 days of feeding time) levels, respectively.

The model was as follows:

$$y_{ijk} = \alpha_i + \tau_j + (\alpha\tau)_{ij} + e_{ijk}$$

where

y_{ijk} is the dependent variable
α_i is the effect of treatment ($i = 1, 2$)
τ_j is the effect of time ($j = 1–4$)
$(\alpha\tau)_{ij}$ is the interaction between ith effect of treatment and jth effect of time
e_{ijk} is the residual error

The homogeneity of variances was checked by means of Levene's test, establishing a level of significance of 0.05. The statistical analysis was done using SPSS® software (version 12, Illinois, United States).

37.3 Results and Discussion

In Table 37.1, color parameters are presented. As regards to b^* color component, raw milk corresponding to the alfalfa-based diet showed significant differences only 60 days after implementing the diet. These samples presented higher b^* values with respect to the other diet.

No significant differences ($p > 0.05$) were observed for L^* parameter. As a tendency, samples that corresponded to ALF diet exhibited lower values at 20 and 40 days compared to those reported at the beginning and end of the assay.

In the case of a^*, no significant differences ($p > 0.05$) were observed for SS diet, even though samples of ALF diet showed an increase (in absolute value) until 40 days and a decrease at 60 days.

As it was said, a^* and b^* color components are influenced by factors related to natural pigment concentration of milk. A significant correlation coefficient of 0.36 ($p < 0.05$) was found between b^* parameter and β-carotene content, even though the differences were observed in the behavior of these variables along feeding time.

Table 37.1 L^*, b^*, and a^* Values in Raw Milk Samples of Holstein Cows Fed on a Silage (SS Diet) and Alfalfa (ALF Diet) Diets (Means and Standard Deviations)

Feeding Time (Days)	Color Component Values					
	L^*		b^*		a^*	
	SS	ALF	SS	ALF	SS	ALF
0	89.4 ± 0.7	89.6 ± 0.6	8.5 ± 0.7 b	8.0 ± 0.5 b	−1.9 ± 0.3 ab	−1.7 ± 0.3 bc
20	89.3 ± 0.7	88.8 ± 0.8	8.6 ± 0.4 b	9.1 ± 0.8 b	−2.2 ± 0.3 ab	−2.1 ± 0.3 ab
40	89.6 ± 0.4	88.9 ± 1.0	8.6 ± 0.5 b	9.1 ± 0.9 b	−2.0 ± 0.2 ab	−2.4 ± 0.1 a
60	89.4 ± 0.2	89.1 ± 0.5	8.7 ± 0.5 b	10.6 ± 0.3 a	−2.0 ± 0.2 ab	−1.2 ± 0.2 c

Note: Different letters show significant differences ($p < 0.05$) in diet*feeding effect for each variable.

Table 37.2 Effect of Treatments and Time in Raw Milk β-Carotene Content (Means and Standard Deviations)

Feeding Time (Days)	β-Carotene (μg/g Milk Fat)	
	SS	ALF
0	2.4 ± 0.4 b	2.9 ± 1.1 b
20	1.1 ± 0.2 c	5.8 ± 0.9 a
40	0.9 ± 0.3 c	5.0 ± 0.8 a
60	1.4 ± 0.4 bc	5.6 ± 0.9 a

Note: Different letters show significant differences ($p < 0.05$) in diet*feeding effect for each variable.

As stated by Priolo et al. (2003), one of the major challenges is to develop a method to trace grass feeding in ruminant products in order to satisfy the demands of farmers, consumers, and certifications. In addition, Langman et al. (2009) demonstrated that reflectance spectrum pattern in the range of 450–510 nm could be a representative index of the feeding system in raw milk of cows.

In this context, b^* values had been proposed to trace grass feeding in sheep or cow milk, even though this parameter could only discriminate between diets up to 60 days of feeding under the trial conditions described in this study.

A significant increase of the β-carotene content was observed after 20 days, a tendency that was maintained at least up to 60 days after the change of diet (Table 37.2), with dissimilar values in the range of 5–6 μg/g of fat (in milk obtained from animals fed on alfalfa) vs. 1–1.4 μg/g of fat (in milk obtained from animals fed on silage). These data are consistent with the results reported by Calderón et al. (2006) where similar values of β-carotene concentration were observed in milk obtained from Montbéliarde dairy cows fed on diets rich in carotenoids (67% on dry basis of grass-based silage).

37.4 Conclusion

In dairy products, color assessment is an important part of the product quality and process management. Several authors proposed indexes that involve the study of the main pigments of milk in order to trace grass-feeding characteristics. Present results show that more research is needed in the application of b^* parameter to comprehend its response under different feeding conditions.

Abbreviation

*a** CIELAB red-green parameter
*b** CIELAB blue-yellow parameter
*L** CIELAB lightness-darkness parameter

Acknowledgments

We thank laboratory technicians Mónica Pecile and Luis Sanow for skillful assistance. This work was granted by National Institute for Agricultural Technology (INTA) through the projects AETA 2681 and PNLEC 1101.

References

Burrows, A. 2009. Palette of our palates: A brief history of food coloring & its regulation. *Comprehensive Reviews in Food Science and Food Safety* 8: 394–408.

Buttriss, J. L. and A. T. Diplock. 1984. HPLC methods for vitamin E in tissues. In *Methods in Enzymology*, vol. 105, eds. L. Parker and A. N. Glazer. New York: Academic Press, pp. 131–138.

Calderón, F., G. Tornanbé, B. Martin, P. Pradel, B. Chauveau-Duriot, and P. Nozière. 2006. Effects of mountain grassland maturity stage and grazing management on carotenoids in sward and cow's milk. *Animal Research* 55: 533–544.

Celestino, E. L., M. Iyer, and H. Roginski, H. 1997. The effects of refrigerated storage of raw milk on the quality of whole milk powder stored for different periods. *International Dairy Journal* 7: 119–127.

Guirao, M. 1980. El sistema visual. In *Los sentidos. Bases de la percepción*. Madrid, Spain: Alambra Universidad, pp. 235–280.

Kalač, P. 2011. The effects of silage feeding on some sensory and health attributes of cow's milk: A review. *Food Chemistry* 125: 307–317.

Kaya, S. 2002. Effect of salt on hardness and whiteness of Gaziantep cheese during short-term brining. *Journal of Food Engineering* 52: 155–159.

Langman, L. 2009. Calidad organoléptica en leche expresada en su color y perfil de olor. Relación de estos parámetros con la incorporación de antioxidantes naturales en la dieta implementada en las vacas. La Plata, Argentina: Facultad de Ciencias Exactas, UNLP, Thesis.

Langman, L., A. M. Sancho, E. Comerón, A. M. Descalzo, and G. Grigioni. 2009. Reflectance spectrum pattern of raw milk from silage and alfalfa-fed cows. In *IDF/DIAA Functional Dairy Foods 2009*. Melbourne, Victoria, Australia.

Morales, F. J. and M. A. van Boekel. 1998. A study on advanced Maillard reaction in heated casein/sugar solutions: Colour formation. *International Dairy Journal* 8: 907–915.

Nozière, P., B. Graulet, A. Lucas, B. Martin, P. Grolier, and M. Doreau. 2006. Carotenoids for ruminants: From forages to dairy products. *Animal Feed Science and Technology* 131: 418–450.

Priolo, A., M. Lanza, D. Barbagallo, L. Finocchiaro, and L. Biondi. 2003. Can the reflectance spectrum be used to trace grass feeding in ewe milk? *Small Ruminant Research* 48: 103–107.

Rossetti, L., L. Langman, G. M. Grigioni, A. Biolatto, A. M. Sancho, E. Comerón, and A. M. Descalzo. 2010. Antioxidant status and odour profile in milk from silage or lucerne-fed cows. *The Australian Journal of Dairy Technology* 65 (1): 3–9.

PART V

Food Environment: Color in Packaging, Sensory Evaluation, and Preferences

CHAPTER 38

Study and Analysis of Consistency of Color from the Piece of Food to the Virtual Representation on Screen and in Packaging

MARCELA MARÍA B. ROJAS and JULIA INÉS FOSSATI

Contents

38.1 Introduction

The objective of this chapter is to show the main results of the applied research project performed by Gutenberg in association with the Federación Gráfica Argentina (FAIGA) and the Unión Industrial Argentina (UIA), with the support of the Consejo Federal de Ciencia y Tecnología (COFECyT).

This work focuses on the technical and nontechnical variables that interactively impact in the consistency of color during the analysis of a graphic workflow for packaging.

Herewith appear the main steps of the graphic workflow, which is critical when it comes to obtaining color consistency.

38.2 Technical Printing Process for Packaging and Its Regulations

The selection of the printing support is the first step that initiates the technical printing process of the packaging of food, from the printing point of view. This support is selected together with the total production required, and it will determine which is the best printing system for its production.

Once the printing system has been selected, it should be established what kind of standards will take place during the technical evaluation of the printing production. Therefore, there could be American, European, or Japanese regulations or even the standards defined by the company, for instance. This chapter deals with the ISO European regulations, whose printing standard is ISO 12647. Considering the choice of the printing standard, the designer should base all of his/her decisions to create the digital file accordingly with the selected standard. As in every other phase of the graphic workflow, the design phase has critical points directly affecting the consistency of color from the conception of the file, like the formats and modes of color when the designer receives the images that were obtained through scanners or professional photographers, the monitor visualization of the work files, the parameterization of the design programs in accordance with the printing system, the color conversion from RGB (red, green, blue) to CMYK (cyan, magenta, yellow, black), and the type of format of the final file, among other things. All of the aforementioned are critical points. However, the critical points depending from the selected printing system should be distinguished from the critical points not depending from the selected printing system. There are many depending critical points in the graphic design phase, but we can mention the parameterization of the design programs and the color conversion from RGB to CMYK, which has to be performed through ICC profiles. We can also mention the transformation of the RGB and CMYK images through the Fogra 39 profile to convert images from RGB to CMYK when an offset printing process using coated support is selected. By doing this, the ISO 12647 offset printing standard is followed.

Moreover, we can mention two critical points which are independent of the design printing system: the technical conditions that should be followed by the monitors to offer a good visualization of the images in accordance with ISO 12646 regulations and the visualization conditions established by the ISO 3664 regulations. The visualization of images in the monitor and the comparison between monitor and test, or the comparison of the color test with the printed material, is perfectly established by these regulations. This means that there are some conditions clearly specified of how the processes of visualization of files in monitors or comparisons are performed. In this manner, the fact that a designer should correct images in an inappropriate monitor or under poor lighting conditions is excluded. It is also excluded the fact that a graphic salesman would show the color test to his/her client in his/her office without considering these conditions in cases where the color test is technically correct.

The designer could create the final file in PDF format, which is also accepted by the regulations. Through preflight programs, this format allows the detection of other variables that impact on the color consistency, like resolution and bits quantity of the images.

In the prepress area, after the file goes to prepress, the ripping of the file is one of the key points when it comes to maintaining the color consistency. In this process, the required technical data to color separation are specified for the selected printing system. The ISO printing regulations suggest parameters for line, angle, point gaining, etc., which are defined during the ripping phase. The generation of the CTPs (computer to plates), the photopolymer plates, or the rotogravure cylinders, which are produced after this process, are controlled through color strips that determine if they have been obtained in accordance with the quality standards required to achieve consistency.

The color tests by contract, which are also performed during this workflow phase, should agree with the ISO 12647-7 standard. The test by contract that does not agree with this condition should be considered as a reference test. This regulation establishes requirements such as the obligation of the test to have an ID tag that records the printing system simulated by the test, the printing system for tests that was used, the date of the test printing, the line registries of the digital printer that was used, the name of the support and of the inks that were used, the color rendering used to obtain the test, and if it complies with the colorimetric tolerances that link it with the printing conditions that were selected through the demand of the color test with a control stripe.

In this phase, the independent critical points are the monitors, the image setter of the CTP or polymers plates, and the digital test printer. All of these points, as well as other equipments, should be calibrated, profiled, and in line. In order to carry out these processes, densitometers and spectrophotometers are used, and their parameterizations, although they are independent of the printing system used for the production of the packaging, should also comply with the graphic standards of the ISO 12647-1 regulations which are based on the ISO 5-3 and ISO 5-4 regulations. Consequently, before going to the printing phase, there are many critical points that should be checked to obtain consistency. When the previous points are not considered, the work performed no longer falls under the standardization concept of the color consistency. In the everyday use, the test printing is used as an approximation leaving the color definition in the hands of the printer during the printing process. In order to exclude this situation, that is, the printing color test not being adequate, the regulations establish that if the workflow is standardized by the regulations, the printing material should comply with the colorimetric and densitometric technical specifications established by the regulations.

In this phase of the printing process, the critical points depending on the printing system are the choice of the ink, specified by the ISO 12648 regulations, and the selection of the substrate whose technical specifications are described in the ISO 12647 printing regulations.

Up to this point, we have mentioned the most important points affecting the color consistency in a selected printing system, and, as we have already established, all of the phases of the graphic workflow have critical elements that should be controlled for the color consistency to be successful. In the packaging of food, the color consistency is required in more than one printing system. The packaging of the same food could require an offset printing and flexography. Working in a standardized manner is the only possible methodology that could be applied to achieve color consistency between printing systems in the packaging of food. By doing this, the color of the brand of food is preserved, which facilitates its unique identification.

This research project emphasizes the study of these critical points because, in accordance with what we have already shown, not every professional that participates in this process has the same level of interest in color consistency. The following comparative chart summarizes the state of the art at the beginning of the project. This chart relates all of the players of the graphic workflow chain for the packaging of food with their respective production tools, environments, evaluation tools, color measurement, and color consistency relevance granted in their job. For instance, a publicist cares about the color of the brand. The color consistency could be granted with great importance, but when it comes to advertising, the requirements of the monitor or of the visualization conditions are not considered, at least, most of the time (Table 38.1).

38.3 Associated Costs of Achieving Color Consistency

In the study, it was also determined that the loss of interest in color consistency is originated in the associated costs that imply color measurements in the actions that have to be performed in the diverse corrections. Performing color management implies that the company has to pay costs such as the ones detailed in Table 38.2.

Even though the costs of the test printer ink, the color specialist, and the production test can be considered to be the most typical ones, other unexpected costs appeared. The higher costs come from the following:

1. Difficulties for the provision of normalized consumables
2. Equipment and software highly sensitive to problems in the electric network
3. The need of building adequate conditions compatible with the comfort of workers (light conditions, neutral gray color in walls, and clothes)
4. Impossibility of achieving certification and calibration of foreign equipments

Table 38.1 Relationship between the Actor, the Tool of Reproduction, the Color Mode, the Tools Used for Measurement, and the Importance Attributed to the Color Consistency

Actor	Color Object Observed-Reproduction Tool	Medium: Physical-Real Color Mode	Tools Most Commonly Used for Color Measurement	Importance Attributed to the Consistency of Color in All the Workflow
Producer	Food	Physical sample	Visual observation	Low
Publicist	Screen	Digital RGB	Visual comparison with pantone or with a real piece	High
Photographer	Image in digital camera or screen	Digital RGB	Color chart, spectrophotometer	Medium
Designer	RGB screen Contract digital proof	RGB digital CMYK digital	Photoshop information tool (info), pantone, spectrophotometer	Very high
Operator	Screen + rip + imagesetter Proof printer	Digital: RGB Digital: RGB	Spectrodensitometer	Medium
Printer	Box-production printing machine	Physical CMYK-pantone	Spectrodensitometer	Medium
Distribution	Packaging	Physical CMYK-pantone	None	Low
Buyer	Box-box Box-food	Physical CMYK	Visual comparison	High Low

Table 38.2 Percentage Distribution of the Additional Costs in order to Achieve Color Consistency

Necessary Item	Annual % Incidence
Color specialist	23%
Spectrodensitometer w/soft	10%
Normalized light camera	1%
Test paper	6%
Adequate screen	16%
D50 light installation	1%
Maintenance of building conditions	1%
PC or Mac and printing server	5%
Test printer	5%
Original inks	21%
Calibration	2%
Rip for the creation of proof and linearization	9%
Printer with control strip	
Tests in production printers	19%

5. Repetition of plates and tests for the different behaviors in the RIPs (compensations occurred in the control strips when it was not expected)
6. Fluctuations in the material specifications: inks, plates, cartons, and papers
7. The need of the external supplier to determine the properties, such as brightness and rub resistance of the proof
8. The need of adequacy of the printing machines
9. The reproducibility of different spectrodensitometers
10. The resistance to change in the work modality

38.4 Conclusion

There are several factors on each stage of the graphic workflow that may affect negatively the color consistency. In order to achieve the color consistency in the packaging of food, it is necessary that all the work chain operates in line with the working directives and objectives determined beforehand.

References

All references are to the ISO (International Organization for Standardization):
ISO 5-3. 1995. Photography—Density measurements. Part 3: Spectral conditions.
ISO 5-4. 1995. Photography and graphic technology—Density measurements. Part 4: Geometric conditions for reflection density.

ISO 12646. 2004. Graphic technology—Displays for colour proofing—Characteristics and viewing conditions.
ISO 12647. 2004–2007. Graphic technology—Process control for the production of half-tone colour separations, proof and production prints: Part 1: Parameters and measurement methods. Part 2: Offset lithographic processes. Part 3: Coldset offset lithography on newsprint. Part 4: Publication gravure printing. Part 5: Screen printing. Part 6: Flexographic printing. Part 7: Proofing processes working directly from digital data.
ISO 12648. 2006. Graphic technology—Safety requirements for printing press systems.
ISO 3664. 2009. Graphic technology and photography—Viewing conditions.

CHAPTER 39

Influence of Package Color for Mineral Water Plastic Bottle on Consumers' Purchase Motivation

SAORI KITAGUCHI, YUHI YONEMARU, YOJI KITANI, ORANIS PANYARJUN, and TETSUYA SATO

Contents

39.1 Introduction

Tap water is good enough to drink in Japan; however, more and more people purchase bottled water. In 1998, Japanese annual consumption of mineral water was 0.87 million kiloliters. It has increased up to 2.5 million kiloliters in about 10 years (The Mineral Water Association of Japan 2011). Currently, around 1000 brands of bottled water are commercially available in Japan (Hayakawa 2008). In this highly competitive market, package design of products is one of the key issues to attract customers. Nowadays, package design is important for any product. Even a product as simple as mineral water needs careful consideration over the design for its packaging. Water itself is transparent and colorless and so for is the bottle. For this reason, the label on the transparent bottle is a very important aspect in the marketing of this product for the appeal of the consumers. There are lots of elements in a label such as color, shape, letters, and pictures. Since bottled water is not like luxury brand items which consumers examine one by one carefully, it is necessary to attract consumers at that moment when consumers come to a drink section in a shop. In this case, color can be an essential element to catch consumers' eyes rather than such as literal information. Therefore, this study focused on label color of bottled water. Then influences of label color on our impression as well as on consumers' purchase motivation were investigated.

39.2 Colors of the Bottled Water

First of all, label colors of commercially available bottled water were surveyed. A total of 101 bottled water were collected. Fifty-two of them were Japanese products, and forty-nine were foreign products. Colors of their labels were identified in terms of CIELAB (*Commission Internationale de l'Éclairage* $L^*a^*b^*$) values (CIE 2004). To do this, digital images of the bottles were captured via a digital camera Canon EOS 20D. The RGB values read out from each image were converted to CIE *XYZ* values in terms of the CIE 1964 standard colorimetric observer (CIE 2004) using a camera characterization model: a polynomial model using the least square method (Westland and Ripamonti 2004). Then, $L^*a^*b^*$ values were calculated from the *XYZ* values. Two colors of the first and second largest areas on the label were extracted as representative colors of the bottle. Note that if the second largest area was smaller than double the size of the third largest area, only the color of the largest area was used as a representative. Figure 39.1 shows the distributions of the extracted colors on a CIELAB a^*–b^* diagram and an L^*–C^*_{ab} diagram. It can be seen that around 60% of the colors were bluish colors; namely, the colors were concentrated on the hue area of greenish to purplish blue. The differences between the Japanese and the foreign products were that yellowish and reddish colors were used more for the foreign products than the Japanese products; around 30% of the foreign products surveyed used yellowish and reddish colors, but only around 10% of the Japanese products used yellowish and reddish colors. In addition to the label colors, colors of the caps of the bottles were also obtained. Figure 39.2 shows the distributions of the cap colors on a CIELAB a^*–b^* diagram and an L^*–C^*_{ab} diagram. Most of the Japanese products had white caps. On the other hand, various colors were utilized for the foreign products.

39.3 Purchase Motivation

Bottled water producers are trying to make their products stand out from others in some way. But what is important for consumers' purchase decision

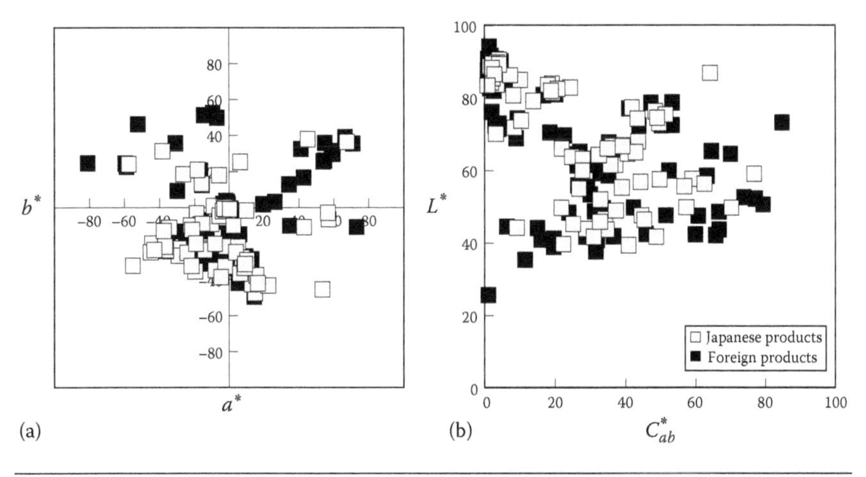

Figure 39.1 The label colors of the bottled water plotted on (a) the a^*–b^* diagram and (b) the L^*–C^*_{ab} diagram.

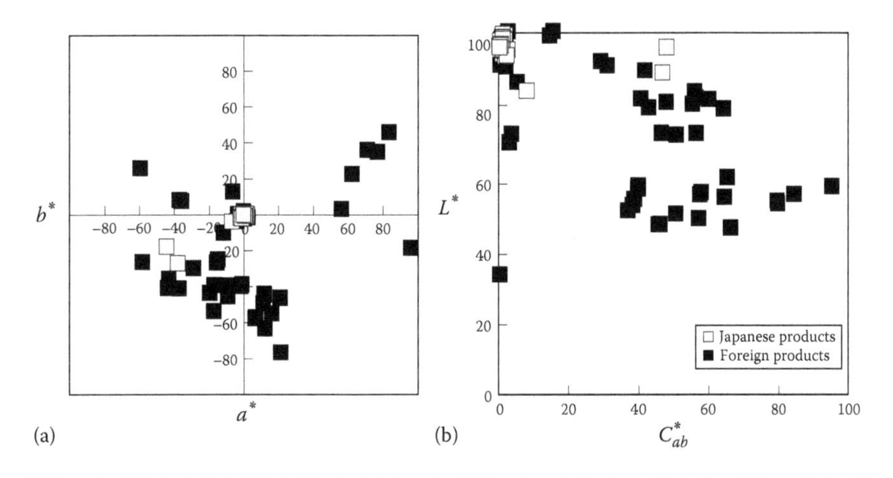

Figure 39.2 The cap colors of the bottled water plotted on (a) the a^*–b^* diagram and (b) the L^*–C^*_{ab} diagram.

process? To see the consumers' purchase motivation, a questionnaire survey was carried out. In this survey, 50 Japanese university students including 25 females and 25 males participated. Their average age was 21.6 years old. The participants were asked to select any from eight factors which would stimulate their purchase motivations. Note that multiple answers were allowed. As a result, 73% of the participants answered "reasonable price" as the most important purchase motivation followed by "taste" (61%); "shape, design, and color of label/bottle" (55%); "brand" (50%); "still/sparkling water" (32%); "soft/hard water" (32%); "country of origin" (12%); and "mineral content" (7%).

39.4 Color Suitability

The label colors of the commercially available products have been seen in the previous section. Bluish colors seem very popular. Here, consumers' opinions about label color were investigated. The investigation was done by asking subjects to scale the suitability of color samples as a label of bottled water.

39.4.1 Experimental Details

In this experiment, 199 Practical Color Coordinate System (PCCS) color papers were used as samples. The colorimetric values of the samples were obtained by measuring the reflectance of the samples using a spectrophotometer Minolta CM-3600d and then computing CIELAB $L^*a^*b^*$ values from the reflectance assuming the D65 illuminant and using the CIE 1964 standard colorimetric observer. The size of the samples was 175 × 120 mm. As shown in Figure 39.3, the sample was presented in a viewing cabinet Judge II with a light source of a CIE D65 simulator. A seven-point semantic differential (SD) method (Union of Japanese Scientists and Engineers 1999) was applied so that a subject was asked to assign a category for each sample in terms of "Unsuitable–Suitable" as a label color of bottled water on a −3 to 3 scale. The categories were −3, "extremely unsuitable"; −2, "quite unsuitable"; −1, "slightly unsuitable"; 0, "neither suitable nor unsuitable"; 1, "slightly suitable"; 2, "quite suitable"; and 3, "extremely suitable." A total of 30 Japanese university students with normal color vision, including 15 females and 15 males, took part in the experiment. Their ages ranged from 18 to 25 years old with an average of 21.6 years.

39.4.2 Results

The average of the categories assigned by the subjects was used as a measure of "Unsuitable–Suitable." Figure 39.4 shows the colors of the samples in

Figure 39.3 The experimental arrangement.

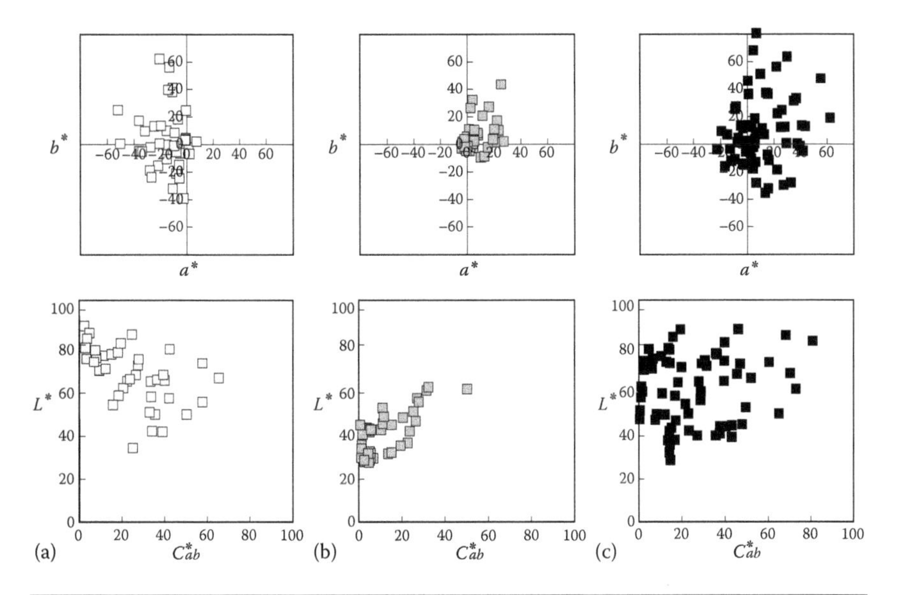

Figure 39.4 The colors of the samples in a CIELAB a^*–b^* diagram (top) and an L^*–C^*_{ab} diagram (bottom) with the indications of the color suitability. (a) The top 1–40 suitable colors, (b) the worst 1–40 suitable colors, and (c) the others.

a CIELAB a^*–b^* diagram and an L^*–C^*_{ab} diagram with the indications of the color suitability. The samples were divided into three groups according to their suitability scores: the top 1–40 suitable colors, the worst 1–40 suitable colors, and the others. The result indicates that the bluish colors are found to be most suitable and the dark and less saturated colors are not suitable for a label of bottled water.

39.5 Impression of Label Colors

In the previous experiment, only the flat papers were used as samples. In order to investigate more various impressions received from label color and the impressions evoked by more realistic samples, an experiment was carried out using bottled water with colored labels.

39.5.1 Experimental Details

Thirty-three PCCS papers were used as labels. The 33 colors were 28 chromatic colors (7 hues × 4 tons: red, orange, yellow, green, blue green, blue, and purple hues, and pale, soft, vivid, and dark tones) and 5 achromatic colors (white, light gray, middle gray, dark gray, and black). The size of the paper wound around a bottle was 55 × 175 mm. As shown in Figure 39.5, similar to the previous assessment using the paper samples, the samples were presented in the viewing cabinet *Judge II*, and a seven-point SD method was applied to scale the impressions of the samples. The impressions evaluated in this

Figure 39.5 The experimental arrangement.

Table 39.1 SD Word Pairs and Correlation Coefficient of *r* between "Unsuitable–Suitable" or "Would Not Buy–Would Buy" and the Other Impressions

SD Word Pairs	Unsuitable–Suitable	Would Not Buy–Would Buy
Would not buy–Would buy	0.86	—
Not ecological–Ecological	0.89	0.80
Not refreshing–Refreshing	0.96	0.90
Hard to drink–Easy to drink	0.92	0.90
General–Unique	−0.84	−0.76
Unhealthy–Healthy	0.87	0.87
Bad taste–Good taste	0.91	0.95
Unclean–Clean	0.92	0.91
Cheap–Expensive	−0.44	−0.09
Low mineral–High mineral	0.48	0.60
Warm–Cool	0.58	0.68
Not fashionable–Fashionable	0.47	0.68

study (SD word pairs) are given in Table 39.1. A subject was asked to assign a category for each sample in terms of the impressions on a −3 to 3 scale. A total of 40 Japanese university students with normal color vision, including 20 females and 20 males, took part in this experiment. Their ages ranged from 18 to 25 years old with an average of 22.3 years.

39.5.2 Results

The average of the categories assigned by the subjects was used as a measure of the impression. Figure 39.6 shows the colors of the samples with the indications of the subjects' answer to the question, "Would not buy–Would buy" ("would buy"). The samples were divided into three groups: the top 1–7, the worst 1–7, and the other colors. The bluish colors found to give the positive

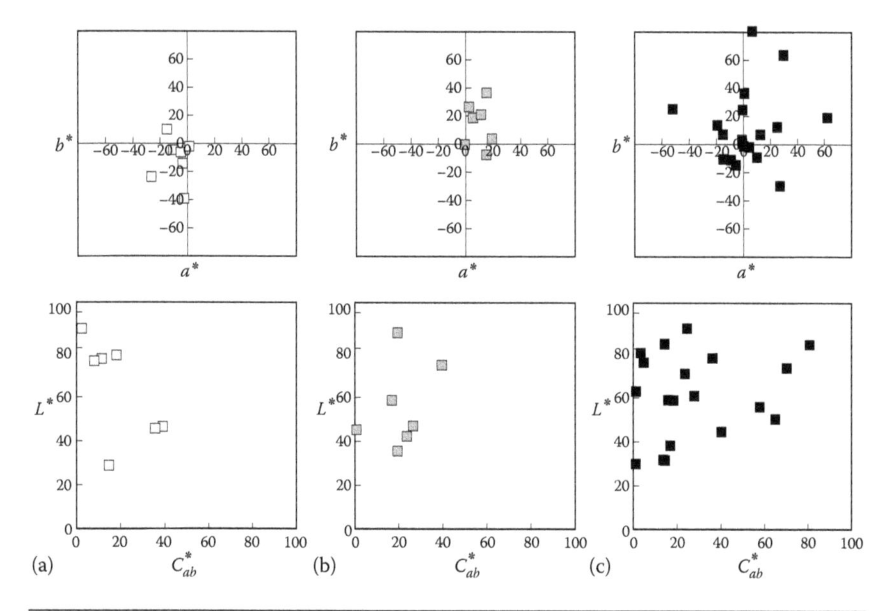

Figure 39.6 The colors of the samples in a CIELAB a^*–b^* diagram (top) and an L^*–C^*_{ab} diagram (bottom) with the indications of the impression of "Would not buy–Would buy." (a) The top 1–7, (b) the worst 1–7, and (c) the others.

impression. This is the same tendency as the result of the color suitability obtained from the previous experiment. The correlation coefficient of $r = 0.86$ between the "would buy" and the suitability results of the corresponding samples also indicates the similarity between them.

The correlations between "would buy" or the color suitability and the other impressions were summarized in Table 39.1. The high correlations ($r > \pm 0.9$) are found from the impressions of "good taste," "clean," "refreshing," and "easy to drink." The colors which gave these positive impressions showed high correlation with "would buy." It means that these positive impressions of colors stimulate consumers' purchase motivation.

From the result of the questionnaire in Section 39.3, it was found that "reasonable price" was the most important purchase motivation. Therefore, the result of "would buy" was compared with that of "expensive." There is no definite correlation between them; however, as it can be seen from Figure 39.7, "expensive" correlated inversely with the L^* values ($r = 0.79$). It indicates that the subjects associated "expensive" with the dark colors and "cheap" with the light colors. Thus, lighter colors stimulate purchase motivation rather than darker colors.

The impression of "fashionable" is usually positive for many commercial products such as fashion items, and it is considered to stimulate purchasing interest. However, in the case of bottled water, the subjects seem not to associate "fashionable" with the purchasing interest of "would buy." The reddish

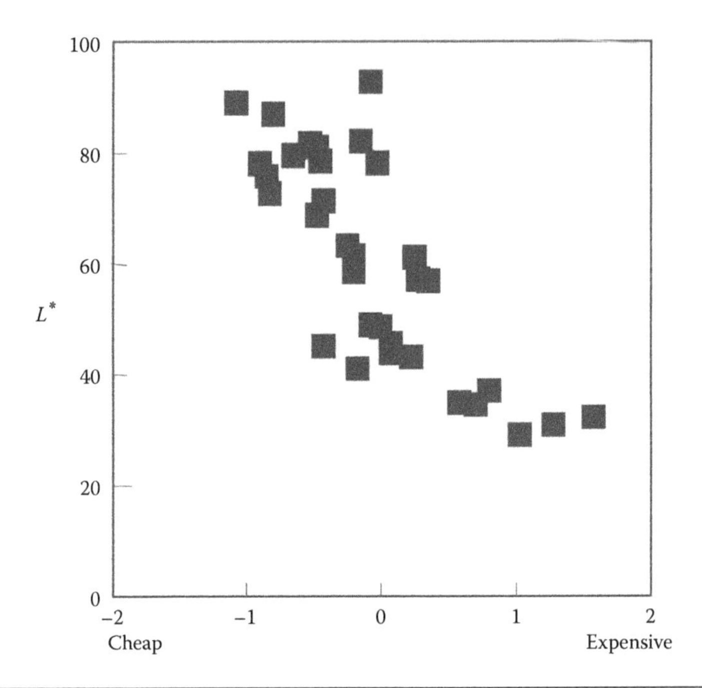

Figure 39.7 The relationship between the impression of "Cheap–Expensive" and the L^* values of the samples.

colors were associated with "fashionable." These results suggested that the subjects were not really concerned about fashionability of bottled water.

39.6 Conclusions

The survey was carried out about consumers' purchase motivation regarding bottled water. The impressions evoked by various label colors were also investigated. According to the result of the questionnaire, although the most important motivation was "reasonable price," more than half of the participants answered that "shape, design, and color of label/bottle" stimulated their purchase motivation. The experimental results showed that the colors give the positive impressions such as "good taste," "clean," "refreshing," and "easy to drink," which were similar to the colors which stimulate purchase motivation; however, unlike fashion items, there is no strong relationship between the impressions of "would buy" and "fashionable." The subjects associated the impressions of "would buy," "good taste," "clean," "refreshing," and "easy to drink" mostly with the bluish colors. The bluish colors were also chosen as suitable label colors for bottled water. Therefore, it can be concluded that bluish colors of labels tend to give positive impressions and stimulate purchase motivation. However, many commercial products have already used bluish colors. Thus, for eye-catching bottled water, it would be better not to use bluish color labels, but to use saturated reddish labels which give the impressions of "fashionable" and "unique," and for expensive-looking bottle, it would be better to use darker colors.

References

CIE (Commission Internationale de l'Éclairage). 2004. *Colorimetry*, 3rd edn. Vienna, Austria: CIE Central Bureau, Publ. 15.

Hayakawa, H. 2008. *Mineral Water Guide Book*. Tokyo, Japan: Shinchosha Publishing Co.

The Mineral Water Association of Japan. 2011. Mineral water kokusan yunyuu no suii (Annual record of domestic/imported mineral water sales). In http://minekyo.net/public/_upload/type017_5_1/file/file_13004318674.pdf. Accessed April 12, 2011.

Union of Japanese Scientists and Engineers. 1999. *Sensory Evaluation Hand Book*. Tokyo, Japan: JUSE Press Ltd.

Westland, S. and C. Ripakmonti. 2004. *Computational Colour Science Using Matlab*. West Sussex, U.K.: John Wiley & Sons.

CHAPTER 40

Legal Value of Color and Form in the "Small Print"

DARDO G. BARDIER

Contents

40.1 Introduction

"Segregation of visual information begins at the retina" said Kandel (2000). The retina has a depression called fovea. The foveola is in the center of the fovea, the cone cells here are thinner and in higher density than in the rest of the fovea. The immediate processing layers of the retina devote to these cones a great proportion of neural cells. They also retire for the cones to get light more directly (Hart 1994) (Figure 40.1).

For a standard observer, the foveola (whose diameter is about 0.25 mm) by its distance to the pupil, determines a high-resolution geometric solid cone of almost 1° (54′). What is seen in this field of maximum acuity (FMxA) will be at its maximum perceptible detail (Hart 1994) (Figure 40.2).

At common reading distance (about 30 cm), that cone geometry has a circular base of about 6 mm diameter on the read text. In order to read, we need to move our eyes, so at the most common reading distance, the band traveled by the reader watching is of about 6 mm. It is composed of three strips of 2 mm each: what was written in the separate line by two leadings (Bardier 2001). Within the diameter of the foveola, there are 60–75 cone cells. Our maximum resolution power is, therefore, of 1/60–1/75 the diameter of the field maximum resolution. This determines the minimum resolution angle (MRA), approximately 1 minute degree (Hart 1994). At common reading distance, this is almost 0.1 mm.

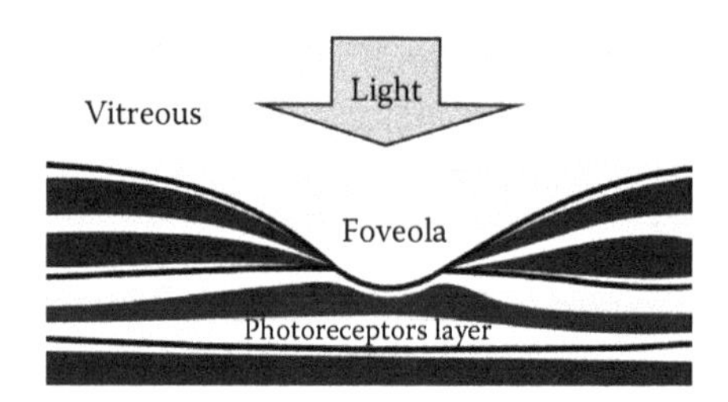

Figure 40.1 At the foveola, the photoreceptors are thinner and compacted; layers of neurons are retrieved. (Adapted from Hart, W.M. Jr., *Adler. Fisiología del ojo*, Mosby/Doyma Libros, Madrid, Spain, 1994.)

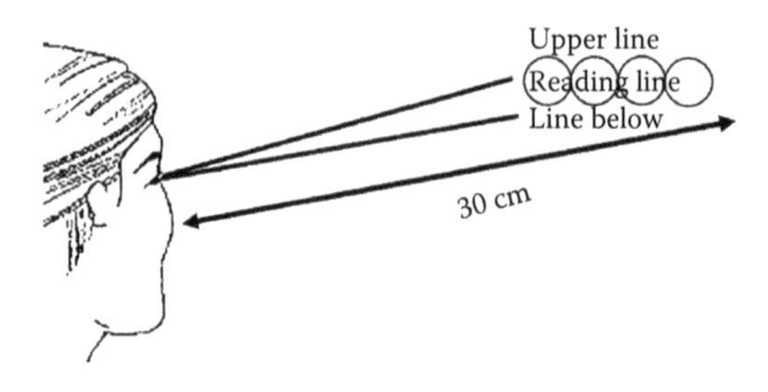

Figure 40.2 When reading, we cross, with our circle of maximum acuity, a strip 6 mm wide, centered on the writing line.

Based on this information of the human eye, the main formal characteristics that a text should fulfill to be clearly readable can be inferred.

The aim of this chapter is to elaborate, based on the human eye properties, a list of recommendations for people and institutions responsible for the control of provisioning food, drugs, and all kinds of products that must alert on compounds and effects.

40.2 Methodological Approach

Formal characteristics of a printed text were analyzed through the human eye properties described earlier. Common fonts such as Times New Roman and Arial were used as illustrative models. The text characteristics that were analyzed are height, width, thickness, contrast, and printing quality. Following this analysis, a list of recommendations for the format of fonts in foods was elaborated.

40.3 Height of Letters

40.3.1 Policy about the Height of the Letters in Some Latin-American Countries

To highlight the importance of this characteristic of printed text, regulations of some countries are cited as follows:

Argentina: "Article 1: ... typographic characters must not be less than one point eight (1.8) millimeters in height." That article and the resolution number 906/98 (Ministry of Economy of the Nation) set the minimum height the letter used for consumption, offers, budgets, and contracts must have: "one point eight (1.8)," because what the law says is true in those words (Figure 40.3).

Chile: "... with a letter size not less than 2.5 mm". Clauses that do not comply with these requirements will not produce any effect on the consumer (*sensu* Act 19496, 14 July 2005).

Figure 40.3 This printing meets Argentine law because the uppercase has 1.8 mm height. Note the circle of 6 mm.

Uruguay: "...Typographical characters used in contracts of accession may not be in any case smaller than 10 points of size". (*sensu* Circular 2016 Banco Central del Uruguay, Article 195: Conditions of contracts).

40.3.2 Minimum Height yet Optimal

The Times New Roman 12 font has a lowest height (e.g., a, n, z) of 2 mm. To complete the band read, we add a 2 mm completely free net leading on the written line as follows. We must also add 2 mm free space to the top written line, carefully taking this measure from the bottom of the lower letters (e.g., g, p, y).

In that type and font size, net reading band meets the 6 mm required for reading without confusion. In such text, when lines are only common letters without outbound, up or down, each leading measures 2.8 mm. The total gross of the reading band is 7.6 mm, which makes it even more comfortable for reading.

Capital letters (e.g., A) of that type and font size measure 2.8 mm. If you mix with lowercase, it is the same as when all are lowercase.

Although uppercase is larger and more readable, they differ less from one another, and reading becomes less understandable.

40.4 Width of Letters

40.4.1 Optimal Width

For the mentioned biological reasons, letters should have a layout with an average width of 1.5 mm, being the thinnest of about 0.8 mm (e.g., l, t), until the thicker, with 2.8 mm (m, w). Most (e.g., a, n, z) are very close to the average width. If we add separations between letters, the measure becomes 2 mm: each time we take a look with our FMxA, we cover two to four letters. This coincides with a syllable.

We need to place separator points or commas for each three figures: 1,000,000. The fonts used in press, such as Times New Roman 12 or Arial 12, satisfy either this optimum standard.

40.4.2 Minimum Width Visible

The width reduction should be, at best, proportional to the reduction of height of the letter. Therefore, common letters, such as "o," should be not smaller than 1.3 mm width. The usual letters Arial 9 and Times New Roman 10 meet this minimum. Thinner letters are unreadable to the average reader because these letters (1) diminish the visualization of each letter and (2) introduce into the FMxA more than one syllable and more than one line. The font Cooper Black 10 meets the minimum standards of height, but is clearly uncomfortable when reading. The same applies to Bodomi MT Poster Compressed 9.

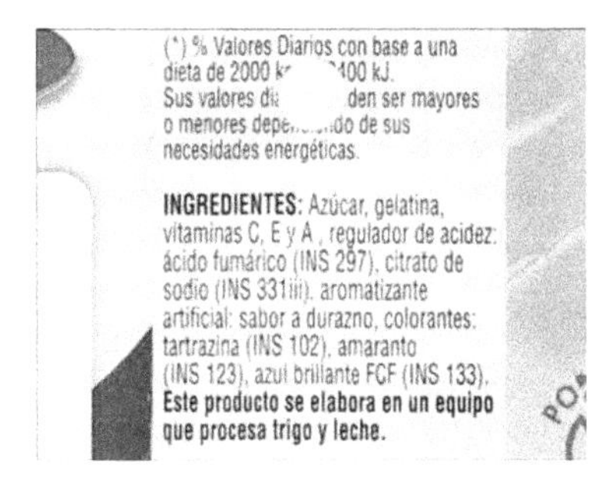

Figure 40.4 Text of acceptable height but unacceptable bandwidth. Circle of 6 mm.

Those fonts are still more difficult to read for readers below the average (Figure 40.4).

40.5 Contrast

40.5.1 Optimal Contrast

All the above specifications are for text in black ink on white paper or vice versa.

40.5.2 Minimum Contrast

In the case of colored letters and/or colored background, the font size shall be increased because between colors the contrast is always lower than between black and white (Hart 1994). This is true even for the best contrasting colors (purple, blue, or red) on yellow, yellow-green, orange, or green background (or vice versa).

Mimetic contrasts: It is not possible to read when the letter and the background colors are isovalent (orange in green, purple or blue in dark red, yellow in orange, yellow in white). If you should warn of dyes harmful to health, there is a double damage (Figure 40.5).

Nonexistent contrasts: Letters of equal color as the background color are still less legible. White against white and red against red are invisible, although they actually can be printed.

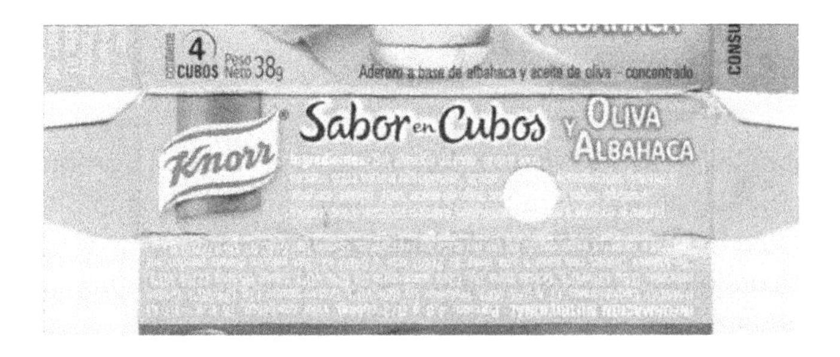

Figure 40.5 White letters on clear background prevent reading. Circle of 6 mm.

Changing contrasts: When the background and letters produce camouflage. Letters that change shape, texture, and/or color are not legible.

40.6 Thickness of Letters

40.6.1 Optimal Thickness

We can write thick or thin letters. However, if the thickness of the sticks is less than the visible minimum, reading becomes impossible because the thickness of the cells limits the discernment of the parts of the letters.

This thin thickness (0.1 mm) is valid only if we stop to look at or remain in the horizontal travel view. However, for mobile reading, it is necessary for the thickness of the vertical strokes to be three times greater (on average): 0.3 mm.

40.6.2 Minimum Thickness

The thickness of letter components is not reducible because they become almost unreadable with normal view at normal distance. However:

1. The Arial 9 font has uniform thickness of approximately 0.3 mm everywhere. In theory, if the reading eye makes arrests of its movements, as the FMxA is circular, all sides should be of equal thickness. If this size is reduced, the ability to distinguish between letters gets lost (McLean 1993).
2. Times New Roman 10 maintains the thickness of the vertical sticks (0.3 mm) and refines the horizontal (0.1 mm). As in the beginning and end of the stop between saccadic movements, there still remains mobile vision; this suits best to rapid reading than the constant thickness (Hart 1994).

Bold: A letter thicker than the rest of the text is more readable almost all the times, so for the most part of the fonts, there is a bold type. Arial 9 and Times New Roman 10 improve when they get bold.

Separation between letters: 0.1 mm in letters with outgoing horizontal and more than 0.3 mm in fonts of uniform thickness (e.g., Arial 9).

40.7 Printing Quality

The text should be placed where it can be easily readable. Locations perpendicular to the rest of the text in the background, in the bottom of the product, not integrated to the main text or hidden in foldings, are inconvenient.

Texts that do not meet standards of good printing, punctuation, and understandable lexicon are difficult or impossible to read. Although the size should be increased, these shortcomings may not be offset. If those

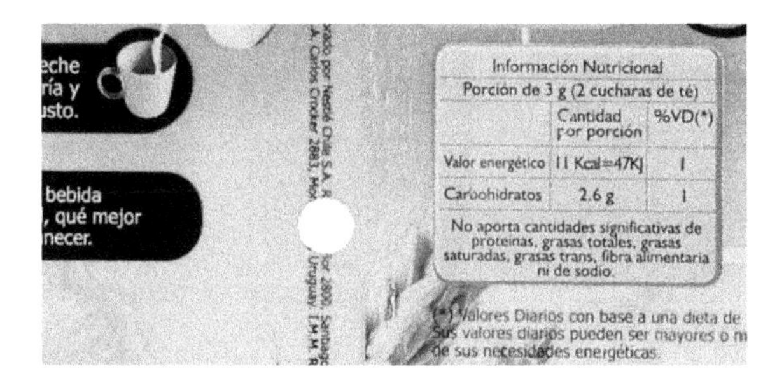

Figure 40.6 Vertical texts on secondary sites, with low contrasts and unreadable letter, are inadequate. A circle of 6 mm is represented in the figure.

standards are not met, texts should be corrected or given by unwritten (Figure 40.6).

Additionally, small print suffers the printing defects more than normal text. Warnings legally required should not be taken as accomplished when letters are badly printed, stained, grassed, or discolored. The printing quality of warnings should be better than the rest.

40.8 Recommendations

It should not be taken as written what may not be easily readable. A text is comfortably readable at common reading distance when it meets all the following minima:

1. Letter height: 1.7 mm. Height of uppercase letters, numbers, or high letters: 2.3 mm. Average height of the letters [(common + high)/2]: 2 mm.
2. Net leading, completely empty: 1 mm. Gross spacing: 2 mm.
3. Nonhorizontal elements' thickness: 0.3 mm. Horizontal elements' thickness: 0.1 mm. Separation between letters with outbound: 0.1 mm. Without outbound: 0.3 mm.
4. Average letter width: 1.5 mm. Separation between words: 1 mm.
5. Contrast between text and background: white and black. Only homogeneous and heavily contrasting colors if all the dimensions of the letters are increased. Camouflage colors should not be used (changing colors, similar colors in letters and background).
6. Printing quality of warnings should be equal to or better than the rest of the form.
7. The text should meet the usual rules of the language and the typography.
8. The text should not be located in disadvantaged or uncomfortable areas, vertical, slanted, or in hidden positions.

Conditions 1–4 are fulfilled by Arial 9 and Times New Roman 10 letters. Competent legal bodies should publish lists of fonts and acceptable sizes of letters.

Acknowledgments

Thanks to Cecilia Bardier and Virginia Suárez for valuable contributions and English translation.

References

Bardier, D. G. 2001. *De la visión al conocimiento*. Montevideo, Uruguay: The Author.
Hart, W. M. Jr. 1994. *Adler. Fisiología del ojo*. Madrid, Spain: Mosby/Doyma Libros.
Kandel, E. 2000. *Neurociencia y conducta*. Madrid, Spain: Prentice Hall.
McLean, R. 1993. *Manual de tipografía*. Madrid, Spain: Blume.

CHAPTER 41

Color Temperature Variation for Fresh Food Lighting

A Test on User Preferences

MAURIZIO ROSSI and ALESSANDRO RIZZI

Contents

41.1 Introduction

The way food is presented and how it appears to the final user has a great effect on his or her will to buy or consume it, especially in the case of fresh food. Color and glossiness appearance (Hunter 1975) can be easily manipulated through choosing the proper lighting, varying its geometry, luminance level, and spectral content. The final visual appearance to the brain (Zeki 1993) can make the food appear more or less fresh, ripe, juicy, appetizing, etc. This chapter presents an experiment designed as follows. Five light booths have been prepared with five different light sources and filters (see Table 41.1). The chosen food, from a big retail store, is usually exposed under artificial light directly or through transparent wrapping. The test

Table 41.1 Configuration of the Five Lighting Fixture Prototypes

No.	Lamp	Power (W)	Lamp CCT (K)	CRI	Filter
1	Halogen	150	2900	100	Neutral anti-UV
2	White sodium	100	2550	83	Neutral anti-UV
3	Metal halide	150	4200	96	Neutral anti-UV
4	Metal halide	150	4200	96	Warm white dichroic
5	Metal halide	150	4200	96	Cold white dichroic

aims at verifying the changes in consumers' buying intentions as caused by changes in CCT illumination and consequently appearance. Test has been held in a temperature-controlled room without natural lighting. The subjective preference test has been carried out on 124 subjects (62 males and 62 females) subdivided into two age groups (20–30 years and above 30 years). The chosen types of food have been red meat, bakery products, fish, fruits, and vegetables.

The presented research aims at verifying the performances of five lighting system prototypes with different light sources and dichroic filters. The performances have been verified both quantitatively and qualitatively. The five prototypes have been realized arranging different luminous sources and different anti-UV and dichroic filters. The configuration choice has been done in order to use high-efficiency discharge lamps with high color rendering index (CRI), using dichroic filters to tune the light spectral properties on the exposed food.

41.2 Chosen Fresh Food

The fresh food has been chosen among those normally distributed in large supermarkets which are directly exposed under light or through a transparent cover. To choose them, customers highly rely on their visual appearance. The five chosen food types are the following:

1. Red meat: filet, ground meat, and three different kinds of sliced ham
2. Selection of Mediterranean blue fish
3. Products from the oven: round- and stick-shaped bread, brown toast bread, and three different kinds of pastry and biscuits
4. Green, green-yellow, and yellow vegetables and fruits: green and yellow peppers, courgettes, green lettuce, lemons, bananas, pumpkins, grapefruits, etc.
5. Orange, red, and other colors of vegetables and fruits: pink grapefruits, mandarins, oranges, carrots, potatoes, red peppers, red apples, eggplants, and red lettuce

Figure 41.1 Lighting fixture "Minisosia" used in the tests. (Photo by Maurizio Rossi.)

41.3 Light Sources

Five different prototypes of lighting fixtures have been set up in collaboration with the factory Ing. Castaldi Illuminazione SpA, using a commercial product like "Minisosia" (Figure 41.1) with Philips lamps with different color temperatures and CRIs. Prototypes are equipped with neutral anti-UV filters and dichroic filters as in Table 41.1.

Light sources have been chosen for their energetic efficiency and halogens to have a high CRI source to compare with. Their spectra are visible in Figure 41.2.

Neutral anti-UV filter has been used to eliminate UV component that can damage food. The two dichroic filters have been used to modify the CCT coming from the lighting fixtures.

41.4 Measures on the Light Sources

Spectral and chromatic characteristics of the five prototypes have been measured in the temperature-controlled dark room of Lighting Lab at Politecnico di Milano, set on a temperature of 25°C and humidity at 50%. Measures have been taken after an aging period, as prescribed by standards, using a stabilized power supply Elettrotest—TPS/M/6KW. The instrument used is a

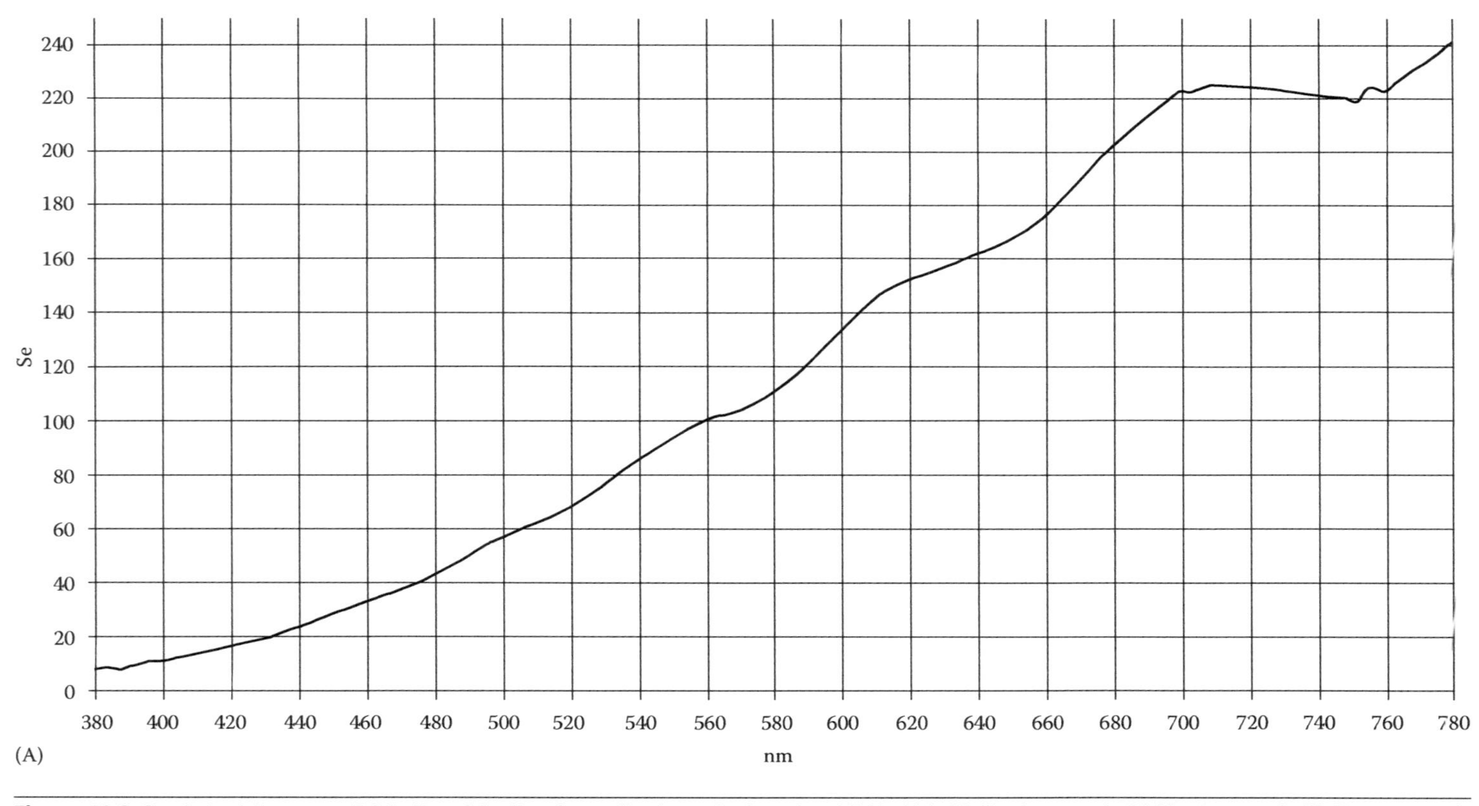

Figure 41.2 Spectral relative power distribution of the five chosen illuminants in the order of Table 41.1. (A) Illuminant no. 1. (B) Illuminant no. 2. (C) Illuminant no. 3. (D) Illuminant no. 4. (E) Illuminant no. 5.

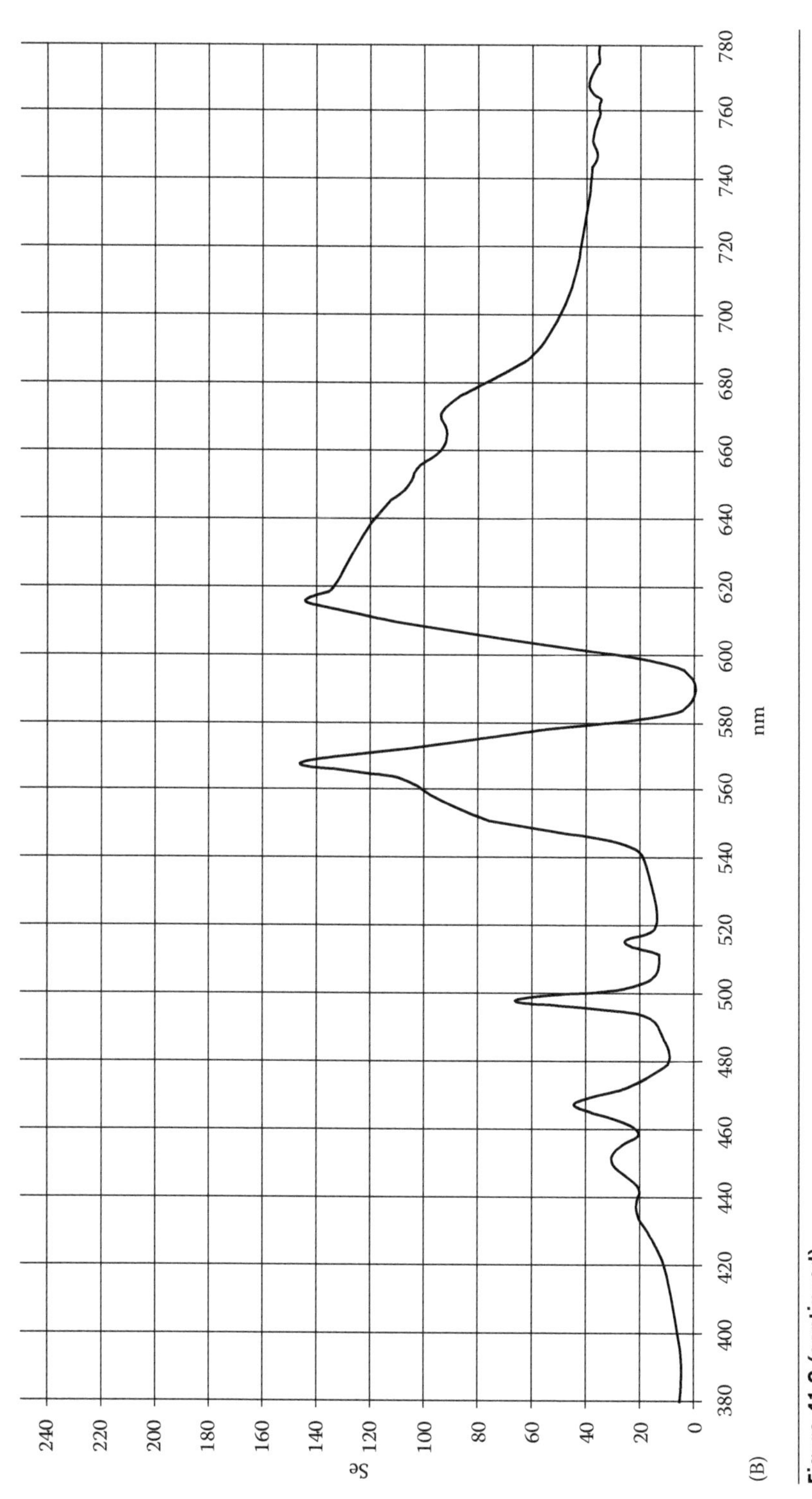

Figure 41.2 (continued)

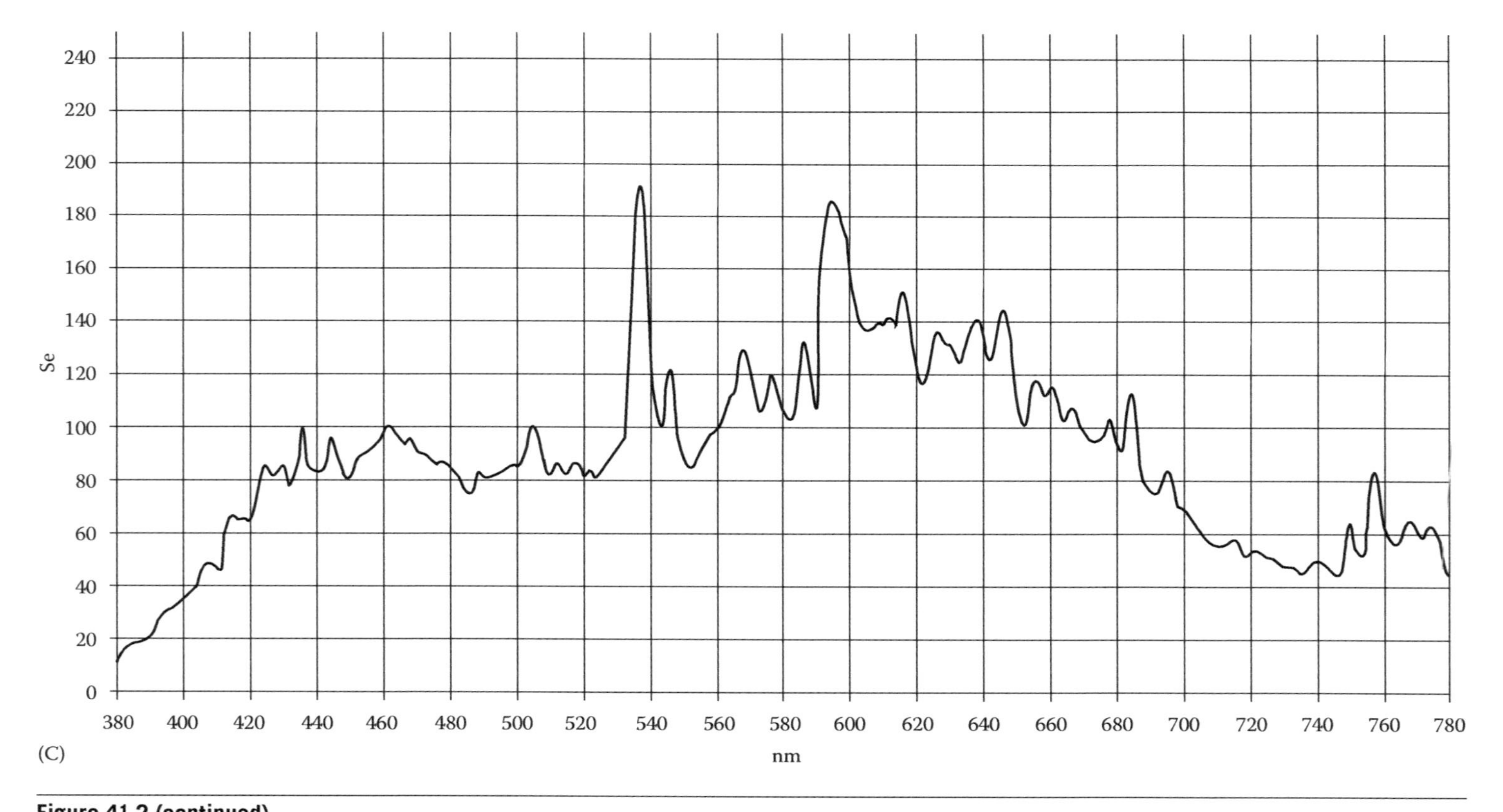

Figure 41.2 (continued)

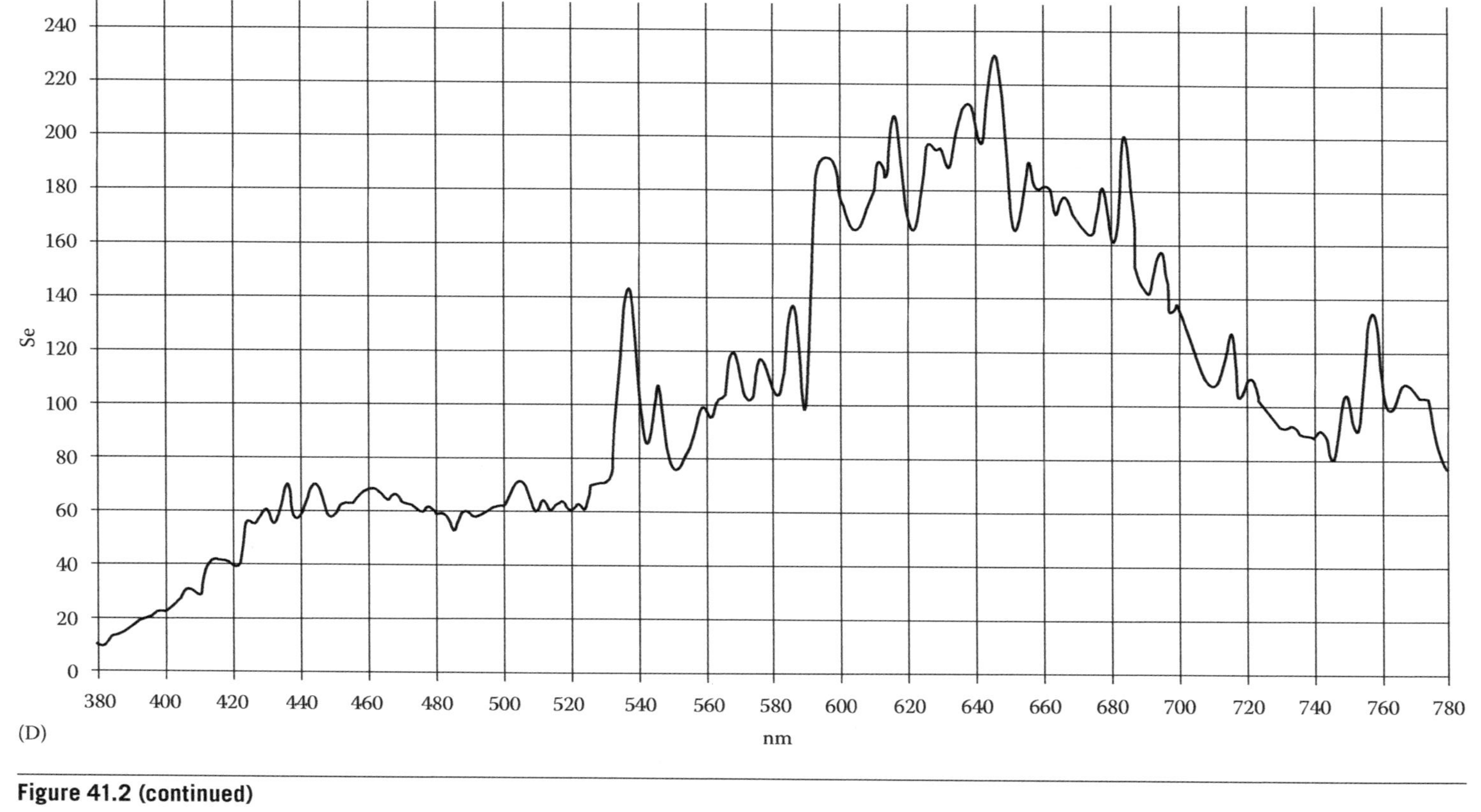

Figure 41.2 (continued)

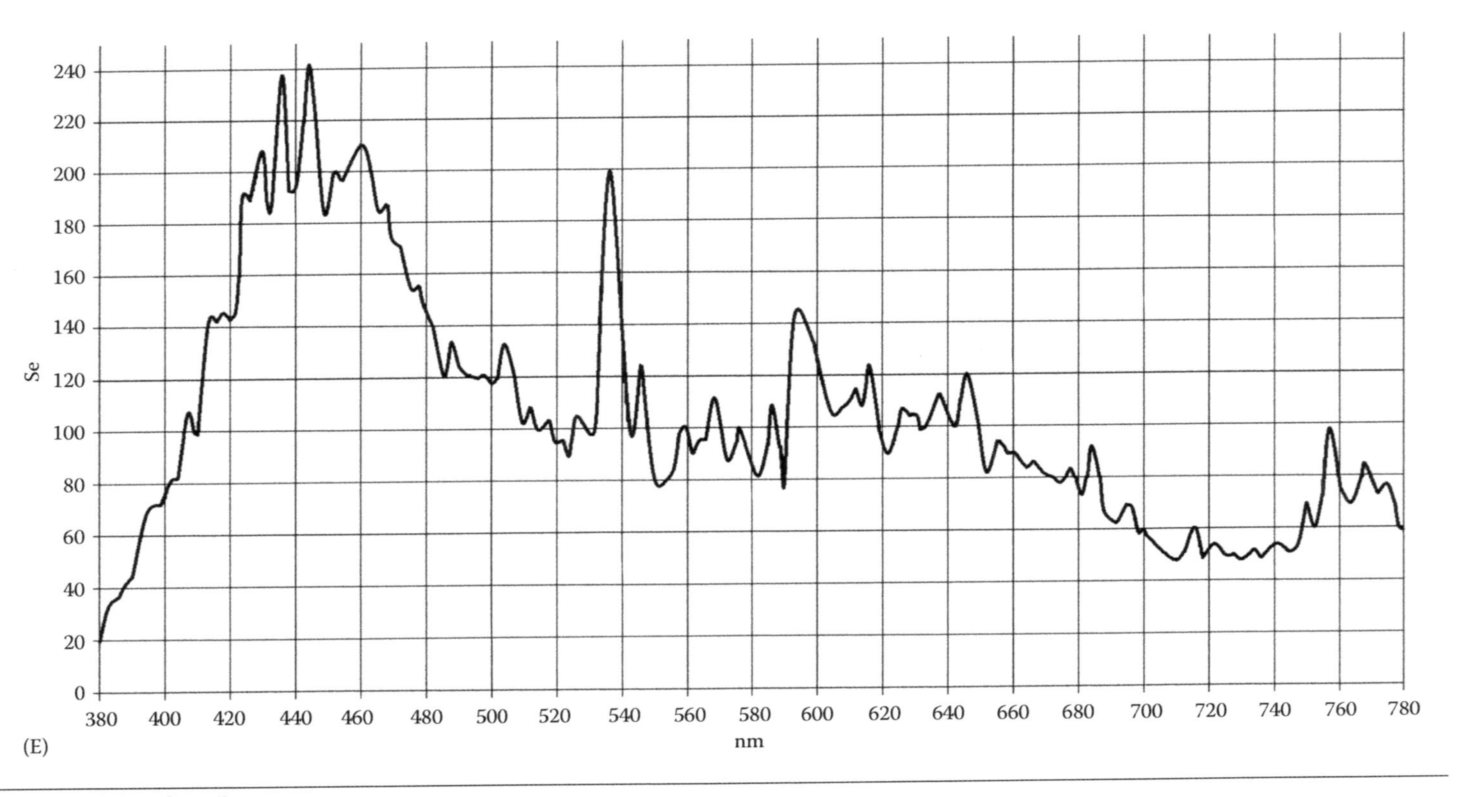

Figure 41.2 (continued)

spectroradiometer SpectraScan (Photo Research)—PR701S, with a ±2% accuracy on spectral radiance and ±20 K for the CCT, certified by the National Institute of Standards and Technology.

We have measured (CIE 2004) the following:

- *X*, *Y*, and *Z* tristimulus values, CIE 1931 standard observer
- *x* and *y* chromatic coordinates, CIE 1931 standard observer
- *u* and *v* chromatic coordinates, CIE 1960 used to compute CCT
- u' and v' chromatic coordinates, CIE UCS 1976
- CCT
- CCT Δuv chromatic distance from the Plankian locus

From the spectral radiance $L_e(\lambda)$, we have determined the relative spectral power distribution $S_e(\lambda)$ (ISO-CIE 1999).

It can be noticed that halogen and white sodium lamps and metal halide lamp with warm white dichroic filter are characterized by a warm light lower than 2800 K. In particular, the warm white dichroic filter can lower by around 1500 K the metal halide lamp emission.

On the other side, the metal halide lamp with cold white dichroic filter is characterized by a 10,228 K color temperature, with a strong blue dominant.

Finally, the anti-UV filter lowers the metal halide lamp emission, from 4200 to 3977 K.

We have also measured the radiance reflected from the food in relation to the different illuminant spectra. □E_{ab}^{*} chromatic distance in the CIELAB space is always much greater than unity. Even if no color adaptation transform has been computed, due to the ongoing scientific debate on the topic, users were always able to perceive the difference due to illuminant.

41.5 User Preference Tests

Perceptual tests have been performed on 124 persons equally distributed between males and females and subdivided equally in two age groups: 20–30 years and above 30 years. The subject had simply to indicate his or her buying choices.

Five preference tests have been performed with three multiple light booths. The light sets have been chosen following the principle that with warm colors, warm lights are preferred and, in the same way, with cold colors, colder lights. For the first three types of food (red meat, oven products, Mediterranean fish), two identical trays (in the content and the disposition) have been shown under two different lighting system. Both kinds of fruits and vegetables have been shown on identical trays under three different types of lighting systems. The trays have been placed at 85 cm height, while the lighting systems have been installed at 1.2 m height over the food, in a way to avoid unwanted light mix. A medium achromatic gray has been chosen as background.

For red meat and oven products, subjects had to choose between the following two lighting systems:

A. Metallic halide lamp 150 W-G12-4200 K—CRI = 96, with warm dichroic filter
B. White sodium lamp 100 W-PG12-1-2550 K—CRI = 83, with anti-UV neutral filter

For the fish, subjects had to choose between the following two lighting systems:

A. Metallic halide lamp 150 W-G12-4200 K—CRI = 96, with anti-UV neutral filter
B. Metallic halide lamp 150 W-G12-4200 K—CRI = 96, with cold dichroic filter

Finally, for fruits and vegetables of the two kinds, subjects had to choose between three possible lighting systems:

A. Metallic halide lamp 150 W-G12-4200 K—CRI = 96, with anti-UV neutral filter
B. White sodium lamp 100 W-PG12-1-2550 K—CRI = 83, with anti-UV neutral filter
C. Halogen lamp 150 W-E27-2900 K—CRI = 100, with anti-UV neutral filter

41.6 Results

The objective has been to test the changes in visual preference, in relation to various types of lighting system, for the three food categories (meat and oven products, fish, fruits and vegetables). Subjects had to compare the same category of products under various illuminations.

Every subject has been asked to choose which product they would have bought, observing the products for no more than 10 ÷ 30 s at a distance of approximately 0.8 ÷ 2 m. A supporting person has filled the questionnaire in order to control the test timing. The question asked was "which of the presented products would you buy?"

Measures have been confirmed by perceptual tests. In fact, in the red meat test, 56% of users' preferences (61% for women) go for solution (A). For the products from the oven test, we have a similar result with a little increment to 64%, see Figure 41.3. For the Mediterranean blue fish, preferences are definitely for solution (A) (81%), but women over 45 years preferred solution (B); this is the case of bigger tested difference between sexes (see Figure 41.4). In the green-yellow vegetables and fruits test, the majority of users (55%) preferred solution (A); while 39% preferred solution (B), and 6% solution (C). Different results have been obtained in the orange-red vegetables and fruits test; in this case, solution (B) won with 60% of the preferences.

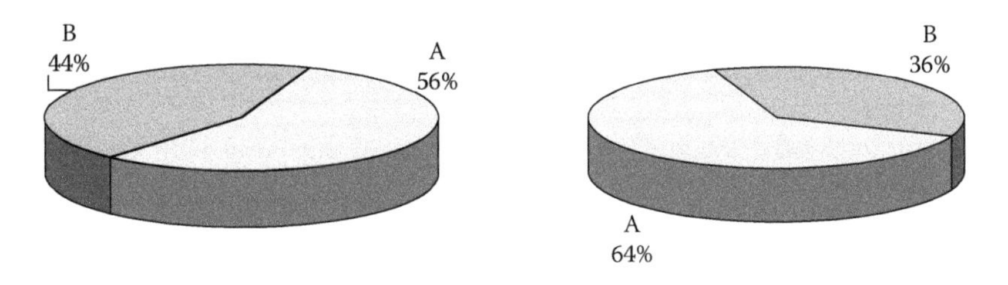

Figure 41.3 Chosen lighting for red meat (left) and oven products (right). Letter A refers to the metallic halide lamp 150W-G12-4200K—CRI = 96, with warm dichroic filter; letter B refers to the white sodium lamp 100W-PG12-1-2550K—CRI = 83, with anti-UV neutral filter.

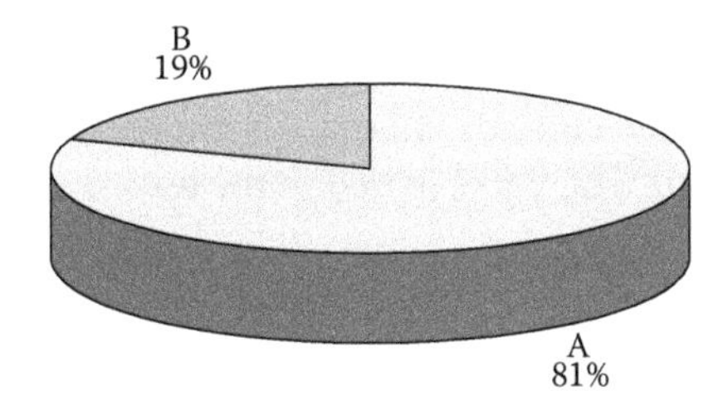

Figure 41.4 Chosen lighting for fish. Letter A refers to the metallic halide lamp 150W-G12-4200K—CRI = 96, with anti-UV neutral filter; letter B refers to the metallic halide lamp 150W-G12-4200K—CRI = 96, with cold dichroic filter.

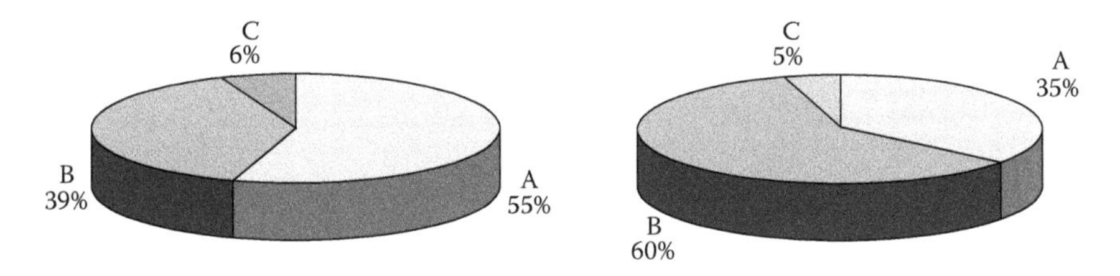

Figure 41.5 Chosen lighting system for red yellow-green fruits and vegetables (left) and other colors of fruits and vegetables (right). Letter A refers to the metallic halide lamp 150W-G12-4200K—CRI = 96, with anti-UV neutral filter; letter B refers to the white sodium lamp 100W-PG12-1-2550K—CRI = 83, with anti-UV neutral filter; letter C refers to the halogen lamp 150W-E27-2900K—CRI = 100, with anti-UV neutral filter.

The results, shown in Figures 41.3 through 41.5, can be used to develop practical criteria for the choice of food lighting.

41.7 Conclusions

A test on the effect of different lighting systems on food choice has been presented. The data reported here are necessarily a synthesis from a more detailed set. The overall preferences are the following:

- Red meat: metal halide lamp with warm dichroic filter
- Oven products: metal halide lamp with warm dichroic filter

- Fish: metal halide lamp with neutral filter
- Green-yellow fruit and vegetables: metal halide lamp with neutral filter
- Orange-red fruit and vegetables: white sodium lamp with neutral filter

Results show that the visual effects of high-efficiency discharge lamps with dichroic filters are comparable to halogen lamps with higher chromatic index but with a lower energetic efficiency.

Acknowledgments

We would like to thank Engr. Giorgio Castaldi, the technicians of the Laboratorio Luce at Politecnico di Milano, Cristina Fallica, Andrea Siniscalco, and all the 124 persons who participated in the test.

References

CIE (Commission Internationale de l'Éclairage). 2004. *Colorimetry*, 3rd edn. Vienna, Austria: CIE Central Bureau, Publ. 15.

ISO-CIE. 1999. ISO 10526/CIE S005:1999. Joint ISO/CIE standard: CIE standard illuminants for colorimetry.

Hunter, R. S. 1975. *The Measurement of Appearance*. New York: John Wiley & Sons.

Zeki, S. 1993. *A Vision of the Brain*. Oxford, U.K.: Blackwell Scientific Publications.

CHAPTER 42

Psychological Effects of Tablecloth Color and Tray Color on Diners

KEIKO TOMITA, TERUMI AIBA, JANGMI KANG, MOTOKO MATSUI, and KIMIKO OHTANI

Contents

42.1 Introduction

Due to the decrease in the quality of human relationships not only in society widely but also within families under a lot of stress, the number of Japanese subjects with depression is increasing without regard to age and sex. More than 30,000 Japanese commit suicide each year. Our previous paper showed that mealtime was the most important chance for Japanese to communicate and cultivate mutual understandings among family members (Ohtani et al. 2003; Tomita et al. 2005a,b), but many Japanese spend mealtimes focused only on eating. If a daily comfortable dining environment can be created by using color effectively to promote conversation and increase energy, the number of persons who feel relieved may increase and then those who commit suicide may decrease.

We generally use all five senses to enjoy our meals, although information obtained through sight is the most important factor for enhancing the diner's appetite (Birren 1963; Hutchings 2003). In particular, the colorful dishes and foods on the dining table stimulate diners to feel various emotions and can increase appetite. Although some reports have examined the psychological effects of the color in the food business, such as package design color, the interior colors of a restaurant, and so on (Hutchings 1999; Ohtani et al. 2000), the psychological effects of color on the dining table have not been investigated.

The tablecloth color occupies the largest color area on the dining table, and we can easily change it for refreshment depending on time, place, and occasion. In addition, the psychological effects of the color are thought to change according to the strength of illumination. We investigated the psychological effects of tablecloth color under ordinary lighting conditions (400–600 lx) toward the university students, as shown in Section 42.2. In addition, we investigated the effects under 850–1050 lx (bright condition) or 12–22 lx (dark condition) toward the university students, as shown in Section 42.3.

In Japan, small tableware such as meal trays and place mats are more familiar than tablecloths. In particular, fiber-reinforced plastic (FRP) trays are used frequently for school lunches and meals at a food court and in institutions. The number of persons living in nursing homes is increasing as society ages in Japan. Improvements in the dining space environment of nursing homes may be important in maintaining good nutritional status and increasing activities of daily living (ADL) and quality of life (QOL). It is possible that if an appropriate color of meal tray is used, the QOL of residents in nursing homes may increase. Thus, we also investigated the psychological effects of a small color area on the dining table provided by FRP trays for both university students as well as elderly persons living in nursing homes and in their own homes, as shown in Section 42.4.

42.2 Psychological Effects of Tablecloth Color on Diners under 400–600 lx in the Dining Hall

42.2.1 Materials and Methods

The questionnaire consisted of three parts (Table 42.1): physical and mental condition on the day of testing, the effects of the colored tablecloth on the psychology of the diner, and the impact of six colored, cotton tablecloths (red, yellow, beige, blue, white, and black). The hue, value, and chroma of each tablecloth are shown in Table 42.2. The questionnaire study was performed

Table 42.1 The Questionnaire

Place of Examination	Variety of Questionnaire	Chapter 2	Chapter 3
Questionnaire in the waiting room	Attributions	Age, sex, etc.	
	Physical and mental conditions	How about your todays conditions for sleeping, health, appetite, mood	
	Preference color and recommendable	Which color do you like best[a]?	
		Which color do you recommend for dining tablecloth[a]?	
Questionnaire in the experimental room[b]	Effects on the psychology of diner	With whom and how many persons do you want to eat? (large number of people, small number of people, a couple, alone)	
		Does this table make you want to talk?	
		Does this tablecloth color make you heal?	
		Does this tablecloth color make your appetite whet?	

(continued)

Table 42.1 (continued) The Questionnaire

Place of Examination	Variety of Questionnaire	Chapter 2	Chapter 3
	Color image of tablecloth by using SD methods	Prefer–Dislike, Suitable–Unsuitable, Appetizing–Unappetizing, Healthy–Unhealthy, Hot–Cold, Warm–Cool, Sanitary–Insanitary, Cheerful–Gloomy, Pleasant–Unpleasant, Exhilarated–Melancholy, Happy–Unhappy, Friendly–Unfriendly, Intimately–Lonely, Calm–Irritate, Comfortable–Uncomfortable, Lively–Unlively, Active–Inactive, Relax–Tense, Refined–Vulgar, Sophisticated–Unsophisticated, Open–Closed, Spacious–Confined, Adult–Childlike, Feminine–Masculine, Classical–Modern, Formal–Casual, Romantic–Realistic, Natural–Artificial, Decorative–Simple, Individual–General, Popular–Gorgeous, Emotional–Unemotional (32 adjective antonym pairs)	Prefer–Dislike, Appetizing–Unappetizing, Healthy–Unhealthy, Hot–Cold, Warm–Cool, Tensed–Relaxed, Cheerful–Gloomy, Pleasant–Unpleasant, Exhilarated–Melancholy, Happy–Unhappy, Friendly–Unfriendly, Intimately–Lonely, Calm–Irritate, Comfortable–Uncomfortable, Lively–Unlively, Conversational–Silent, Emotional–Unemotional, Healing–Impatient, Gaily–Depressing, Refined–Vulgar, Sophisticated–Unsophisticated, Open–Closed, Spacious–Confined, Adult–Childlike, Feminine–Masculine, Classical–Modern, Formal–Casual, Fantastic–Realistic, Deep–Shallow, Dazzling–Dark, Private–Public, Vivid–Dull, Soft–Hard (33 adjective antonym pairs)

[a] Subjects answered this question by observing the color chart made by a snip of the tablecloth (15 mm × 25 mm).
[b] Subjects answered these questions by sitting in front of the table covered with color tablecloth.

Table 42.2 Hue, Value, and Chroma for Each Tablecloth Color Based on Munsell Values

Tablecloth Color	Hue	Value	Chroma
Beige	2.0 Y	8.5	1.4
Red	5.8 R	4.6	13.4
Yellow	7.5 Y	8.7	9.4
Blue	1.1 B	5.6	7.2
White	2.1 PB	9.2	0.2
Black	9.0 PB	2.3	0.4

Source: Data from Tomita, K. et al., *J. ARAHE,* 13, 173, 2006.

for female university students (mean age, 22.1 ± 2.3 years) by using the SD method with 32 antonym adjective pairs. The number of subjects for each tablecloth color study was 25–28. All subjects took part in the experiment more than 1.5 h after the meal.

Six tables (900 mm wide × 1800 mm deep × 700 mm tall) were covered with different colored tablecloths and separated from each other by a beige partition. The position of the tablecloth color was randomly changed every day. The subject answered the questionnaire at the table on which two goblet-shaped water glasses (transparent glass) and two dinner plates with similar colors to the tablecloth were set.

The dining hall was illuminated by a ceiling light which was the fixed fluorescent lamps (FLR 40S EX-N/M-X36; National Co. Ltd., Japan, straight-tube type and rapid-tube type) with three wavelengths (435 nm [44.5%], 545 nm [100%], and 611 nm [59.5%]).

The brightness was measured by an illuminance meter (type 51005, Yokogawa Electric Corporation, Japan) on the center of the table (30 cm from the front side, 90 cm from both the right and left edges). The room temperature and the humidity were about 25°C and 50%, respectively.

Data analyses were performed using the SPSS (10.0 version for Windows). A principal component analysis was performed in which the following five points were confirmed: (1) the eigenvalue must be 1.0 or more; (2) the cumulative contribution rate must be 50% or more; (3) the reliability coefficient, Cronbach's alpha, must be 0.65 or more; (4) the number of components was verified by Scree plot; and (5) adjectives with factor loadings of 0.4 or on more than one component were omitted.

42.2.2 Results

42.2.2.1 Image of the Tablecloth Color The experiment under 400–600 lx was performed as the model of dining spaces at home and at a restaurant according to the JIS and IEIJ (CSAJ 1998; IEIJ 2004). By principal component analyses together with a Varimax rotation technique from the questionnaire distributed

Table 42.3 Four Components ("Activity," "Elegance," "Softness," and "Ordinary") Were Extracted from 32 Adjective Pairs by Principal Component Analysis under 400–600 lx in the Dining Hall

	Components			
Antonym Adjective Pairs	**1**	**2**	**3**	**4**
Cheerful–Gloomy	0.850	−0.195	0.176	−0.071
Pleasant–Unpleasant	0.845	−0.098	0.234	−0.016
Open–Close	0.835	−0.185	−0.046	−0.055
Healthy–Unhealthy	0.817	0.080	0.006	0.195
Exhilarated–Melancholy	0.809	−0.054	0.274	−0.011
Lively–Unlively	0.791	−0.212	0.162	−0.084
Spacious–Confined	0.788	0.074	−0.124	0.088
Happy–Unhappy	0.754	0.056	0.353	0.155
Appetizing–Unappetizing	0.699	0.169	0.102	0.261
Intimately–Lonely	0.649	−0.126	0.219	0.280
Friendly–Unfriendly	0.614	0.108	0.203	0.440
Emotional–Unemotional	0.595	0.094	0.431	0.046
Refined–Vulgar	−0.034	0.836	0.090	−0.009
Sophisticated–Unsophisticated	0.194	0.752	−0.086	−0.226
Adult–Childlike	−0.356	0.721	0.105	0.002
Sanitary–Insanitary	0.236	0.688	−0.260	0.065
Modern–Classical	0.161	−0.568	−0.078	−0.021
Calm–Irritate	−0.153	0.514	0.251	0.380
Warm–Cool	0.434	−0.185	0.666	−0.018
Hot–Cold	0.455	−0.313	0.592	−0.105
Feminine–Masculine	0.003	0.266	0.570	0.298
Romantic–Realistic	0.256	0.286	0.530	−0.267
Gorgeous–Popular	0.039	0.203	0.242	−0.748
Relax–Tense	0.257	−0.004	0.154	0.744
Natural–Artificial	0.479	0.319	−0.026	0.553
Eigenvalue	7.940	3.482	2.260	2.219
Cumulative contribution rate (%)	31.8	45.7	54.7	63.6
α-Coefficient	0.942	0.544	0.658	−0.620

Source: Data from Tomita, K. et al., *J. ARAHE*, 13, 173, 2006.
The gray shades show the range of antonym adjective pairs having the feature of each component.

by the SD method, four components were extracted: "activity," "elegance," "softness," and "ordinary" (Table 42.3). The image of each tablecloth color is shown in Figure 42.1. Chromatic colors were in the positive zone for "activity" and "elegance" except for yellow. Yellow was negative for "elegance."

42.2.2.2 Atmosphere Produced by the Tablecloth Color The warm-colored tablecloths (red and yellow) were more likely than other colors to produce an

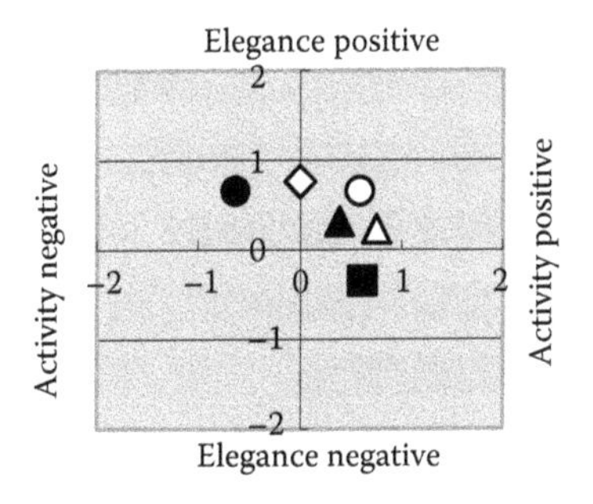

Figure 42.1 The factor score of each tablecloth color on a graph with vertical axis representing "elegance" and the horizontal axis representing "activity" as extracted by principal component analysis from 32 antonym adjective pairs under 400–600 lx in the dining hall. (Data from Tomita, K. et al., *J. ARAHE*, 13, 173, 2006.) △ Red, ■ yellow, ○ beige, ▲ blue, ◇ white, ● black.

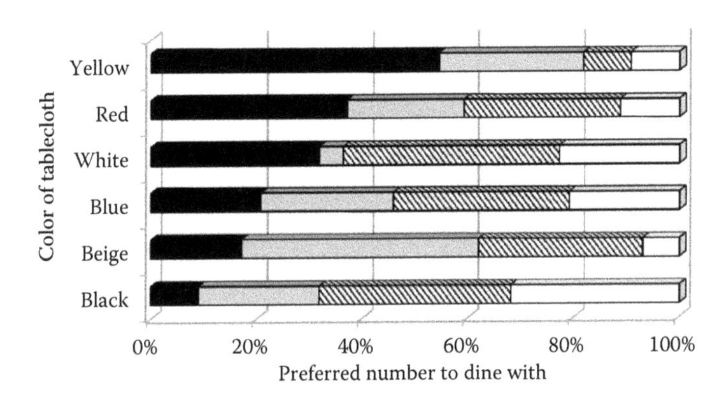

Figure 42.2 Effects of tablecloth color on preferred number of people to dine under 400–600 lx in the dining hall. ■ A large number, □ a small number, ▨ as a couple, □ alone.

atmosphere in which diners wanted to dine with a large number of people (Figure 42.2). The beige tablecloth was most likely to promote a conversation with a small number of people (Figure 42.3). Although Birren (1963) reported that orange, red, and yellow enhanced one's appetite and yellow-green and purple decreased one's appetite, the results of this study showed that beige increased one's appetite (Figure 42.4). In addition, the beige tablecloth led to feelings of healing (returning to a healthy mental and emotional state after a tired, sad, stressful, tense experience, etc.) (Figure 42.5). The beige tablecloth was likely to make one feel like being with one's family (Tomita et al. 2007).

42.2.2.3 Table Style Produced by the Tablecloth Color The table style produced by the tablecloth color is shown in Figure 42.6. Chromatic tablecloth colors such as yellow, red, and blue were associated with modern and casual styles, whereas the beige tablecloth was associated with a classical style.

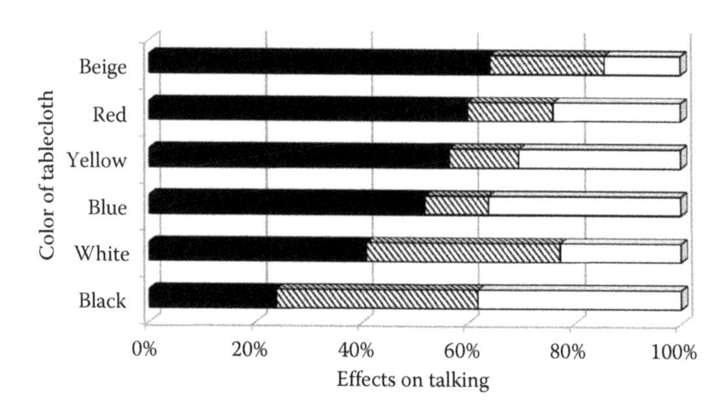

Figure 42.3 Effects of tablecloth color on desire to talk with others under 400–600 lx in the dining hall. ■ Increased, ▨ unchanged, □ decreased.

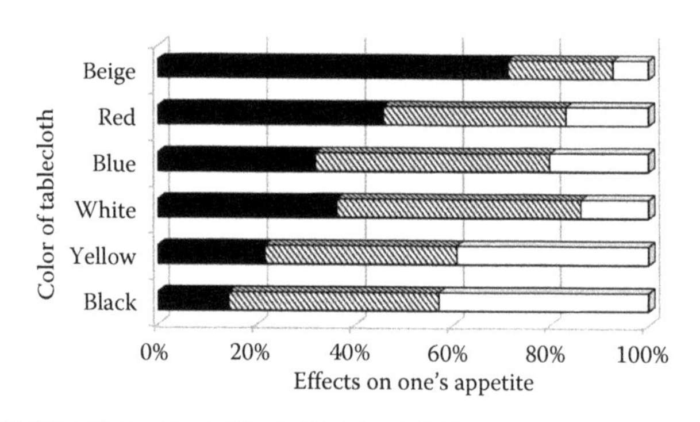

Figure 42.4 Effects of tablecloth color on one's appetite under 400–600 lx in the dining hall. ■ Increased, ▨ unchanged, □ decreased.

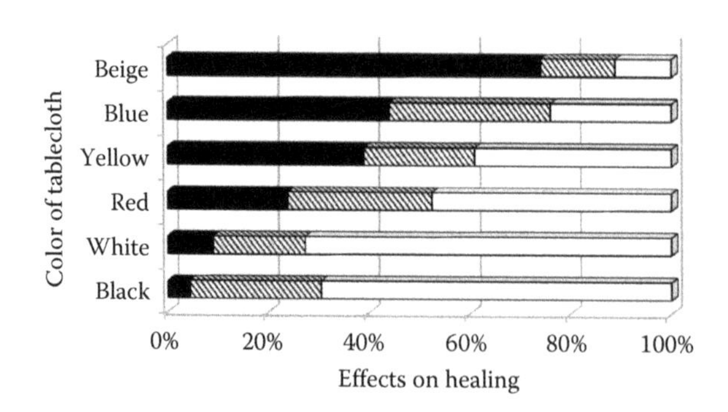

Figure 42.5 Effects of tablecloth color on healing under 400–600 lx in the dining hall. ■ Increased, ▨ unchanged, □ decreased.

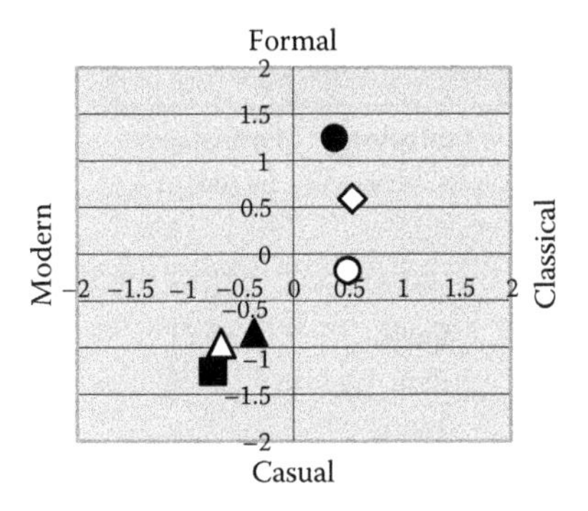

Figure 42.6 Relationship between table style and tablecloth color under 400–600 lx in the dining hall. △ Red, ▲ blue, ■ yellow, ◇ white, ○ beige ● black. (Data from Tomita, K. et al., *J. ARAHE*, 13, 173, 2006.)

Achromatic tablecloth colors (i.e., white and black) were associated with formal and classical styles.

42.3 Psychological Effects of Tablecloth Color on Diners under 850–1050 or 12–22 lx in the Climatic Chamber

42.3.1 Materials and Methods

The questionnaire is shown in Table 42.1. This experiment was performed according to the method described in Section 42.2 except for the following four points: (1) the illuminant condition was 12–22 lx or 850–1050 lx; (2) the experiment was performed in a climatic chamber, where we can easily change not only temperature but humidity and brightness; (3) in the case of the experiment under 12–22 lx, an indirect lighting lamp (AT-3805 white sand <PTH-10> φ80xH156 with circle type <PTH-05> 140x105xH65; TKG Co. Ltd., Japan) put on the center of the table was turned on; and (4) 33 antonym adjective pairs were applied. The brightness and the color temperature were measured by an illuminance meter (same in Section 42.2) and luminance meter (CS-200, Konica Minolta, Japan), respectively, on the center of the table (30 cm from the front side, 90 cm from both right and left sides) (Table 42.4).

42.3.2 Results

42.3.2.1 Image of the Tablecloth Color The experiment under 850–1050 lx (bright condition) was meant to represent a model of dining near a window with blinds under 10,000 lx of sunlight (CSAJ 1998; IEIJ 2004). The experiment under 12–22 lx (dark with dim light condition) was meant to represent a model of a relaxing and comfortable space such as a bar, a lounge, or a tea ceremony performed at night, known as *Yobanashi no Chaji* in Japan, which is held commonly under 5 lx (CSAJ 1998; IEIJ 2004).

Table 42.4 Luminance and Color Temperature of Each Tablecloth in the Climatic Chamber Measured by Using the Luminance Meter (CS-200 Made from Konica Minolta in Japan)

Bright Condition	Illuminance (lx)	Tablecloth Color	Luminance Lv (cd/m²)	Color Temperature (K)	Δ*uv*
Dark with dim light condition	12–22	Red	1.19	—	—
		Yellow	6.46	3165	0.0256
		Beige	5.31	3569	0.0089
		Blue	2.29	7011	0.0362
		White	7.29	3805	0.0054
		Black	0.20	3986	0.0057
Bright condition	850–1050	Red	50.34	—	—
		Yellow	277.30	3274	0.0277
		Beige	244.41	3739	0.0121
		Blue	92.01	7043	0.0401
		White	312.52	3963	0.0083
		Black	8.55	4298	0.0121

In the dark with dim light and bright conditions, two components were extracted from the questionnaires by principal component analyses together with a Varimax rotation technique: "activity" and "elegance" (Table 42.5). The image of each tablecloth color is shown in Figure 42.7. In the dark with dim light condition, the images of all tablecloth colors were in the negative zone for "activity." In the bright condition, all tablecloth colors except black were in the positive zone for "activity." Thus, the images produced by a black tablecloth were barely changed by the strength of brightness, whereas the images produced by a yellow tablecloth were shown to be most affected by the strength of the brightness (Tomita et al. 2006).

42.3.2.2 Atmosphere Produced by the Tablecloth Color With all tablecloth colors, as the brightness increased, the desire to dine with more people also increased, which indicated that strength of the brightness leads to an increase in affection of diners more than the color of the tablecloth (Figures 42.8 and 42.9). In the bright condition, the beige tablecloth was most likely to enhance one's appetite and lead to feelings of healing (Figures 42.10 and 42.11) (Tomita et al. 2007).

42.3.2.3 Table Style Produced by the Tablecloth Color The table style produced by a tablecloth color is shown in Figure 42.12. In the dark with dim light condition, all tablecloths colors were associated with formal and classical styles. In the bright conditions, all colors except white were associated with a modern style. In particular, the property of the yellow tablecloth was strengthened by the strength of the illumination (Tomita et al. 2006).

Table 42.5 Two Components ("Activity" and "Elegance") Were Extracted from 33 Adjective Pairs by Principal Component Analysis under 12–22 lx (Dark with Dim Light Condition) and 850–1050 lx (Bright Condition) in the Climatic Chamber

	Components	
Antonym Adjective Pairs	**1**	**2**
Cheerful–Gloomy	0.9014	−0.0670
Pleasant–Unpleasant	0.8896	−0.0083
Healthy–Unhealthy	0.8643	0.1430
Open–Close	0.8579	−0.1129
Gaily–Depressing	0.8538	−0.1697
Dazzling–Dark	0.8506	−0.1950
Exhilarated–Melancholy	0.8482	−0.0014
Happy–Unhappy	0.8360	0.2059
Vivid–Dull	0.8293	−0.0556
Lively–Unlively	0.7881	0.0097
Spacious–Confined	0.7759	0.0923
Intimately–Lonely	0.7417	0.0671
Lively–Unlively	0.7349	0.1053
Refined–vulgar	0.0339	0.8152
Vivid–Dull	0.1371	0.7735
Healing–Impatient	0.0799	0.7692
Eigenvalue	9.182	2.601
Cumulative contribution rate (%)	54.0	15.3
α-Coefficient	0.962	0.733

Source: Data from Tomita, K. et al., *J. ARAHE*, 13, 173, 2006.
The gray shades show the range of antonym adjective pairs having the feature of each component.

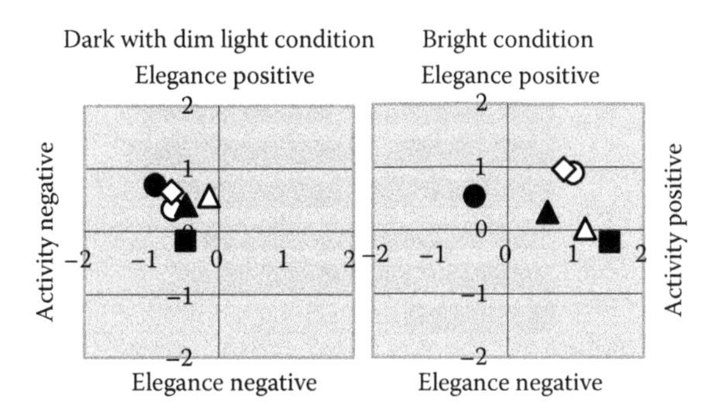

Figure 42.7 The factor score of each tablecloth color on a graph with the vertical axis representing "elegance" and the horizontal axis representing "activity" as extracted by principal component analysis from 33 antonym adjective pairs in the climatic chamber. △ Red, ■ yellow, O beige, ▲ blue, ◇ white, ● black. (From Tomita, K. et al., *J. ARAHE*, 13, 173, 2006.)

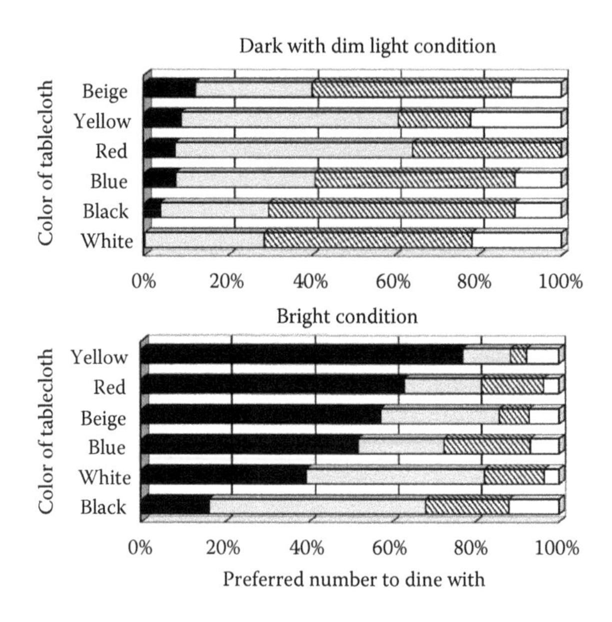

Figure 42.8 Effects of color of tablecloth on preferred number of people to dine with in the climatic chamber. ■ A large number of people, ▫ a small number of people, ▨ as a couple, □ alone.

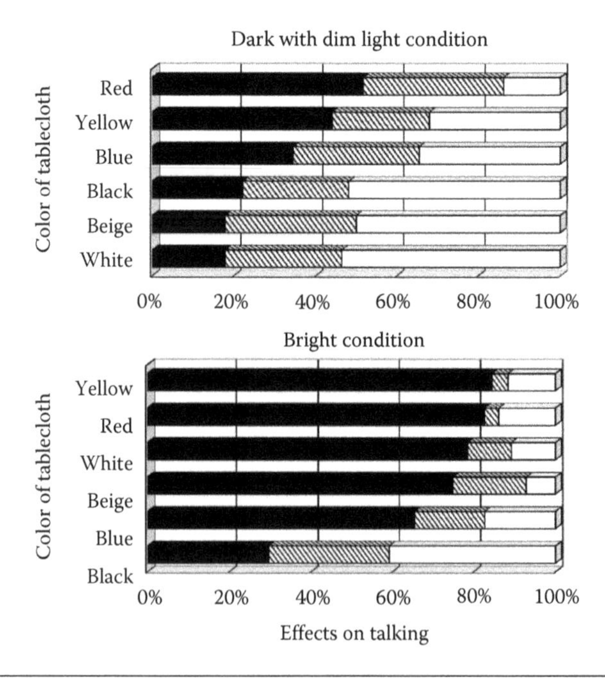

Figure 42.9 Effects of tablecloth color on desire to talk with others in the climatic chamber. ■ Increase, ▨ unchanged, □ decreased.

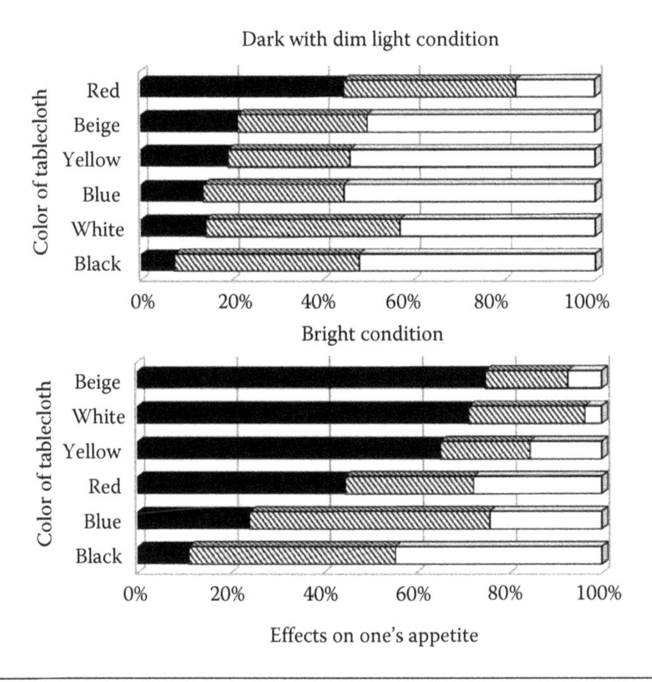

Figure 42.10 Effects of tablecloth color on one's appetite in the climatic chamber. ■ Increased, ▨ unchanged, □ decreased. (Data from Tomita, K. et al., *J. Integr. Study Diet Habits*, 18, 48, 2007.)

42.4 Psychological Effects of FRP Tray Color on Diners

42.4.1 Materials and Methods

Questionnaires consisted of general questions regarding the physical and mental situation on the day of the study and images of six colored FRP trays; questionnaire studies were performed by using the SD method, which consisted of 11 antonym adjective pairs. Subjects were elderly persons living in nursing homes (nursing-elderly; mean age, 83.6 ± 8.3 years; $n = 22$/color), elderly persons living in their own homes (home-elderly; mean age, 68.4 ± 6.6 years; $n = 20$/color), and university students (mean age, 20.6 ± 1.5 years, $n = 34$/color). In the case of elderly persons, a few supplementary questions about their eye conditions such as the presence of cataracts and how good their vision was were asked. For elderly persons in nursing homes, interview surveys were performed instead of having subjects fill out a questionnaire. Subjects who had good vision and no serious cognitive impairment that would interfere with answering the questionnaire took part in the experiment. All experiments took place more than 1.5 h after a meal. The questionnaire studies were performed in the dining room under 400–600 lx. The room temperature and the humidity were around 25°C and 50%, respectively. Subjects answered the questionnaire

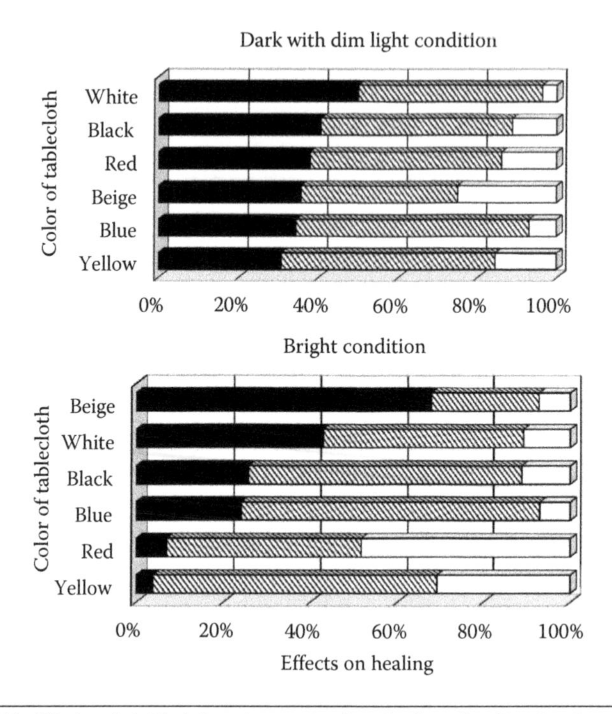

Figure 42.11 Effects of tablecloth color on healing in the climatic chamber. ■ Increased, ▨ unchanged, □ decreased. (Data from Tomita, K. et al., *J. Integr. Study Diet Habits*, 18, 48, 2007.)

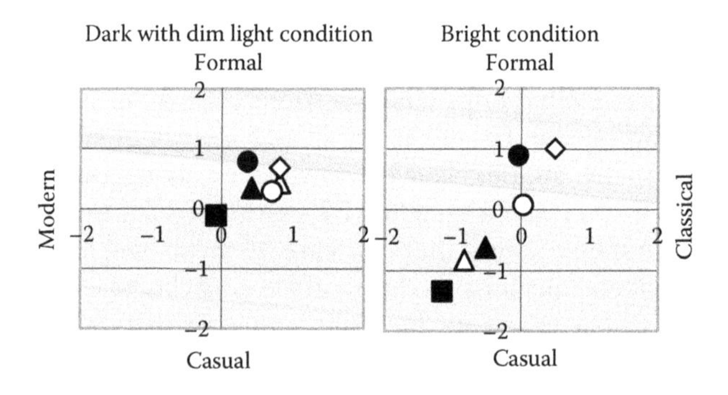

Figure 42.12 Relationship between table style and tablecloth color in the climatic chamber. △ Red, ■ yellow, O beige, ▲ blue, ◇ white, ● black. (Data from Tomita, K. et al., *J. ARAHE*, 13, 173, 2006.)

at the table covered with a beige tablecloth (Table 42.2) which was shown to be most useful for mealtime as shown in Section 42.3, on which a randomly selected colored tray was set with dish plates and bowls made from melamine. The colors of FRP trays (430 mm deep × 310 mm wide × 20 mm tall) are shown in Table 42.6. Data analyses were performed using PASW statistics 18.

Table 42.6 L^*, a^*, b^* Values of Each Tray Color Measured by Using the Spectrophotometer (CM-2600d Made from Konica Minolta in Japan)

		Yellow*	Pink*	Green*	Blue	Brown	Orange
L^*(C)	SCI	87.27	82.23	78.51	58.97	38.45	73.94
	SCE	84.93	79.59	75.87	54.55	30.52	70.87
a^*(C)	SCI	−5.09	6.26	25.22	2.01	4.73	30.12
	SCE	−5.29	6.55	27.36	2.17	6.19	31.85
b^*(C)	SCI	28.21	10.01	−15.77	−18.96	10.64	28.92
	SCE	30.15	10.98	−16.68	−20.41	16.75	32.19

Source: Data from Tomita, K. et al., Psychological effects of the tray-color on diners—Comparison between young persons and elderly persons, in *Proceedings of AIC 2010*, Mar del Plata, Argentina, 2010, pp. 383–389.

42.4.2 Results

42.4.2.1 Image Profiles of Tray Colors Image profiles of the six tray colors are shown in Figure 42.13. Although the image profile of the blue tray was significantly different in university students compared with elderly persons, the image profiles of the yellow, green, orange, and pink trays were almost the same between groups. Pastel colors such as pink, yellow, and green were shown to promote the diner's appetite and to lead to feelings of comfort, cheerfulness, and happiness.

42.4.2.2 Psychological Effects of Tray Color Two components were extracted by factor analyses of the images of tray colors obtained by the SD method: "activity" and "relaxation" (Table 42.7). The scores for each component for the six colors are shown in Figure 42.14.

In the case of elderly persons living in their own homes and university students, the pastel colors such as pink, yellow, and green were in the positive zone for both components. University students and elderly persons living in their own homes answered that the pastel colors were good for breakfast and lunch and that brown was good for dinner (Figure 42.15).

42.5 Conclusion and Summation

In Section 42.2, the images produced by six tablecloth colors were investigated under 400–600 lx. The beige tablecloth was shown to be useful for family dining, as it led to the production of a comfortable atmosphere and a desire to have conversations with a small number of people.

In Section 42.3, the images produced by six tablecloth colors under 850–1050 and 12–22 lx were investigated. In the bright condition, the yellow and red tablecloths were shown to be useful for a large number of people to enjoy a mealtime together. The black tablecloth in the dark condition promoted intimate conversation among a small number of people.

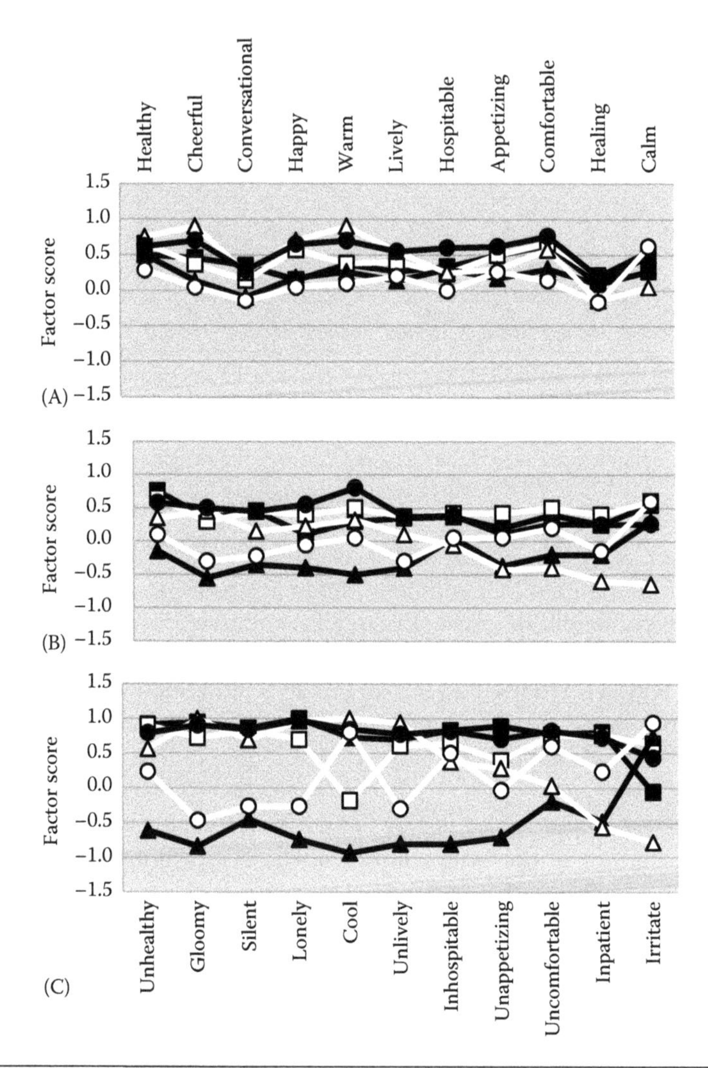

Figure 42.13 Image profiles of each tray color by the SD method using 11 antonym adjective pairs. (A) Elderly persons living in nursing homes. (B) Elderly persons living in their own homes. (C) University students. Yellow, green, blue, orange, pink, brown.

In Section 42.4, the psychological effects of FRP tray colors were investigated under 400–600 lx. The pastel colors were best for promoting appetite without regard to age and living location. In addition, changing the tray color depending on the mealtime showed a possibility of contributing to the daily life rhythm and increasing QOL.

Depending on the diner's situation (physical and mental condition) and time, place, and occasion, the effective usage of tablecloth color or tray color, which is easily changeable, may change the diner's mood and increase QOL.

Table 42.7 Two Components ("Activity" and "Relaxation") Were Extracted from 11 Antonym Adjective Pairs by Factor Analysis

	Components	
Antonym Adjective Pairs	**1**	**2**
Cheerful–Gloomy	0.912	0.053
Happy–Lonely	0.892	0.081
Lively–Unlively	0.805	0.105
Healthy–Unhealthy	0.701	0.273
Conversational–Silent	0.679	0.203
Warm–Cool	0.573	0.035
Healing–Impatient	0.368	0.706
Comfortable–Uncomfortable	0.416	0.671
Calm–Irritate	−0.342	0.65
Eigenvalue	4.605	1.638
Cumulative proportion (%)	44.228	60.991
Cronbach's α-Coefficient	0.895	0.685

Source: Data from Tomita, K. et al., Psychological effects of the tray-color on diners—Comparison between young persons and elderly persons, in *Proceedings of AIC 2010*, Mar del Plata, Argentina, 2010, pp. 383–389.

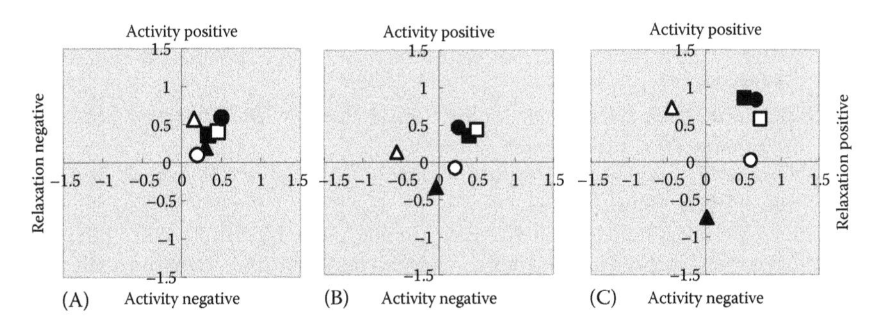

Figure 42.14 The factor score of each tray color on a graph with vertical axis representing "activity" and horizontal axis representing "relaxation" as extracted by factor analysis from 11 antonym adjective pairs. (A) Elderly persons living in nursing homes. (B) Elderly persons living in their own homes. (C) University students. ■ Yellow, □ green, ▲ blue, △ orange, ● pink, O brown.

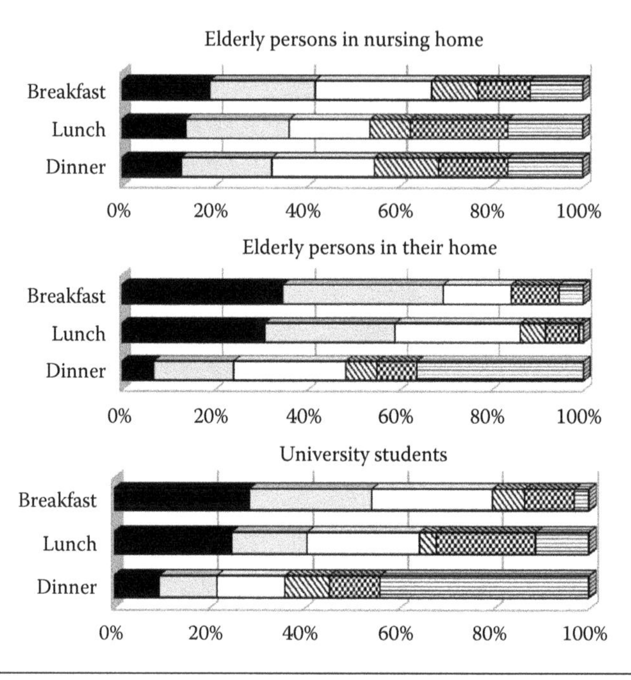

Figure 42.15 Suitable tray color for each meal. ■ Yellow, ▫ green, □ pink, ▨ blue, ■ orange, ▤ brown.

Acknowledgments

The work in Section 42.3 was supported by a Grant-in-Aid for Scientific Research (C) (21500760) in Japan.

References

Birren, F. 1963. Color and human appetite. *Food Technology* 17: 553–555.

CSAJ (Colour Science Association of Japan). 1998. *Handbook of Colour Science*. Tokyo, Japan: University of Tokyo Press.

Hutchings, J. B. 1999. *Food Colour and Appearance*, 2nd edn. Gaithersburg, MD: Aspen Publishers.

Hutchings, J. B. 2003. *Expectations and the Food Industry. The Impact of Color and Appearance*. New York: Kluwer Academic.

IEIJ (Illuminating Engineering Institute of Japan). 2004. *Lighting Handbook*, 2nd edn. Tokyo, Japan: Ohm-sha Co.

Ohtani, K., R. Nakakita, T. Aiba, J. Kang, K. Tomita, and T. Minamide. 2003. Effects of dietary experiences in their home and parents-child interaction on their self-independence in the late adolescence. *Journal for the Integrated Study of Dietary Habits* 14: 14–27.

Ohtani, K., A. Ozaki, O. Rhee, J. Kang, M. Matsui, and T. Minamide. 2000. Effects design colour of canned beverage on the various senses—For Japanese and Korean Women's students. *Journal of the Color Science Association of Japan* 24: 223–231.

Tomita, K., T. Aiba, J. Kang, M. Matsui, and K. Ohtani. 2010. Psychological effects of the tray-color on diners—Comparison between young persons and elderly persons. In *AIC 2010 Color and Food, Proceedings*, eds. J. Caivano and M. López. Buenos Aires, Argentina: Grupo Argentino del Color, pp. 383–389.

Tomita, K., T. Aiba, J. Kang, and K. Ohtani. 2005b. Effect of the estimation for their father on their sense of eating with families in course of the development — Toward for junior and senior high school, and university students. *Journal for the Integrated Study of Dietary Habits* 16: 230–241.

Tomita, K., R. Nakakita, T. Aiba, and K. Ohtani. 2005a. Effect of the estimation toward their mother on their sense of eating with families in course of the development—Toward for junior and senior high school, and university students. *Journal for the Integrated Study of Dietary Habits* 15: 229–239.

Tomita, K., M. Ono, T. Aiba, and K. Ohtani. 2006. The atmosphere of the dining produced by the various kinds of tablecloth colours under three kinds of illuminant condition. *Journal of ARAHE* 13: 173–184.

Tomita, K., M. Ono, T. Aiba, and K. Ohtani. 2007. Psychological effects of tablecloth color on diners under different brightness. *Journal for the Integrated Study of Dietary Habits* 18: 48–55.

CHAPTER 43

Color Evaluation in Association with Consumer Expectations of Green Tea Drinks in Thailand

SUCHITRA SUEEPRASAN and CHAWIKA TRAISIWAKUL

Contents

43.1 Introduction

Second only to water, tea is the most widely consumed beverage in the world, owing to its benefit of relaxation effect (Rogers et al. 2008). Three major types of tea—green, black, and oolong—are classified by their processing. Green tea is unfermented, while black tea is fully fermented, and oolong tea is half fermented. The differences in their fermentation process result in different amounts of chemical content among tea types (Alcázar et al. 2007; Keenan et al. 2011). Green tea is reportedly rich in chemical constituents attributed to health benefits (Alcázar et al. 2007; Ying et al. 2005). Due to the propagation of its potential health benefits to regular green tea drinkers (Kuriyama et al. 2006), green tea has recently become one of the most popular drinks in Thailand. As a natural product with various methods of preparation, green tea could vary greatly in color.

It is well known that appearance properties of food products have an effect on the perception of food quality (Francis 1995; Jaros et al. 2000). No food would be accepted without the acceptable appearance. Studies on consumer response to food and sensory evaluation have thus become of great interest, and the number of published papers regarding the subject has increased over the past decades (Tuorila and Monteleone 2009). Previous studies showed that color played an important role in consumers' preference for beverages (Duangmal et al. 2008; Tang et al. 2001). Consumers have expectations of taste based on the color of the beverages (Wei et al. 2009). Liang et al. (2008) found that sensory preference of green tea was dependent on its color, taste, and aroma.

Since commercial green tea drinks are usually contained in a clear bottle, the color of green tea drinks is one of the major attributes that affects consumers' impression on the product. The aim of this study was to investigate a color range of green tea drinks in Thailand, whereby the aspect of color and its association with consumers' expectation and preference were examined. The amounts of chemical constituents of green tea drinks were not investigated in this study.

43.2 Experimental Methods

43.2.1 Commercial Ready-to-Drink Green Teas

A survey was carried out to investigate the color range of green tea drinks available in the market. It was found that four major brands, namely, Fuji, Kirin, Oishi, and Unif, were readily available in most stores, and the most common flavor was "original" flavor. It is worthy to note that the original flavor of ready-to-drink green tea in Thailand is a sweetened green tea drink. The actual original flavor with no sugar added is clearly labeled with "No Sugar" and is only available in big stores as it is not a popular flavor. Since the original flavor (sweetened one) was more common in the market, it was chosen to be investigated in this study. For each brand, 15 bottles, original flavor, were randomly sampled from various convenient stores and supermarkets. Spectral transmittance of each sample was measured using a HunterLab ColorQuest XT spectrophotometer using a 50 mm path length cell. CIELAB (*Commission Internationale de l'Éclairage* L^*, a^*, and b^*) color values were then calculated with the CIE standard illuminant D65 and the CIE 10° standard observer. Figure 43.1 shows the distributions of these samples in the CIELAB space. The average values of the green tea drinks for each of the four

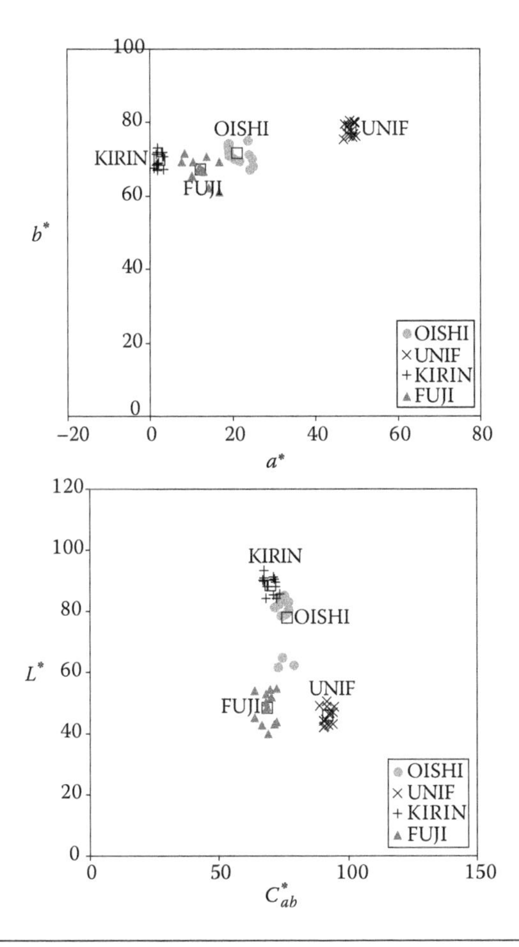

Figure 43.1 The distributions of commercial ready-to-drink green teas in the a^*–b^* plane and the L–C^*_{ab} plane. The open squares represent the average values for each brand.

brands are presented with open squares. It can be seen that there is not much variation within the same brands; however, the colors from different brands are distinctly different. The colors of green tea drinks from different brands were visually distinguished and ranged from yellowish to brownish colors. None of them was green as the name suggested.

43.2.2 Sample Preparation

Liquid color samples were prepared to simulate the color of green tea drinks. Based on the results of the survey of commercial ready-to-drink green teas, the color samples were generated to give a good perceptual coverage of the color range of commercial green tea drinks. These samples were prepared with a mixture of water and food colorants, i.e., tartrazine (providing yellow shade), Sunset Yellow FCF (orange shade), Brilliant Blue FCF (blue shade), caramel (dark brown), and Ponceau 4R (red shade). A total of 23 color samples

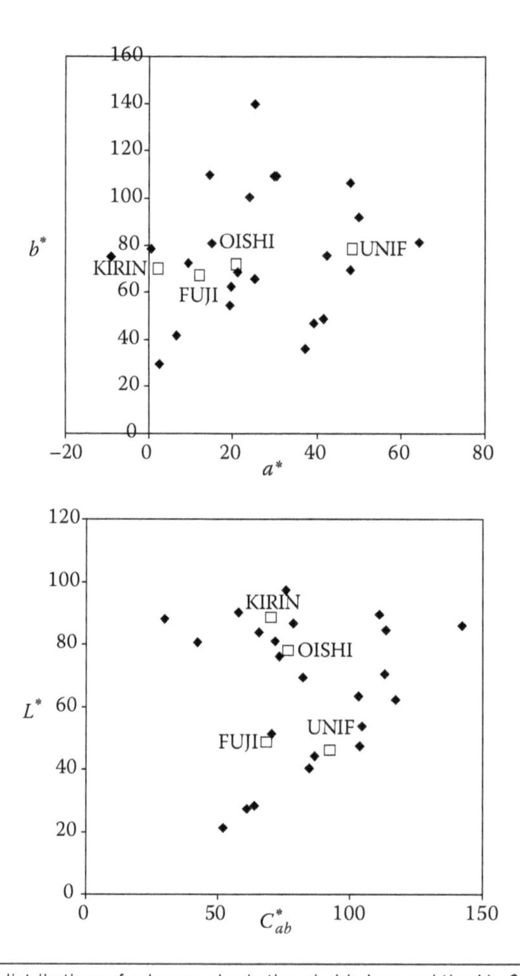

Figure 43.2 The distributions of color samples in the a^*–b^* plane and the L^*–C^*_{ab} plane.

were prepared. Their colors were measured in the same manner as mentioned in the previous section for commercial green tea drinks. The distributions in the CIELAB space of sample colors, together with colors of the commercial green tea drinks, are shown in Figure 43.2. It can be seen that these samples well cover the colors with respect to hue, lightness, and chroma of green tea drinks from the four brands.

43.2.3 Visual Assessment

A panel of 40 Thai observers, including 20 males and 20 females, ranging from 18 to 25 years of age, participated in visual assessments. All observers passed the Ishihara test to ensure normal color vision. The visual assessments were conducted in a darkened room, whereby observers viewed each of the 23 color samples contained in a 250 mL PET bottle which was placed in the middle of a light booth illuminated with D65 simulators (Figure 43.3).

Figure 43.3 Experimental setup.

The distance between the observers' eyes and the samples was approximately 60 cm. The order in which the samples were presented was randomized for each observer. Before carrying out the experiments, observers were given 1 min for adaptation. The tasks of the observers were to regard the samples as green tea drinks and rate their level of acceptability and preference concerning colors of the drinks using a 1 (the least) to 5 (the most) integer scale. Note that in the case of acceptability, observers were also informed that a score of 3 was considered an acceptable level. Observers rated the level of acceptability first, followed by preference, and then provided their expectation of taste in association with the sample colors.

43.3 Results and Discussions

43.3.1 Observer Agreement

The reliability of the experimental results can be determined by the consistency of observers' responses. In this study, two measures, i.e., the correlation coefficient (r value) and coefficient of variation (CV), were used to indicate how well visual scores of each individual observer agreed with the panel results (averaged from all observers). The r value measures the strength and direction of a linear relationship between two variables. Values of r range from −1 (perfect negative correlation) to 0 (no correlation) to 1 (perfect positive correlation). Hence, for a perfect agreement between the visual scores of each individual observer and the panel results, the r value should be 1.

Table 43.1 Observer Agreement

Individual vs. Panel Results	Acceptability		Preference	
	CV	r	CV	r
All observers	21	0.74	20	0.80
Female	21	0.72	20	0.78
Male	20	0.78	19	0.81

In the case of the CV measure, it expresses the root-mean-square deviation of the data points from the 45° line as a percentage of the mean value of the set y, giving the result independent of the size of the set y (Equation 43.1):

$$CV = \frac{100}{\bar{y}}\sqrt{\frac{\sum (x_i - y_i)^2}{N}} \tag{43.1}$$

In this study, set x is the individual results while set y represents the panel results. N is the number of samples and $\bar{y}$ is the mean value of set y. A CV value of 20 means 20% error of set x, i.e., individual results, from set y, i.e., the panel results. For perfect agreement, the CV value should be zero, i.e., 0% error. The larger the value, the poorer the agreement.

Table 43.1 summarizes the results of observer agreement for all observers (individual results against the panel results), female observers (individual female results against the panel results from females), and male observers (individual male results against the panel results from males). It was found that observer agreement was slightly better for preference than acceptability, suggesting that observers tended to like the same color but have different acceptability tolerances for the color of green tea drinks. The mean r values for male and female observers showed that males had better agreement between themselves than females did for both acceptability and preference. Nevertheless, the mean values of CV were not much different for all cases, showing that there was not much disagreement between observers' responses. The results of observer agreement found in this study were not much different from those found in similar studies (González-Miret et al. 2007; Ji et al. 2005; Jung et al. 2009). Furthermore, the agreement between male and female observers was examined using the r value. It was found that the responses from males and females were in agreement with r values of 0.97 and 0.98 for acceptability and preference, respectively.

43.3.2 Acceptability

The visual scores obtained from all observers were averaged, in which samples with an average score of 3 and above were considered acceptable as having the colors of green tea drinks. The results are shown in Figure 43.4. The acceptable samples scattered around the colors of the four brands. Despite green

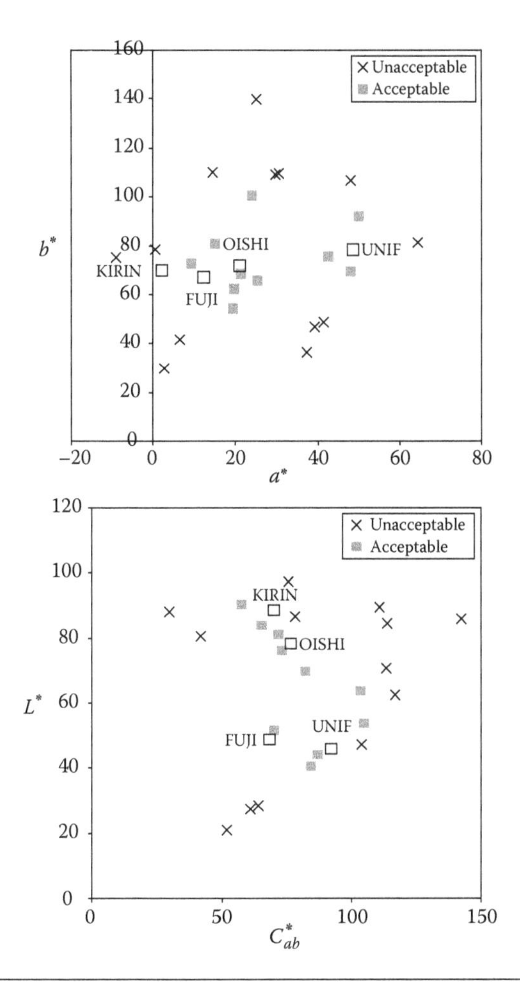

Figure 43.4 The acceptability of colors in the a^*–b^* plane and the L^*–C^*_{ab} plane.

tea having somewhat variety in color depending on chemical compositions, which result from the method of preparation (Keenan et al. 2011; Kodama et al. 2010; Liang et al. 2008), the acceptable range found in this study did not go beyond the range of these commercial green tea drinks. This is possibly due to the fact that observers are familiar with those colors and do not accept any other colors that fall outside the familiar range. It is worthy to note that green tea is not traditionally consumed in Thailand, as opposed to any other Asian countries such as China and Japan. The consumption of green tea has become popular in the recent year owing to an advertising campaign promoting green tea as a health booster. While green tea drinks could actually vary in color from deep green to light yellow, commercial ready-to-drink products had little color variation (from yellowish to brownish color). Thai observers, being familiar with commercially available green tea drinks, thus adopted

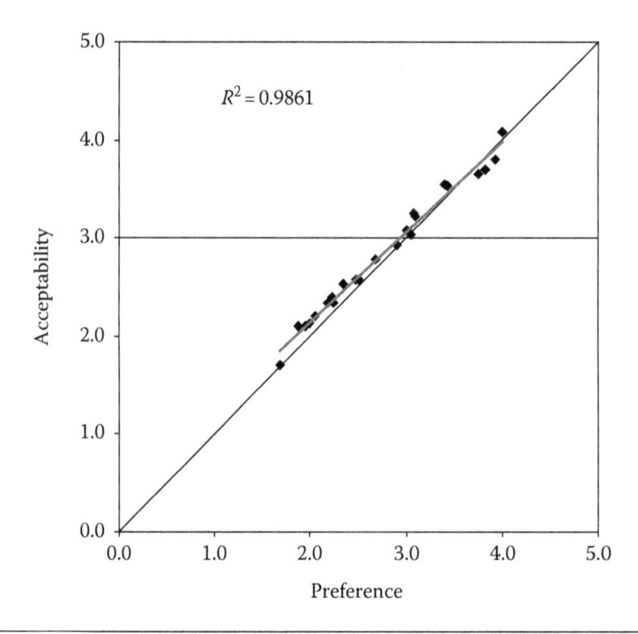

Figure 43.5 Agreement between acceptability and preference.

their acceptability tolerances according to the color range of green tea drinks available in the market.

The visual scores of acceptability were plotted against those of preference to investigate the relationship between them (Figure 43.5). The results showed that below the acceptability level, preference scores were lower than acceptability scores, revealing that observers would accept the colors as green tea drinks only when they liked the colors. The visual scores of acceptability and preference were highly correlated with an *r*-squared value of 0.9861. This suggests that the two aspects can hardly separate from each other.

43.3.3 Preference

Figure 43.6 shows color preference for each sample, whereby the size of each dot represents the magnitude of preference, i.e., the bigger the dot, the higher the preference score. The data were grouped according to tastes that majority of observers associated with the samples. It was found that observers tended to prefer samples having similar colors to green tea drinks of the four brands. The most preferred color was light brown (L^*, a^*, and b^* of 80.96, 21.28, and 68.4, respectively) with a visual score of 4.0, and majority of observers associated it with tastelessness. This color is very close to the color of OISHI, which is the most popular brand for green tea drinks in Thailand. In addition, the results showed that most highly preferred colors were associated with sweetness. This result agrees with the finding of Tang et al. (2001), which revealed that sweetness was a strong promoter of overall pleasantness of soft drinks.

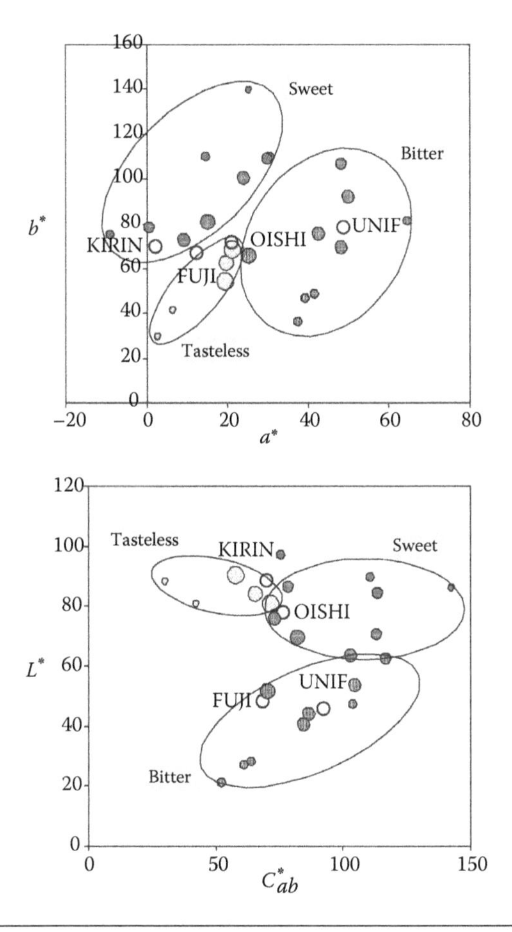

Figure 43.6 The magnitude of color preference in the a^*–b^* plane and the L^*–C^*_{ab} plane.

It should also be noted that the so-called "original" flavor of commercial ready-to-drink green tea in Thailand is sweetened; as Thais are not familiar with drinking tea, sugar is added to promote more pleasing taste. However, the intensity of sweetness for different brands is varied.

The expectation of taste with respect to color could be observed: yellowish was associated with sweetness, high lightness with tastelessness, and low lightness with bitterness. Observers' experience with the tastes of green tea drinks of the four brands could also influence the results. The further investigation regarding taste association is given in the following section.

43.3.4 Association with Taste

In the visual assessments, observers provided the expectation of tastes in association with color of the samples. Observers were free to answer any tastes, and the experimental data were recorded in terms of tastes and the number of observers giving the answer of that taste. The results are summarized in

Table 43.2 Taste Associated with Sample Colors

	Percentage of Observers Expecting the Taste				Color Values		
Sample	Bitter	Sweet	Tasteless	Sour	L^*	a^*	b^*
1	**100**	0	0	0	21.09	37.39	36.12
2	**98**	3	0	0	28.34	41.61	48.61
3	**98**	3	0	0	27.33	39.24	46.91
4	**98**	0	0	3	47.31	64.50	81.21
5	**90**	10	0	0	44.16	42.48	75.60
6	**85**	13	3	0	51.40	25.24	65.51
7	**63**	38	0	0	53.75	49.97	91.83
8	**60**	40	0	0	62.32	48.13	106.51
9	**55**	45	0	0	40.52	48.13	69.42
10	0	**98**	3	0	76.01	9.35	72.48
11	3	**93**	0	5	97.19	−9.06	74.97
12	0	**93**	5	3	86.59	0.52	78.38
13	3	**88**	3	8	89.60	14.53	109.91
14	0	**85**	8	8	84.49	30.50	109.32
15	0	**78**	3	20	85.98	25.19	139.93
16	33	**65**	3	0	69.45	14.99	80.76
17	38	**63**	0	0	70.55	29.81	109.21
18	48	**53**	0	0	63.51	23.93	100.33
19	0	8	**93**	0	88.13	2.56	29.73
20	0	8	**93**	0	80.66	6.59	41.61
21	0	43	**58**	0	80.96	21.28	68.40
22	0	48	**53**	0	90.30	19.32	54.33
23	0	45	**48**	8	83.96	19.61	62.39

The bold figures indicate the taste the majority of observers expected for each sample.

Table 43.2. Four tastes, i.e., sweet, bitter, tasteless, and sour, were collected. No other tastes were reported, and only few observers answered sour. This is possibly because the three tastes (sweet, bitter, and tasteless) are common for tea. It can also be observed that sweetness is more common than the other two tastes as when a number of observers answering either bitter or tasteless decreased, a number of observers answering sweet taste increased. In other words, observers split the answers between sweetness and the other tastes. Since the "original" flavor of commercial green tea drinks is actually sweet, this finding confirms that observers tended to associate the taste with the color based on their experience with the four brands.

To reveal the correlation between the expectation of tastes and colors of green tea drinks, the r values between frequencies of taste responses and colorimetric values, i.e., L^* (representing lightness), a^* (redness-greenness),

Table 43.3 Correlation between Tastes and Colorimetric Values

r Value	L^*	a^*	b^*	h_{ab}	C^*_{ab}
Bitter	**−0.94**	**0.76**	−0.20	**−0.85**	0.00
Sweet	**0.68**	−0.51	0.61	**0.71**	0.47
Tasteless	0.44	−0.43	−0.51	0.29	−0.62
Sour	0.42	−0.15	0.59	0.31	0.54

Bold figures indicate strong correlation.

b^* (yellowness-blueness), h_{ab} (hue angle), and C^*_{ab} (chroma), were calculated, and the results are shown in Table 43.3. It was found that bitterness had a strong negative correlation with lightness, i.e., observers tended to expect a bitter taste when the green tea was low in lightness. A strong negative correlation was also found between hue and bitterness. Since the hue angle spanned approximately between 40° (reddish) and 100° (yellowish), and a positive correlation between bitterness and a^* was found, i.e., expectation of bitterness increased with redness, it could be interpreted that observers associated dark brown green tea drinks with bitterness.

In the case of sweetness, the results showed the tendency of sweetness increasing with an increase of lightness of the samples. It was also found that sweetness correlated with hue in the direction of yellowness, as observed by the positive correlations of the b^* and hue angle values with sweetness. This reveals that the observer associated yellow with sweetness. Observers tended to expect a sweet taste of green tea when its color was light yellow. Tastelessness seemed to be expected when the color of the drinks was low in chroma with high lightness, implying that intensity of hue is related to tastelessness.

43.4 Conclusions

This study investigated the colors of green tea drinks in Thailand with respect to observers' preference and their associations with tastes. The color range of commercial ready-to-drink green teas available in Thailand was first examined. The colors of green tea drinks from four major brands ranged from yellowish to brownish. Not much difference was found within the same brands. Liquid color samples were then generated to cover the range of commercial green tea colors. Visual assessments were conducted using these samples to investigate the acceptability range, color preference, and observer expectations of taste. Since consumption of green tea is not part of traditional culture in Thailand, observers do not have a preconception of colors and tastes of green tea. Drinking green tea has recently become fashionable due to the advertising campaign. Thai observers are acquainted with green tea drinks that are available in the market. Hence, it was found that observers based their judgments on colors and tastes of commercial green tea drinks. The observers accepted color samples as a green tea drink when their colors were similar to those of commercial green teas. They also tended to prefer samples having similar colors to commercial green tea drinks. In addition, the most common and popular flavor of green tea drinks in the market is the "original" flavor, which is in fact a sweetened drink. Thus, sweetness was found to be a common expectation of taste from the observers. Most of highly preferred colors were associated with sweetness. The results showed the tendency of bitterness increasing with a decrease in lightness. On the contrary, samples with high lightness and low chroma were associated with tastelessness, whereas yellow was related to sweetness.

Acknowledgments

The authors acknowledge the financial support from the Faculty of Science, Chulalongkorn University. They also thank all observers who took part in the visual experiments.

References

Alcázar, A., O. Ballesteros, J. M. Jurado, F. Pablos, M. J. Martín, J. L. Vilches, and A. Navalón A. 2007. Differentiation of green, white, black, Oolong, and Pu-erh teas according to their free amino acids content. *Journal of Agricultural and Food Chemistry* 55 (15): 5960–5965.

Duangmal, K., B. Saicheua, and S. Sueeprasan. 2008. Colour evaluation of freeze-dried Roselle extract as a natural food colorant in a model system of a drink. *LWT—Food Science and Technology* 41 (8): 1437–1445.

Francis, F. J. 1995. Quality as influenced by color. *Food Quality and Preference* 6 (3): 149–155.

González-Miret, M. L., W. Ji, M. R. Luo, J. B. Hutchings, and F. J. Heredia. 2007. Measuring colour appearance of red wines. *Food Quality and Preference* 18 (6): 862–871.

Jaros, D., H. Rohm, and M. Strobl. 2000. Appearance properties—A significant contribution to sensory food quality? *Lebensmittel-Wissenschaft und-Technologie* 33 (4): 320–326.

Ji, W., M. R. Luo, J. B. Hutchings, and J. Dakin. 2005. Scaling transparency, opacity, apparent flavour strength and preference of orange juice. In *AIC Colour 2005, Proceedings of the 10th Congress of the International Colour Association*, eds. J. L. Nieves and J. Hernández-Andrés. Granada, Spain: Comité Español del Color, pp. 729–732.

Jung, M., W. Ji, P. Rhodes, M. R. Luo, and J. B. Hutchings. 2009. Colour and appearance of an ideal beer. In *AIC Colour 2009, Proceedings of the 11th Congress of the International Colour Association*. Sydney, Australia: Colour Society of Australia.

Keenan, E. K., M. D. A Finnie, P. S. Jones, P. J. Rogers, and C. M. Priestley. 2011. How much theanine in a cup of tea? Effects of tea type and method of preparation. *Food Chemistry* 125 (2): 588–594.

Kodama, D. H., A. E. Gonçalves, F. M. Lajolo, and M. I. Genovese. 2010. Flavonoids, total phenolics and antioxidant capacity: Comparison between commercial green tea preparations. *Ciência e Tecnologia de Alimentos* 30 (4): 1077–1082.

Kuriyama, S., T. Shimazu, K. Ohmori, N. Kikuchi, N. Nakaya, Y. Nishino, Y. Tsubono, and I. Tsuji. 2006. Green tea consumption and mortality due to cardiovascular disease, cancer, and all causes in Japan. The Ohsaki study. *Journal of the American Medical Association* 296 (10): 1255–1265.

Liang, Y. R., Q. Ye, J. Jin, H. Liang, J. L. Lu, Y. Y. Du, and J. J. Dong. 2008. Chemical and instrumental assessment of green tea sensory preference. *International Journal of Food Properties* 11 (2): 258–272.

Rogers, P. J., J. E. Smith, S. V. Heatherley, and C. W. Pleydell-Pearce. 2008. Time for tea: Mood, blood pressure and cognitive performance effects of caffeine and theanine administered alone and together. *Psychopharmacology* 195 (4): 569–577.

Tang, X., N. Kälviäinen, and H. Tuorila. 2001. Sensory and hedonic characteristics of juice of sea buckthorn (*Hippophae rhamnoides* L.) origins and hybrids. *Lebensmittel-Wissenschaft und-Technologie* 34 (2): 102–110.

Tuorila, H. and E. Monteleone. 2009. Sensory food science in the changing society: Opportunities, needs, and challenges. *Trends in Food Science & Technology* 20 (2): 54–62.

Wei, S. T., L. C. Ou, M. R. Luo, and J. B. Hutchings. 2009. Quantification of the relations between juice colour and consumer expectations. In *AIC Colour 2009, Proceedings of the 11th Congress of the International Colour Association*. Sydney, Australia: Colour Society of Australia.

Ying, Y., J. W. Ho, Z. Y. Chen, and J. Wang. 2005. Analysis of theanine in tea leaves by HPLC with fluorescence detection. *Journal of Liquid Chromatography & Related Technologies* 28 (5): 727–737.

CHAPTER 44

Orange Juice Color

Visual Evaluation and Consumer Preference

ROCÍO FERNÁNDEZ-VÁZQUEZ, CARLA M. STINCO, ANTONIO J. MELÉNDEZ-MARTÍNEZ, FRANCISCO J. HEREDIA, and ISABEL M. VICARIO

Contents

44.1 Introduction

Citrus juices are among the most widely consumed fruit juices because of their combination of desirable flavor, appealing color, and health benefits (Rouseff et al. 2009). The natural bright color of the citrus juices has been considered traditionally as one of their main advantages over other juices (Barron et al. 1967).

It is due to carotenoids which belong to one of the main classes of natural pigments. The carotenoid profile of orange juices (OJs) is one of the most complex in nature, and it comprises carotenes and xanthophylls. Some of these compounds (β-carotene, α-carotene, and β-cryptoxanthin) have provitamin A activity, and they may exhibit other biological activities, like antioxidant and anticarcinogenic activity (Krinsky et al. 2004). The color of citrus fruits in general depends on several factors apart from the stage of maturity, such as climate, species, and variety (Casas and Mallent 1988). In oranges, it ranges from pale yellow at the beginning of the season to red-orange at the end of the season.

Consumers frequently judge food quality based on color, so the relevance of color as a quality attribute in the food industry is undoubtedly. Some studies have revealed that the color of citric beverages is related to the consumer's perception of flavor, sweetness, and other quality characteristics related to these products (Huggart et al. 1977, Pangborn 1960, Tepper 1993). Valencia OJs are worldwide appreciated due to their deep orange color (Francis and Clydesdale 1975, Robards and Antolovich 1995). In this sense, manufacturers may add other juices, which provide a more intense coloration, to improve OJs varieties which are poorly colored. Certain varieties of oranges, whose pulp has a peculiar reddish color, are becoming increasingly studied (Lee 2001, 2002); however, few data can be found in the literature in relation to the color of other varieties.

Color measurement can be done by visual evaluation or instrumental analysis. In the literature, information concerning the sensory evaluation of food color, including guidelines for panel selection, physical requirements for visual assessments, and types of sensory tests, can be found (Hutchings 2011). As a result of the visual analysis, a particular description of the color is obtained, for which there is a certain vocabulary. Nevertheless, this kind of analysis is subjective since it is influenced by several factors (illumination, type of container, volume, etc.). For that reason, implementation of instrumental measurement of color within the OJ industry is very important for quality control purposes (Meléndez-Martínez et al. 2005).

The objectives of this study were to characterize the color of OJs from nine different orange varieties (not previously reported in literature) and to determine the correlation between the color attributes (lightness, chroma, and hue) determined by a digital system and evaluated by a trained panel. The consumers' preference for OJ color was also explored.

44.2 Materials and Methods

44.2.1 Samples

OJs were obtained from nine orange varieties: Navel Foyos (NF), Fisher (F), Navel Powell (NP), Fukumoto (FU), Navel Lane Late (LL), Valencia Midnight (VM), Rohde Late (RL), Navel Rohde (NR), and Valencia Delta (VD).

The samples analyzed in this study were kept at −21°C until their analysis in the laboratory. Thawing was carried out at room temperature (23°C) for 24 h.

44.2.2 Color Measurement

Digital images (DigiEye imaging system, Verivide Leicester, United Kingdom) were obtained in order to evaluate the total appearance of juice at depths observed by consumers (Luo et al. 2001). The latter system includes a digital camera Nikon D-80, a computer (provided with appropriate software), a color sensor for calibrating displays, and an illumination box designed by DigiEye Plc. The computer software included the functions of camera characterization, color measurement, monitor characterization, and various specialized functions such as color texture mapping, color selection, and fastness grading (Hutchings et al. 2002). In these measurements, the samples were illuminated by a diffuse D65 simulator. A GretagMacbeth ColorChecker DC chart was used for calibration purposes (Li et al. 2003). The samples were measured in 75 mL transparent plastic rectangular bottles (5 × 7.5 × 2.5 cm).

44.2.3 Juice Selection for Sensory Analyses and Consumers' Preference Test

After analyzing the results by image analysis, five samples of OJ were chosen for sensory analysis based on their color differences (NF, F, NP, VM, RL). Those samples in which $\square E^{*}_{ab}$ exceeded the threshold for visual discrimination (Martínez et al. 2001) were chosen.

44.2.4 Sensory Analyses

Visual analyses of the samples were carried out within a well-illuminated room provided with two cabins (70 × 70 × 55 cm), with white walls. The 75 mL plastic bottles were used. Each series contained the different samples of OJs, which were placed randomly. Before each visual analysis, all the bottles were vigorously shaken to avoid pulp sedimentation. The trained panel consisted of 18 panelists, aged between 30 and 45 years, with normal color vision (which was verified using the Farnsworth-Munsell 100 hue test) and experienced in both visual assessments of color and tristimulus colorimetry concepts. All of them were asked to classify the samples in increasing order of hue (h_{ab}) (yellowish-reddish), chroma (C^{*}_{ab}) (dull-vivid), and lightness (L^{*}) (clear-dark). They were also asked to score the colorimetric parameters on a continuous scale of 10 cm, anchored at the ends.

44.2.5 Consumers' Preference

The consumer panel consisted of 111 panelists (78 females and 33 males) recruited among students and staff of the University of Seville. The consumers were asked to order the five samples previously selected (NF, F, NP, VM, RL), without giving consumers information about them. OJs were presented according to their color preference. The ranking decision was based only on the color characteristics, without further information. OJs were presented in the same bottles and within the same cabins as in sensory analyses.

44.2.6 Data Analysis

Friedman test (O'Mahony 1986) ($p < 0.05$) and Page test ($p < 0.05$) were used to analyze the sensory data. For consumer preferences, the Fisher test was applied. These analyses were performed using the Statistica program for Windows (version 8) (StatSoft 2007).

44.3 Results and Discussion

44.3.1 Color Measurement

Figure 44.1 shows the distribution of the samples in the CIELAB color space (a^*b^* diagram) to illustrate the color of the orange varieties. The values of the coordinate L^* ranged from 56.09 in the lighter OJ to 61.34 in the darker one; C^*_{ab} ranged from 54.03 in the dullest OJ to 60.29 in the most vivid, and h_{ab} ranged from 66.43 for the most reddish to 81.99 in the most yellow, measured by image analyses.

□E^*_{ab} values were determined for the different OJ varieties, and those with □$E^*_{ab} > 3$ were selected for visual analyses. The selected samples were NF, F, NP, VM, and RL.

Table 44.1 shows the colorimetric parameters of the OJs selected for consumer preference test. All of the samples selected presented significant differences ($p < 0.05$) for hue being the most reddish sample RL and the most

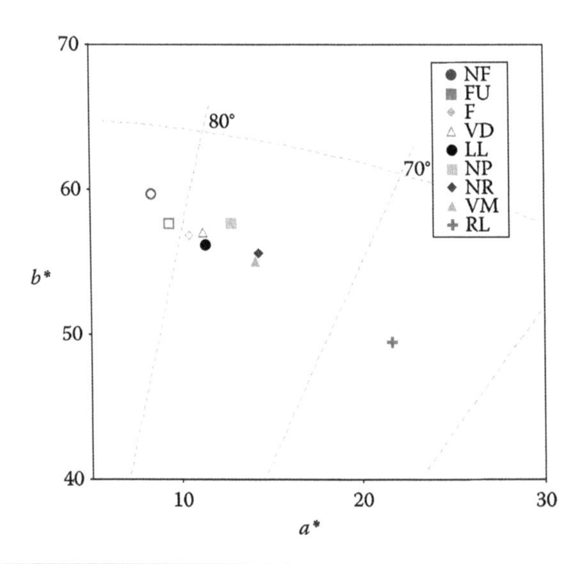

Figure 44.1 *a*b** diagram of orange varieties.

Table 44.1 Color Coordinates (Mean Value ± Standard Deviation) of OJ Samples Selected for Sensory Analysis

Varieties	L^*	a^*	b^*	C^*_{ab}	h_{ab}
NF	61.3 ± 0.3^a	8.4 ± 0.5^a	59.7 ± 0.5^a	60.3 ± 0.6^a	82.0 ± 0.5^a
F	60.4 ± 0.3^a	10.5 ± 0.1^b	56.9 ± 0.5^b	57.8 ± 0.6^{bc}	79.6 ± 0.8^b
NP	59.8 ± 0.3^{ab}	12.8 ± 0.4^c	57.7 ± 0.4^b	59.1 ± 0.4^{ac}	77.5 ± 0.4^c
VM	58.8 ± 0.2^b	14.2 ± 0.5^d	55.0 ± 0.7^b	56.8 ± 0.6^b	75.6 ± 0.5^d
RL	56.1 ± 0.4^c	21.6 ± 0.5^e	49.5 ± 0.6^c	54.0 ± 0.6^d	66.4 ± 0.4^e

[a–e] Different superscripts within columns indicate statistically significant differences ($p < 0.05$).

yellowish sample NF. According to chroma and lightness, only some of the samples presented significant differences.

44.3.2 Sensory Analyses

44.3.2.1 Order Test The trained panel was able to order the OJs correctly based on hue (h_{ab}) and lightness (L^*); in fact, correlation between panel scores and DigiEye data were significant for L^* and h_{ab} ($r^* = 0.96$ and $r^* = 0.92$, respectively). For hue (yellowish-reddish), all samples were correctly ordered by panelist, excepting F and NP which Δh_{ab} was 2.04. For L^*, the samples with $\Delta L^* > 0.663$ were correctly ordered by 83% of the panel, while the rest of 17% were able to order all the samples correctly. However, according to chroma (C^*_{ab}), only 17% of the OJs were ordered properly. This may be due to the low chroma differences among samples (□ C^*_{ab} ranged 1.03–2.77).

44.3.2.2 Score test Table 44.2 shows the average scores given by the panelists to the different samples, according to different color parameters. The judges scored significantly differently all the OJs according to h_{ab} (from the most yellowish [NF] to the most reddish [RL]) and L^* (from the highest [NF] to the lowest lightness [RL]), but as expected, they only found significant differences in C^*_{ab} among some varieties.

Table 44.2 Mean Scores for the Colorimetric Parameters Given by the Trained Panel

	RL	NF	F	VM	NP
h_{ab}	7.9^a	1.6^b	4.3^c	5.6^d	3.0^e
L^*	8.0^a	2.3^b	4.4^c	6.1^d	4.0^c
C^*_{ab}	5.7^a	5.9^a	4.7^{ab}	3.6^b	4.7^{ab}

[a–e] Different superscripts within row indicate statistically significant differences ($p < 0.05$).

Table 44.3 Rank Order Tests: Rank Sums

Samples	Rank Sums
Valencia Midnight	407[a]
Fisher	372[ab]
Rhode Late	338[cb]
Navel Powell	321[c]
Navel Foyos	227[d]

Note: Values followed by the different superscripts are statistically different (95% level Fisher test).

44.3.3 Consumers' Preference

The consumers (111 panelists) were asked to order the five samples according to their color preference.

The results of the ordination test are shown in Table 44.3 and the probabilities of the differences between rank sums (Tukey test). A significant preference ($p<0.05$) was observed for the OJs with intermediate h_{ab} and L^* values (more orangish): VM and F (without significant differences between them) (Table 44.1). On the other hand, the least preferred OJ was the one with the highest lightness and hue values, the most yellowish, the NF variety, and clearly different from the rest. Samples RL and NP showed intermediate preference.

44.4 Conclusions

A trained sensory panel was able to order OJs according to hue when Δh_{ab} was higher than 2.04, and 83% of the panel ordered correctly the samples according to lightness when ΔL^* was higher than 0.67. However, the panel was not able to order samples according to their chroma.

Regarding consumers' preference, no clear preference for any of the varieties was observed. However, as a general trend, consumers preferred samples with orangish hues ranging from 76 to 80 CIELAB units.

Acknowledgments

This work was supported by funding from the Consejería de Innovación Ciencia y Empresa, Junta de Andalucía (Spain) by the project P08-AGR-03784.

References

Barron, R. W., M. D. Maraulja, and R. L. Huggart. 1967. Instrumental and visual methods for measuring orange juice color. *Proceedings of the Florida State Horticultural Society* 80: 308–313.

Casas, A. and D. Mallent. 1988. El color de los frutos cítricos. I. Generalidades. II. Factores que influyen en el color. Influencia de la especie, de la variedad y de la temperatura. *Revista de Agroquímica y Tecnologia de Alimentos* 28: 184–202.

Francis, F. J. and F. M. Clydesdale. 1975. *Food Colorimetry: Theory and Applications.* Westport, CN: Avi Publishing Co.

Huggart, R. L., D. R. Petrus, and B. S. Buzz Lig. 1977. Color aspects of Florida commercial grapefruit juices. *Proceedings of the Florida State Horticultural Society* 90: 173–175.

Hutchings, J. B. 2011. *Food Colour and Appearance*. Glasgow, U.K.: Blackie.

Hutchings, J. B., M. R. Luo, and W. Ji. 2002. Calibrated colour imaging analysis of food. In *Colour in Food: Improving Quality*, ed. D. B. MacDougall. Cambridge, U.K.: Woodhead Publishing, pp. 352–366.

Krinsky, N., S. T. Mayne, and H. Sies. 2004. *Carotenoids in Health and Disease.* New York: Marcel Dekker.

Lee, H. S. 2001. Characterization of carotenoids in juice of red Navel orange (Cara Cara). *Journal of Agricultural and Food Chemistry* 49 (5): 2563–2568.

Lee, H. S. 2002. Characterization of major anthocyanins and the color of red-fleshed Budd blood orange (*Citrus sinensis*). *Journal of Agricultural and Food Chemistry* 50 (5): 1243–1246.

Li., C., G. Cui, and M. R. Luo. 2003. The accuracy of polynomial models for characterising digital cameras. In *AIC 2003 Color Communication and Management, Proceedings*. Bangkok, Thailand: The Color Group of Thailand, pp. 166–170.

Luo, M. R., G. Cui, and C. Li. 2001. Apparatus and method for measuring colour (DigiEye System), British patent.

Martínez, J. A., M. Melgosa, M. M. Pérez, E. Hita, and A. I. Negueruela. 2001. Visual and instrumental color evaluation in red wines. *Food Science and Technology International* 7 (5): 439–444.

Meléndez-Martínez, A. J., I. M. Vicario, and F. J. Heredia. 2005. Instrumental measurement of orange juice colour: A review. *Journal of the Science of Food and Agriculture* 85 (6): 894–901.

O'Mahony, M. 1986. *Sensory Evaluation of Food: Statistical Methods and Procedures.* New York: Marcel Dekker.

Pangborn, R. M. 1960. Influence of color on the discrimination of sweetness. *The American Journal of Psychology* 73 (2): 229–238.

Robards, K. and M. Antolovich. 1995. Methods for assessing the authenticity of orange juice. A review. *Analyst* 120 (1): 1–28.

Rouseff, R. L., P. Ruiz Perez-Cacho, and F. Jabalpurwala. 2009. Historical review of citrus flavor research during the past 100 years. *Journal of Agricultural and Food Chemistry* 57 (18): 8115–8124.

StatSoft. 2007. Statistica (data analysis software system), version 8.0.

Tepper, B. J. 1993. Effects of a slight color variation on consumer acceptance of orange juice. *Journal of Sensory Studies* 8 (2): 145–154.

Author Index

T

U

V

Y

Z

Subject Index

A

B

C

D

E

F

M

N

O

P

Q

R

S

T

U

V

W

Y

For Product Safety Concerns and Information please contact our EU representative GPSR@taylorandfrancis.com
Taylor & Francis Verlag GmbH, Kaufingerstraße 24, 80331 München, Germany

www.ingramcontent.com/pod-product-compliance
Ingram Content Group UK Ltd.
Pitfield, Milton Keynes, MK11 3LW, UK
UKHW021024180425
457613UK00020B/1046

* 9 7 8 1 1 3 8 1 9 9 6 4 4 *